P9-CBT-213

ADVANCES IN CHEMICAL PHYSICS

VOLUME XC

Advances in
CHEMICAL PHYSICS

Edited by

I. PRIGOGINE

University of Brussels
Brussels, Belgium
and
University of Texas
Austin, Texas

and

STUART A. RICE

Department of Chemistry
and
The James Franck Institute
The University of Chicago
Chicago, Illinois

VOLUME XC

AN INTERSCIENCE® PUBLICATION
JOHN WILEY & SONS, INC.
NEW YORK • CHICHESTER • BRISBANE • TORONTO • SINGAPORE

This text is printed on acid-free paper.

An Interscience® Publication

Library of Congress Catalog Number: 58-9935

ISBN 0-471-04234-X

10 9 8 7 6 5 4 3 2 1

CONTRIBUTORS TO VOLUME XC

D. Alonso, Service de Chimie Physique and Centre for Nonlinear Phenomena and Complex Systems, Université Libre de Bruxelles, Brussels, Belgium

I. Burghardt, Service de Chimie Physique and Centre for Nonlinear Phenomena and Complex Systems, Université Libre de Bruxelles, Brussels, Belgium

P. Gaspard, Service de Chimie Physique and Centre for Nonlinear Phenomena and Complex Systems, Université Libre de Bruxelles, Brussels, Belgium

Edward R. Grant, Department of Chemistry, Purdue University, West Lafayette, Indiana

B. Vincent McKoy, Arthur Amos Noyes Laboratory of Chemical Physics, California Institute of Technology, Pasadena, California

Klaus Müller-Dethlefs, Institut für Physikalische und Theoretische Chemie, Technische Universität München, Garching, Germany

Edward W. Schlag, Institut für Physikalische und Theoretische Chemie, Technische Universität München, Garching, Germany

Kwanghsi Wang, Arthur Amos Noyes Laboratory of Chemical Physics, California Institute of Technology, Pasadena, California

INTRODUCTION

Few of us can any longer keep up with the flood of scientific literature, even in specialized subfields. Any attempt to do more and be broadly educated with respect to a large domain of science has the appearance of tilting at windmills. Yet the synthesis of ideas drawn from different subjects into new, powerful, general concepts is as valuable as ever, and the desire to remain educated persists in all scientists. This series, *Advances in Chemical Physics*, is devoted to helping the reader obtain general information about a wide variety of topics in chemical physics, a field which we interpret very broadly. Our intent is to have experts present comprehensive analyses of subjects of interest and to encourage the expression of individual points of view. We hope that this approach to the presentation of an overview of a subject will both stimulate new research and serve as a personalized learning text for beginners in a field.

I. PRIGOGINE
STUART A. RICE

CONTENTS

ZEKE SPECTROSCOPY: HIGH-RESOLUTION SPECTROSCOPY WITH PHOTOELECTRONS

KLAUS MÜLLER-DETHLEFS AND EDWARD W. SCHLAG

Institut für Physikalische und Theoretische Chemie, Technische Universität München, D-85747 Garching, Germany

EDWARD R. GRANT

Department of Chemistry, Purdue University, West-Lafayette, IN 47907

KWANGHSI WANG AND B. VINCENT McKOY

*Arthur Amos Noyes Laboratory of Chemical Physics, California Institute of Technology, Pasadena, CA 91125**

CONTENTS

* Contribution No. 8844.

Advances in Chemical Physics, Volume XC, Edited by I. Prigogine and Stuart A. Rice.
ISBN 0-471-04234-X

I. INTRODUCTION

Zero-kinetic-energy (ZEKE) photoelectron techniques [1–4] have now become new tools of physical chemistry that go far beyond the measurement of the spectra of molecular ions [5–9]. The unprecedented resolution of the ZEKE method [8, 9], which is essentially limited by the laser bandwidth, has already led to many new and exciting applications in chemical physics and chemistry within the last few years [10–23]. The rapid expansion of this method to a number of very active laboratories clearly suggests that this trend can be expected to continue [24–47].

In this section, selected examples will be presented demonstrating the much superior resolution of ZEKE spectroscopy compared to conventional photoelectron spectroscopy (PES) (two to three orders of magnitude). These examples also give a flavor of the perspectives of the method. The first molecule investigated by ZEKE spectroscopy in 1984 was nitric oxide (NO) [1, 2], which was chosen because it had already been studied extensively by vacuum ultraviolet (VUV) and resonant enhanced multiphoton ionization techniques. In addition, at that time, there was much interest in the Rydberg states in NO and substantial progress was being made through the application of multichannel quantum defect theory (MQDT) [10, 48–50] to molecules as a means of understanding the structure of their Rydberg states and their couplings in a manner similar to what had been done for molecular hydrogen (H_2) [49]. Nitric oxide also seemed to be a good starting point to further develop theories of photoionization processes, in particular, by using *ab initio* theories [12, 51] to explain the ionization dynamics. In classical VUV photoelectron techniques [52], information about such dynamics is usually lost since rotational resolution is not achieved (particularly at threshold); the reader is referred to [4] for further discussion. Figure 1 shows the result of the greatly improved resolution of the ZEKE method compared to conventional PES. The VUV–PE spectrum [52] (Fig. 1, top) shows the vibrationally resolved structure of the lowest electronic state of the NO^+ cation. In contrast, the fully resolved **rotational** structure within a single vibrational band [1] (in this case the lowest vibrational band with $v^+ = 0$) is seen in the bottom of Fig. 1 due to the much higher resolution of the ZEKE technique. Details of the photoionization dynamics will be discussed later in this chapter. Note here that for ZEKE spectroscopy different rotational selection rules from those observed for bound–bound transitions are to be expected. For rotational transitions observed in ZEKE spectroscopy higher angular momentum transfers from molecule to ion transitions than $\Delta J = \pm 1$, as for bound–bound transitions, are to be expected and are seen. This apparent difference in rotational selection rules arises because the ejected electron carries an angular momentum ℓ

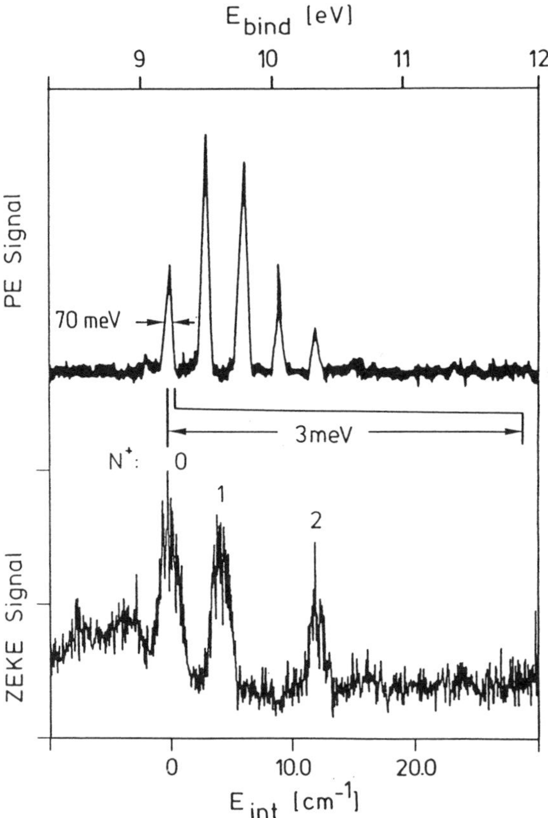

Figure 1. Comparison of VUV–PES [52] and the first rotationally resolved ZEKE spectrum [2] of NO. Here N^+ denotes the total angular momentum quantum number (excluding electronic spin). (Reproduced from [2] and [52] with permission.)

that couples to the total angular momentum J^+ of the ion to yield a state of total angular momentum J_{tot}. For ionization from a neutral state with rotational quantum number J it is this total angular momentum J_{tot} bounded by the triangular condition $J - 1 \leq J_{tot} \leq J + 1$, that is preserved. It can be seen that different rotational states J^+ arise for a given ℓ when J_{tot} is decoupled according to the triangular condition $J^+ - \ell \leq J_{tot} \leq J^+ + \ell$.

The state-of-the-art of ZEKE spectroscopy is best illustrated in its ability to resolve the question of the structure of the benzene cation. The benzene cation, due to its electronically degenerate state, is the textbook example [53] of the Jahn–Teller effect [54, 55]. The strength of this Jahn–Teller interaction has been predicted theoretically [56]. Experimentally, matrix isolation electron spin resonance (ESR) spectra seemed to support a D_{2h} structure at low temperatures (4 K) and a D_{6h} structure at higher temperatures (100 K) [57]. However, matrix perturbations could not be excluded as the cause of the observed effects. Efforts to analyze the free cation spectroscopically in the gas phase have failed since methods of ion spectroscopy [58] are not applicable. With ZEKE spectroscopy it has become possible to completely resolve the rotational structure of the benzene cation, and hence to deduce its molecular structure [8]. Figure 2 shows one of these rotationally resolved spectra, corresponding to the transition from the $J' = 2$, $K' = 2$ rovibronic state in the intermediate resonant $S_1 6^1\ {}^{ev}E_{1u}$ state to the ground state $D_0\,{}^{e}E_{1g}$ of the benzene cation [8]. The observed resolution is essentially laser-limited. The rotational structure shows certain symmetry selection rules [6, 8, 9] that will be discussed later. By applying these symmetry selection rules, the observed ZEKE transitions can be interpreted and the structure of the benzene cation and its molecular symmetry group can be unambiguously determined. The results for the benzene cation are fully consistent with the D_{6h} molecular symmetry group. This leads to the

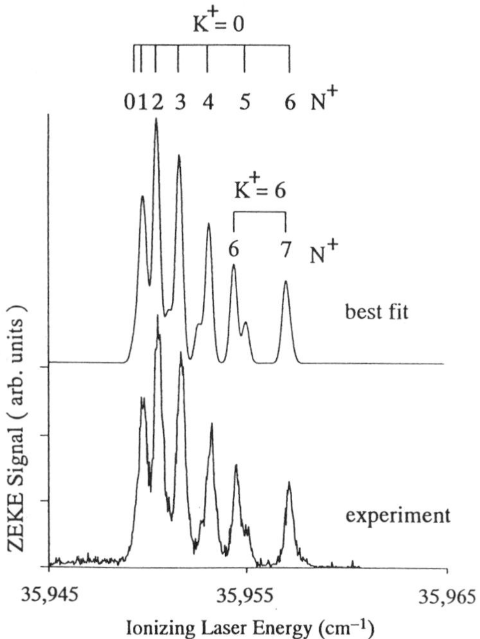

Figure 2. Recorded (bottom) ZEKE spectrum of the ionizing transition $D_0\ {}^2E_{1g} \leftarrow S_1$ $6^1\ {}^1E_{1u}$ ($J' = 2$, $K' = 2$) of benzene. The fully resolved rotational structure, assigned as N^+ progressions in $K^+ = 0$ and $K^+ = 6$, was fitted (top) with the experimentally determined rotational constants and the experimental intensities using a fwhm of $0.3\,\text{cm}^{-1}$. (Reproduced from [8] with permission.)

conclusion that the structure of this fluxional molecule is governed by the *dynamic* Jahn–Teller effect.

To demonstrate the relationship between ZEKE spectroscopy and photoionization efficiency (PIE) measurements, Fig. 3 shows a comparison of a PIE measurement of nitrogen dioxide (NO_2) with its ZEKE spectrum. The PIE measurement shows an obscure onset of ionization. It is obscured because the ionization intensity is not only governed by the threshold signal, but is dominated by transitions into autoionizing states. Such autoionizing states are members of Rydberg series of lower principal quantum numbers converging to higher ion limits. In the photoionization signal these states show up as more or less sharp, resolved structures. Though the dynamics of autoionization is interesting in its own right, here it obscures the observation of the true ionization threshold (i.e., the true states of the corresponding cation). Clearly, in the ZEKE measurement the corresponding structure of the cation is fully resolved. There is essentially no interference from autoionizing states. However, there are certain interactions between Rydberg states (also termed final-state interactions), which have attracted considerable interest [7, 19, 25, 26]. Some of these aspects and the apparent intensity distributions and their influence on ZEKE intensities will be discussed in a later section in this chapter. The spectrum of Fig. 3 shows that the ZEKE method filters out the relevant ionic states. It is already apparent that this technique offers the possibility of state-selected production of ions, that is, with selection of the *rovibronic* ionic state.

A manifold of information on vibrational states of molecular ions has been provided by the use of conventional PE techniques. For many applications in chemistry vibrational resolution is sufficient. However, vibrations of low frequency, and many combination bands cannot be resolved by conventional PES. Even with resonance enhanced multiphoton ionization–photoelectron spectroscopy (REMPI–PES, also termed excited state PES) the resolution is only partly enhanced due to the selection of the intermediate state (see [4] for a discussion of this problem and additional references). Figure 4 shows such a state-of-the-art time-of-flight (TOF) photoelectron spectrum of *para*-difluorobenzene obtained with the $S_1 6^1$ vibronic state as intermediate resonance [59]. The experimental PE resolution in this experiment is better than 10 meV (80 cm^{-1}). Vibrations in the cation are resolved in the TOF–PE spectrum and have been assigned. Comparison of the TOF–PE spectrum with the ZEKE spectrum (Fig. 4, bottom) [60, 61] makes the apparent gain in resolution obvious. The vibrational structure seen in the PES is congested by many more vibrational states that are fully resolved in the ZEKE spectrum. Some of the peaks observed in the ZEKE spectrum are

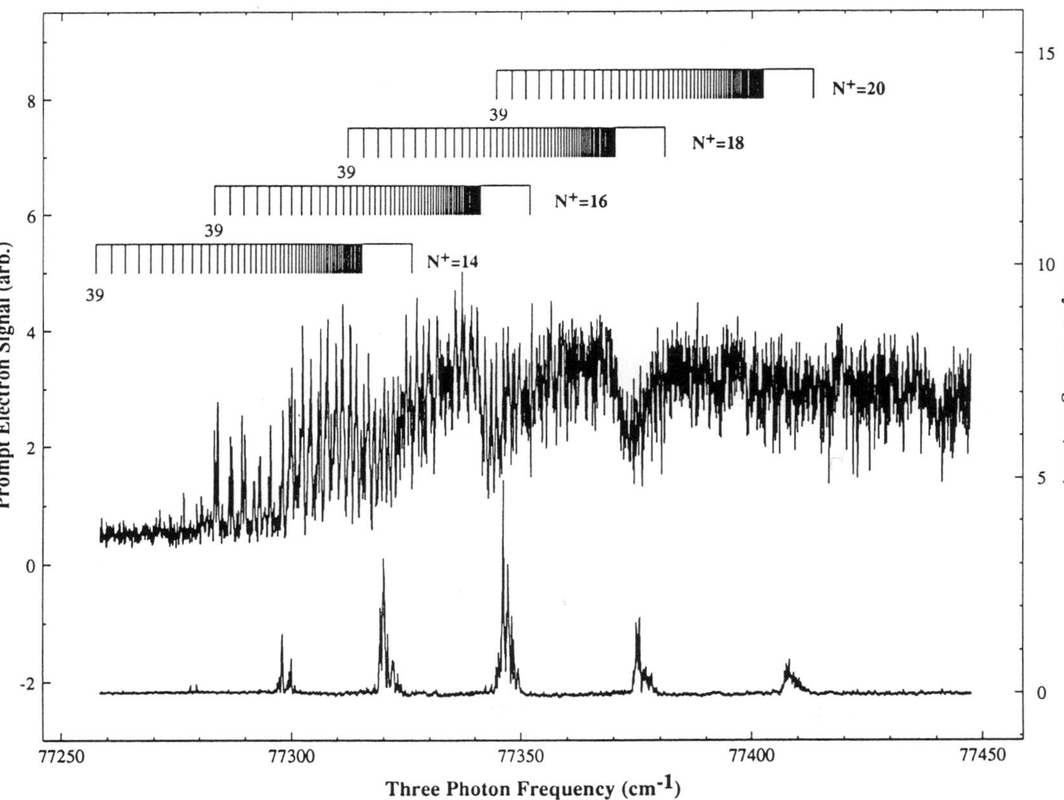

Figure 3. Comparison of (top) photoionization efficiency and rotationally resolved ZEKE spectrum of NO_2.

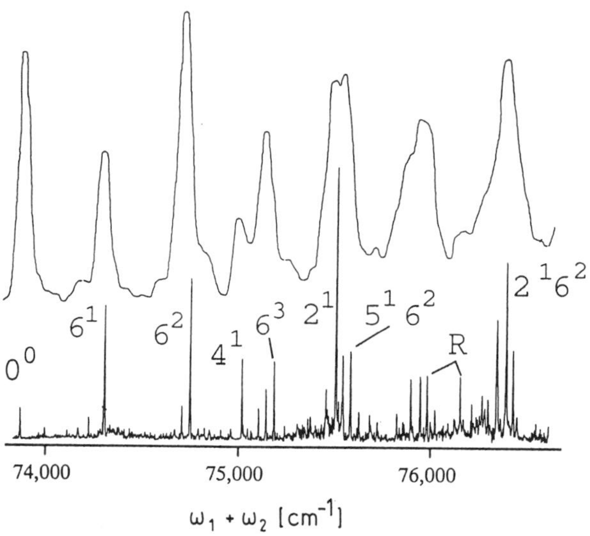

Figure 4. Bottom: ZEKE photoelectron spectrum of *p*-difluorobenzene through the $S_1\,6^1$ intermediate state [61]. Top: TOF photoelectron spectrum [59]. (Reproduced from [59] and [61] with permission.)

combination bands while others are fundamentals. Most of these spectra can now be assigned and understood using *ab initio* methods, which allows comparison of calculated with measured frequencies. Low-frequency fundamentals have also been observed for the first time. Another interesting feature of such vibrationally resolved spectra is that vibrational symmetry selection rules derived from the Franck–Condon principle do not always completely hold. Furthermore, symmetry-forbidden transitions have been unambiguously identified in the ZEKE spectra of several molecules [62, 63].

Another class of chemical compounds addressed in this chapter are molecular clusters. Molecular clusters represent model systems of microsolvation. Interest here has moved very strongly towards a detailed understanding of structural and dynamical properties of such model systems. Molecular clusters are bound by weak **inter**molecular forces and spectroscopic results can be expected to provide a better understanding of the intermolecular potentials involved and of processes such as charge and proton transfer, which are very important in biological systems. The ZEKE technique can yield new information on such systems that actually provides the structure and vibrational frequencies when these clusters are photoionized [18]. An example is given in Fig. 5, which shows the ZEKE

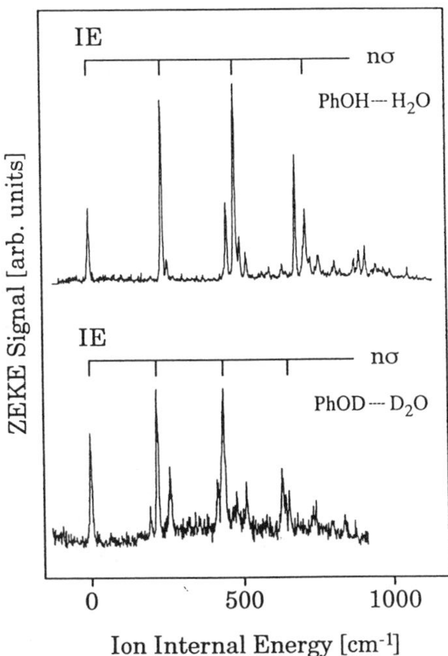

Figure 5. ZEKE spectrum of the h_3 (top) and d_3 (bottom) 1:1 phenol–water complex via the intermediate S_1 0^0 level. The spectra have been corrected for the dye laser output intensity variation of the ionizing laser. (Reproduced from [64] with permission.)

spectra of the protonated Phe–OH$\cdots$$H_2O$ (h_3, top), where Phe is phenol and the deuterated Phe–OD$\cdots$$D_2O$ (d_3, bottom) phenol–water complexes [64]. In these spectra one can see fully resolved vibrational peaks, most of which must be interpreted as intermolecular vibrations. A thorough analysis of the effects of deuteration of the acidic hydrogen on the observed frequencies, the measurement of ZEKE spectra with selection of different intermediate S_1 vibrations (see Fig. 46), and the comparison with *ab initio* calculations [65] lead to the conclusion that six fundamental frequencies, one strong intramolecular (ν_{18b}) and five *inter*molecular vibrations (out of six), can be identified. The problem of understanding the observed fundamental vibrations relies on the ability to assign molecular movements to these fundamental frequencies. This assignment has been made successfully with the help of *ab initio* computations [65], leading to what is now believed to be a rather definitive assignment [64] (including one antisymmetric, i.e., Franck–Condon forbidden vibration).

In addition, this chapter will trace the development of the ZEKE technique and also describe procedures for obtaining optimized resolution. It will be necessary to introduce the concept of pulsed-field ionization of high *n* Rydberg states [3] and discuss the *ab initio* calculations of matrix elements involved in such rotationally resolved photoionization processes (see Section III on theory). We will also discuss what has now been termed ZEKE Rydberg states [45, 46] (or ZEKE states for short) and their properties that can be exploited experimentally.

A. Evolution of Photoionization Experiments

Zero-kinetic-energy PE spectroscopy should be understood in the context of the historical development of photoionization techniques. Photoionization efficiency measurements, VUV–PES, and X-ray photoelectron spectroscopy (XPS) are related to the development of ZEKE spectroscopy. In particular, the threshold PE techniques that originated in parallel to the development of VUV–PES are important.

Experimentally, PES became a general technique for the investigation of photoionization processes in molecules with the invention by Turner et al. [52, 66] of the stable and background-free high-intensity VUV (21.2 eV) He(I) light source. A somewhat similar experiment was independently described by Vilesov et al. [67]. A different development, using X-ray sources, arose from the group of Siegbahn [68], namely, the XPS technique which, reflecting its well-established application, is also termed ESCA (electron spectroscopy for chemical analysis) [69].

Concepts of *ab initio* methods such as an orbital description of the electronic structure of molecules and insight into the nature of the chemical bond have very much profited from the results obtained by PES. In many cases, not only the electronic but also the vibronic states of molecular ions have been observed by PES. However, conventional PES techniques are limited in their energy resolution, since the kinetic energy of the ejected electron cannot be determined with sufficient accuracy to resolve *rovibronic* levels, that is, to obtain the molecular eigenstates. Hence, the orders of magnitude in resolution between laser spectroscopy and conventional electron spectroscopy are essentially due to experimental, but not conceptual limitations. An almost invariant resolution limit of some 10 meV (80 cm^{-1}) seems to be approached by the plethora of PE analyzers that have been employed (see [4] for references). Such resolution is at best sufficient for electronic states and in some cases for the vibronic levels of the molecular ion; rotational state resolution is, generally, beyond the reach of any PES or ESCA technique with the exception of hydrogen [70–73] and ionization from high rotational states in NO [74–77], where partial rotational resolution is achieved.

Indirect threshold measurements were first tried by employing a 127° electron analyzer set to transmit zero-energy electrons. The first direct threshold experiment consisted of the steradiancy analyzer, which exploited the characteristic of threshold energy electrons to faithfully follow an external extraction field. It allowed for the first state selection of molecular ions and detailed information about the kinetics of molecular ions has emerged from such studies [78]. This method is generally referred to as threshold photoelectron spectroscopy (TPES) [78–82]. The resolution, however, was still limited by the presence of fields and the characteristics of the analyzer which leads to a background signal from autoionization, that interferes with the detection of the true ZEKE electrons [83, 84].

B. Principle of ZEKE Detection

Let us recall that the difference between conventional PES and ZEKE–PES is that for PES the system (i.e., atom, molecule, or cluster) is irradiated with light of fixed wavelength and the kinetic energy of the emitted electrons is measured [52, 66, 67] (Fig. 6, right), whereas for ZEKE–PES (as well as for TPES) the wavelength is scanned and only electrons within a small detection window near threshold are measured (Fig. 6, left). The disadvantages of PES are the need for calibration of

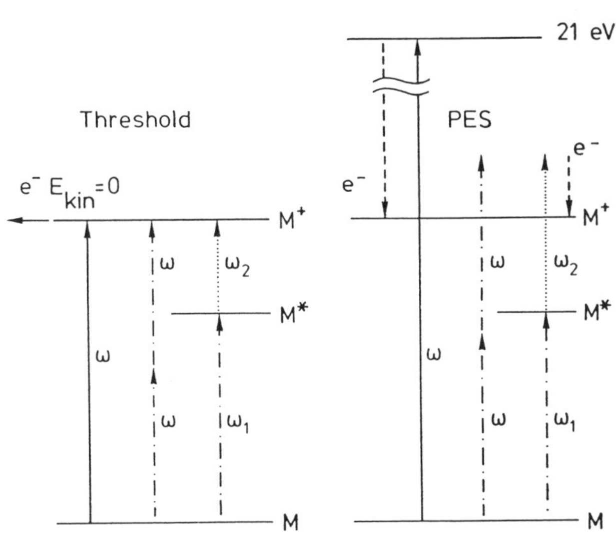

Figure 6. Comparison of ZEKE threshold ionization (left) with PES (right). Excitation, for both cases, is achieved either with one-photon or multiphoton ionization (MPI), nonresonant or resonant MPI.

the PE spectrometer and, more importantly, the limited resolution of about 5–10 meV (i.e., 40–80 cm^{-1}) and the low transmission of the PE analyzer. In contrast, the ZEKE method involves scanning the wavelength of the light source across the ionic states and detection of the electrons that are formed with zero-kinetic energy, whenever the wavelength of the incident light matches the energy of a rovibronic transition between the initial state of the molecule and the ion [4] (Fig. 6, left). The ionization itself takes place under zero-field conditions and the ZEKE electrons (with 100% transmission) are discriminated against kinetic electrons by delayed pulsed-field extraction. Besides the advantage of having the full accuracy of optical spectroscopy, the ZEKE technique offers a spectral resolution of up to 0.2 cm^{-1}, which is sufficient for the rotational resolution of molecules such as NO [2, 7], ammonia (NH_3) [5, 6] and benzene (C_6H_6) [8, 9, 85].

The principal advance in this method came in 1984 when the completely resolved rotational structure, at threshold, of the NO^+ ion was observed in the first true ZEKE experiment [1, 2] (see Fig. 1). This advance came from the delayed pulsed-field extraction principle: ionization under field-free conditions is followed by a delayed extraction pulse, thus only collecting those electrons specific to the ionic eigenstate—**the ZEKE electrons**—and rejecting all other electrons from unspecific channels, the **near-ZEKE** electrons. The signature of a ZEKE electron is its zero velocity. The basic idea is that when the energy of the exciting photon exactly matches the ionization energy of the molecule (i.e., the energy necessary to eject an electron from a molecular orbital) then the electron emitted will have virtually zero kinetic energy. Thus, after some delay the kinetic electrons will have moved away from the ZEKE electrons. This technique is schematically shown in Fig. 7. If the electrons are extracted by a pulsed electric field into a long tube, then the ZEKE electrons will arrive at the other end at a time different from that of the kinetic energy electrons. Such a typical TOF distribution is shown in Fig. 7. Hence, the pulsed-field extraction leads to a strong "time amplification": An initially small energy difference results in a large TOF difference. Therefore, if one sets a detection gate at the right time (marked 1 in Fig. 7), it is possible to see the ZEKE electrons separated from the other electrons. Thus, instead of trying to determine the energy of the photoelectrons, one is looking for signal/no signal at the correct TOF. As the laser energy is scanned, different (ro)vibrational levels in the ion are accessed, and ZEKE electrons will appear each time the energy of the laser matches a transition from a neutral energy level to an ionic level. The ZEKE signal versus wavelength constitutes the ZEKE

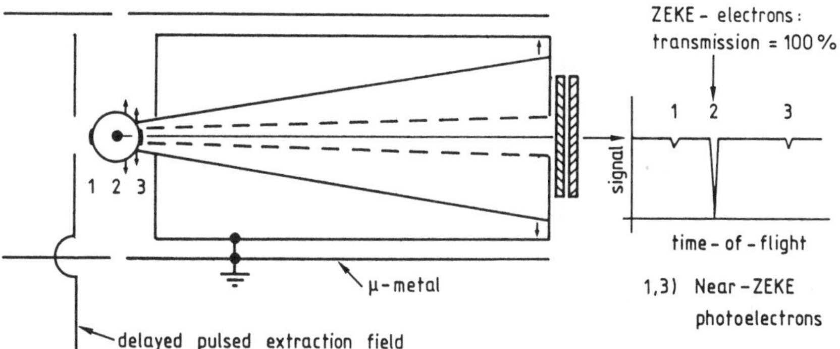

Figure 7. ZEKE photoelectron detection scheme. Left: When ionization occurs ($t = 0$) the electrons with nonzero kinetic energy start to fly away from the ZEKE electrons. Right: After the delay time ($t = t_d$) kinetic electrons flying away (1) or towards (3) the detector are spatially separated from the ZEKE electrons (2); the spatial separation is then transformed into a TOF separation as observed in the electron signal versus TOF curve.

spectrum. Essentially, the resolution of the technique is limited *only by the energy spread of the laser radiation*, which is usually very well known ($<1\,\text{cm}^{-1}$).

C. The Delayed Pulsed-Field Ionization of ZEKE Rydberg States

The experimental observation of long-lived (i.e., up to 100 μs) Rydberg states converging to (ro)vibronic levels of the ion has led to a second variation of the ZEKE technique: the detection of pulsed-field ionized Rydberg electrons [3]. The behavior of the lifetime of molecular Rydberg states is very different from that of atoms (see Fig. 8). Molecular Rydberg states can decay via predissociation or other intramolecular processes that are not open to atoms. Some of these processes are shown schematically in Fig. 9: autoionization and predissociation. Hence, for molecular systems one finds a very short lifetime for those Rydberg states below about $n = 150$, corresponding to about $5\,\text{cm}^{-1}$ below threshold. On the contrary, within the *magic region* of about $5\,\text{cm}^{-1}$ below threshold (see Fig. 8), the lifetime of these highly excited Rydberg states ($n > 150$) can be exceedingly large. These high n, long-lived Rydberg states are called ZEKE Rydberg states. Lifetimes up to tens of microseconds have been observed and it appears that the usual $\tau \propto n^3$ dependence is not valid in this region. Instead, the lifetime of these Rydberg states even seems to be strongly enhanced compared to the $\tau \propto n^3$ law, possibly due to ℓ and m_ℓ

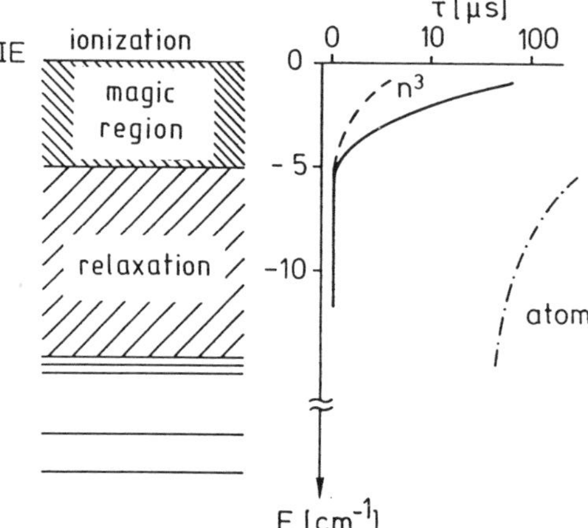

Figure 8. Left-hand side: ionization energy (IE): Adiabatic (field-free) ionization energy; top: Ionization region above threshold; magic region: long-lived high n Rydberg states; relaxation region: short-lived (predissociating) Rydberg states. Right-hand side: approximate lifetime curves for Rydberg states of (1) atoms and (2) molecules following a $\tau \propto n^3$ decay law (dashed line) and with lifetime enhancement.

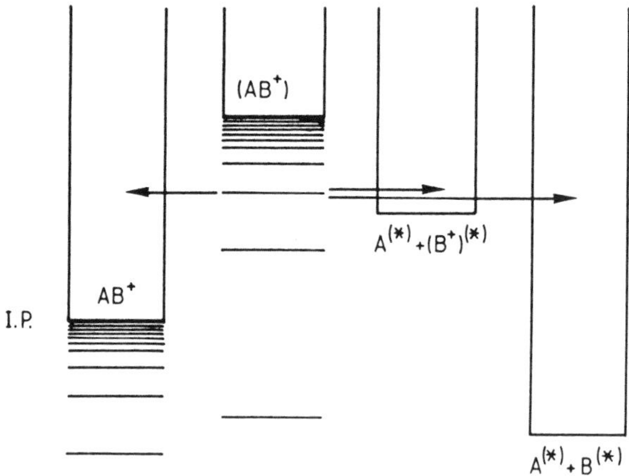

Figure 9. Rydberg couplings: autoionization and predissociation.

mixing by small fields present during the excitation. The longevity of the high-lying Rydberg states is presently understood as being due to the effects of small electric fields in the apparatus [7, 36] and/or effects of higher multipole moments of the core [38] and the presence of ions [46]. The effects of such electric fields have also been investigated [34, 35], for example, in the ZEKE spectrum of NO [34] (exciting via an intermediate $A\ ^2\Sigma^+$ resonant state), and Ar [45] (via a one-photon process).

The ZEKE Rydberg states with their highly excited electrons can be ionized by a very small amount of energy. This energy can be supplied by the pulsed electric extraction field (mentioned above), that is, the electric field pulls the electron away from the ionic core and the molecule is ionized (see Fig. 10). Since the Rydberg states will not move out of the ionization volume, the electrons formed from their ionization will appear at the same TOF as the ZEKE photoelectrons. This result poses no real problem, however, as the spectra obtained are very similar to those that would have been obtained from the ZEKE electrons alone; here both types of electrons will be called ZEKE electrons. (It should be noted, however, that some intensity effects can be observed from the Rydberg–electron spectra and recently, these effects have been reviewed [26].) This technique is now known as ZEKE–PFI spectroscopy (PFI = pulsed-field

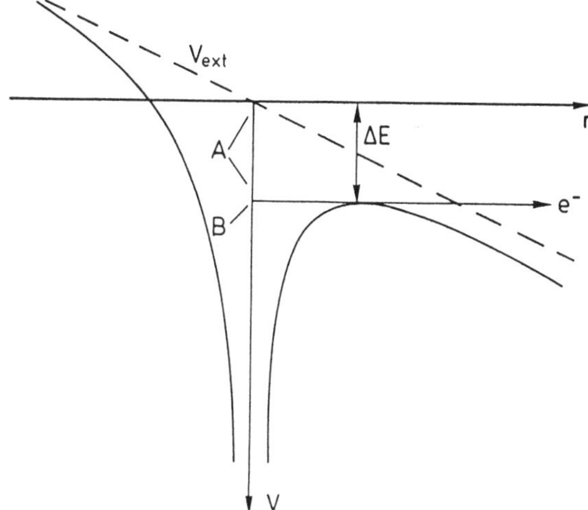

Figure 10. Pulsed-field ionization scheme. In the presence of an electric field F the Coulomb potential V_0 is modified by an additional linear potential V_{ext}, resulting in the final potential V (solid line). The field-free IE is lowered by the Stark shift ΔE, given by the formula: $\Delta E = c \cdot \sqrt{F}$. Hence, by applying a *delayed* extraction field F at time t_d the high-lying, long-lived Rydberg states will be field-ionized.

ionization) and is easier to implement experimentally because stray electric fields (always present in an apparatus) can influence the ZEKE electrons quite easily, making them more difficult to detect; the ZEKE Rydberg states will not be influenced by these stray fields, so these states are more efficiently detected. This method has now been found to be generally applicable to molecular systems. One major advantage of the ZEKE–PFI technique is the insensitivity to stray fields inside the apparatus, which in the case of free ZEKE electrons can cause the electrons to move during the delay time. This result also indicates how PFI electrons and free ZEKE electrons can be detected independently: A small dc electric field across the ionization volume will cause the ZEKE electrons to move slightly, so they will appear at a TOF different from the PFI electrons.

Both types of photoelectrons, either those from pulsed-field ionization of long-lived Rydberg states or the free electrons emitted at or just above threshold, carry the same information about the ionic state since the cross-sections must be continuous across threshold. The spectra obtained are hence very similar, although there is a small energy shift between them that can be empirically calculated and corrected for in order to obtain the true ionization energies with spectroscopic precision. It should be noted, however, that the PFI technique is not applicable to photodetachment studies, because no Rydberg states exist for anions. In this case the original ZEKE method involving the detection of free electrons with negligible kinetic energy must be applied.

D. Details of the ZEKE Experiment: Towards Highest Resolution

The idea and principles of ZEKE spectroscopy have already been discussed in a number of papers [1, 3, 5, 6], review articles [4, 26], and in Nature [20]. Here, we only give a description that focuses on the ZEKE–PFI technique [3] in combination with a slowly rising electric field extraction pulse (i.e., with a linear or multistep slope). The application of this specially shaped pulse has proven to be the key for obtaining the very high resolution (up to $0.2\,\mathrm{cm}^{-1}$), which is essential for rotational resolution in larger molecules, such as benzene [8, 9, 85]. By variation of the slope of the pulse, the spectral resolution of the ZEKE–PFI technique can be adjusted according to the laser bandwidth and to the needs of the system under study, (i.e., whether vibrational or rotational resolution must be achieved). In fact, using a linear or multistep staircase-like extraction pulse provides a technique for the exact determination of the ionization energy under field-free conditions [86, 87].

A typical experimental setup for a ZEKE experiment is shown in Fig. 11. It consists of a laser system and a vacuum apparatus including the

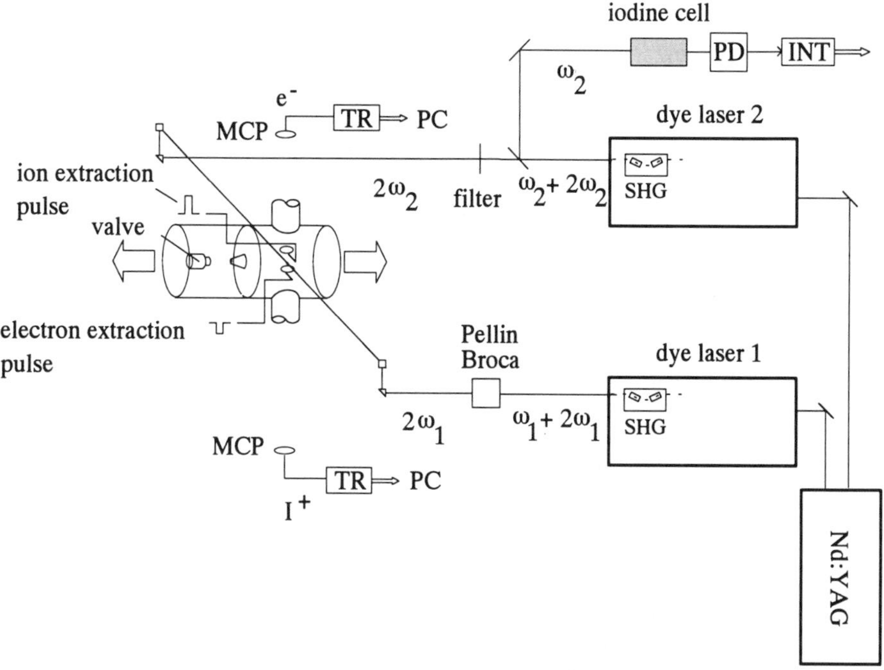

Figure 11. A typical ZEKE apparatus for the two-color ionization scheme employing two synchronously pumped dye lasers.

molecular beam source, the extraction plates, and a μ-metal shielded flight tube with electron detectors (i.e., multichannel plates or a channeltron) at its end. The laser system depends on whether one-color, one-photon VUV [13, 26], one-color, two-photon nonresonant [21, 23, 28], or two-color, two-photon resonant [1, 4] measurements are performed, the latter being the ionization scheme most extensively used. In a typical two-color experiment both lasers are pumped simultaneously by an excimer laser or a Nd:YAG laser. The first dye laser excites a specific vibrational or rotational level of the intermediate resonance and the second dye laser ionizes the molecules or promotes them into long-lived Rydberg states ($n > 150$) converging to (ro)vibronic levels of the electronic ground state or an electronically excited state of the cation (see below). After a delay time of several microseconds, an extraction pulse is applied by either a simple electronic pulsing device or by using an arbitrary function generator (i.e., LeCroy AFG 9100). The electrons are detected with multichannel plates and their TOF signal is recorded with boxcar integrators or a transient digitizer by setting narrow time gates (10–30 ns).

The field ionization of Rydberg states allows the adjustment of the spectral resolution by a variation of slope and magnitude of the delayed extraction pulse. This adjustment can be understood in the following way: Consider a molecule with long-lived Rydberg states in the range of 3–4 cm^{-1} below a particular ionic eigenstate. With the application of a slowly rising pulse, only a small segment of Rydberg states is field ionized at a certain time within the rising slope of the extraction pulse. The TOF of the electrons so produced will hence depend on the energy separation of this segment of Rydberg states from the corresponding ionization limit. This energy separation is defined by the field strength reached at the point in time when the Rydberg state segment is field ionized. Electrons produced at a different point in time within the rising slope of the extraction pulse (i.e., from another Rydberg state segment) will have a different TOF [Fig. 12(a)]. Therefore, only a small region of Rydberg states will be monitored within a certain time gate, thus enhancing the spectral resolution significantly. In contrast, the application of a fast-rising extraction pulse of about 1 V cm^{-1} will cause the field ionization of all Rydberg states in the above-mentioned region and results in the

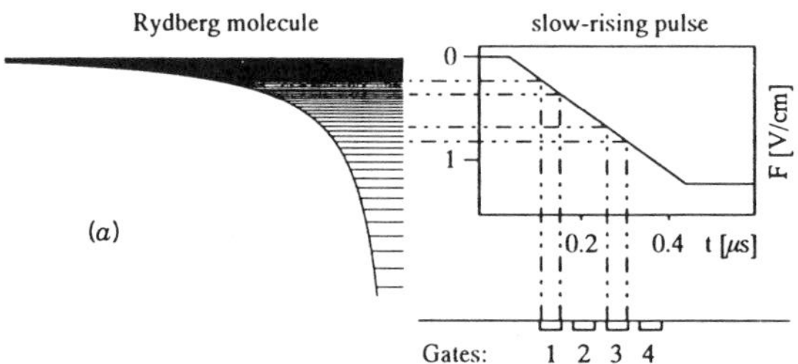

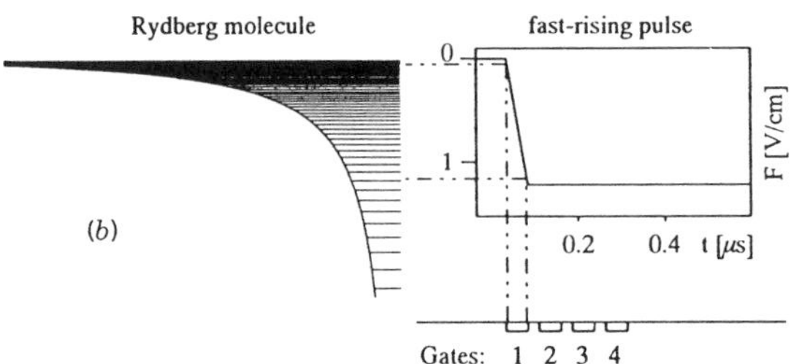

Figure 12. Pulsed-field ionization of long-lived Rydberg states using a slow-rising (*a*) and a fast-rising (*b*) extraction pulse and collection within a certain time gate.

collection of the resulting electrons within a small time gate [Fig. 12(b)]. The scanning of the laser wavelength over this region will therefore produce a ZEKE signal over the entire range, thus limiting the spectral resolution to about 3–4 cm^{-1}. This value is high enough for vibrational resolution, but does not allow for the rotational resolution in most molecules.

II. ROTATIONALLY RESOLVED ZEKE SPECTRA

A. Nitric Oxide

Nitric oxide, (NO), is the molecule most extensively studied in photoionization. It serves as a model for the theoretical description of the photoionization process. The ZEKE spectra from the $C\ ^2\Pi$ state [2], for example, were successfully interpreted by Fredin et al. [10] using multichannel quantum defect theory (MQDT). Ionization from the $A\ ^2\Sigma^+$ state was also investigated. The 6σ orbital is of nearly pure atomic *s* character (94% *s*, 0.5% *p*, and 5% *d* character) [12]. Thus, in the *sudden approximation* the electron is ejected into the odd ℓ continuum channel. Considering parity conservation, $\Delta N^+ = |N^+ - N''|$ should be restricted to even values. Since the *A* state is a Rydberg state with only weak interaction between the core and the electron one would expect no change in core rotation, that is, $\Delta N^+ = 0$. In the corresponding ZEKE spectra [7, 11] (reproduced in Fig. 13), however, strong $\Delta N^+ = \pm 2$ and less intense $\Delta N^+ = \pm 1$ transitions are observed, with the $\Delta N^+ = 0$ transition seen as the most intense peak. The peaks with $\Delta N^+ = \pm 1$ and ± 2 are forbidden in an atomiclike or *sudden approximation* and reflect the molecular character of the ionization process. Rudolph et al. [12, 88] showed in their *ab initio* calculations that scattering of the electron by the non-centrosymmetric core potential into other angular momentum (ℓ) continuum channels (of opposite parity, for instance) causes the appearance of these ΔN^+ peaks which are forbidden in an atomic picture. The molecular nature of the ionization process is clearly demonstrated. The calculation of intensities observed in such ZEKE spectra is fully described in Section III.

B. Benzene

To date, only a few symmetric rotor molecules have been studied with rotational resolution using ZEKE photoelectron spectroscopy [5, 6, 8, 9, 89]. The first example was the ammonia molecule (NH_3) [5, 6]. In this section the discussion is extended to the benzene (C_6H_6) molecule. Both NH_3 and C_6H_6 have been studied using single rovibronic

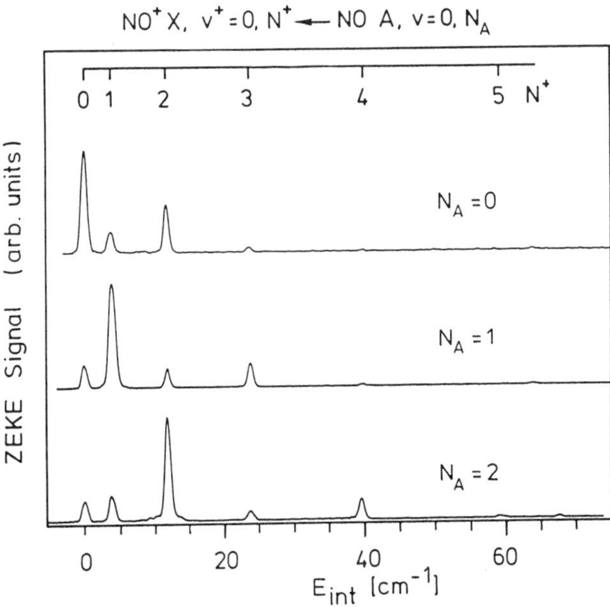

Figure 13. ZEKE spectra of NO via the $A\,^2\Sigma^+\ v^+ = 0$ state with intermediate rotational states selected from $N_A = 0$ to $N_A = 2$. (Reproduced from [7] with permission.)

states as intermediate resonances and full rotational resolution in the cation has been obtained, providing very distinct information about the rotational selective ionization dynamics of these systems. Since the molecular symmetry group is D_{3h} in the case of NH_3 and D_{6h} for C_6H_6, it will be interesting to see how the even- and odd-fold symmetry of the principal axes affects the rovibronic transitions that occur in the spectra and changes in the projection quantum number K of the total angular momentum J on going from the intermediate state to the cation.

1. *The Electronic Ground State of the Benzene Cation*

The neutral C_6H_6 molecule has a hexagonal, planar structure with D_{6h} symmetry [90–93]. In the electronic ground state the outer electron configuration is $(a_{2u})^2\ (e_{1g})^4$. Upon ionization, one e_{1g} electron is removed from the highest molecular orbital, thus leaving an unpaired e_{1g} electron. This results in a doubly degenerate $^2E_{1g}$ electronic ground state of the cation.

The perturbation of the π electron system due to the removal of an electron to form the benzene cation is expected to cause a change in the structure and symmetry of the neutral molecule. Reduction to three-fold symmetry is conceivable, which would correspond to a "localization" of the double bonds, either to D_{3h}, in which the planar structure is maintained, or to a chairlike structure similar to the chair conformation of cyclohexane. However, an in-plane symmetry reduction to D_{2h} is also feasible, as can be concluded from theoretical considerations [56]. The analysis of Rydberg series converging to the electronic ground state of the cation, and also of PE spectra, point more towards the maintenance of a planar structure in the cation [94]. Nevertheless, the ground-state geometry cannot be deduced exactly because of the poor resolution of conventional PE spectroscopy.

The removal of an electron from a degenerate bonding e_{1g} orbital results in the formation of a degenerate state of $^2E_{1g}$ symmetry species. The Jahn–Teller coupling that accompanies the electronic degeneracy causes a distortion along a nontotally symmetric normal coordinate that lifts the degeneracy [53, 54]. Model calculations therefore predict a D_{2h} symmetry for the cation for which there are three equivalent structures [56]. This distortion to D_{2h} can be detected experimentally in a matrix at low temperatures [57]. The ESR spectra of $C_6H_6^+$ in a freon matrix indicate an increased spin density at C1 and C4 at 4.2 K but an equivalent spin density at 100 K. Thus the question arises whether the D_{2h} structure is static and destabilized at higher temperatures by collisions with the matrix or whether a dynamic D_{2h} structure is stabilized by the matrix lattice. For weak Jahn–Teller coupling the stabilization energy for the distorted symmetry is smaller than the zero-point energy of the Jahn–Teller active mode. Isolated (i.e., under collision-free conditions), the three equivalent D_{2h} structures of the cation would rapidly dynamically interconvert, and the ground state of the cation would still be described in the D_{6h} symmetry group [95]. For strong Jahn–Teller coupling the cation would spend much time in either one of the three structures, and would therefore be described in D_{2h}. The knowledge of the structure and the symmetry of the isolated benzene cation is desirable not only for testing quantum mechanical models but is also of fundamental importance for organic chemistry. Rotationally resolved PE spectroscopy allows an unambiguous determination of the symmetry of the cation. The molecular symmetry group can be deduced from the rotational transitions that occur, together with group theoretical considerations [53, 96].

A first attempt to resolve the question was carried out by gas-phase PES [95]. These authors obtained the PE spectrum shown in Fig. 14 by using the standard TOF technique and two-photon, two-color $(1 + 1')$ ionization with the $S_1 6^1$ state as intermediate resonance. The vibrational structure found in that PE spectrum was assigned to the 0° vibrational origin and the 6^1 progression; the $6^1(\frac{3}{2})$ and $6^1(\frac{1}{2})$ states show the linear

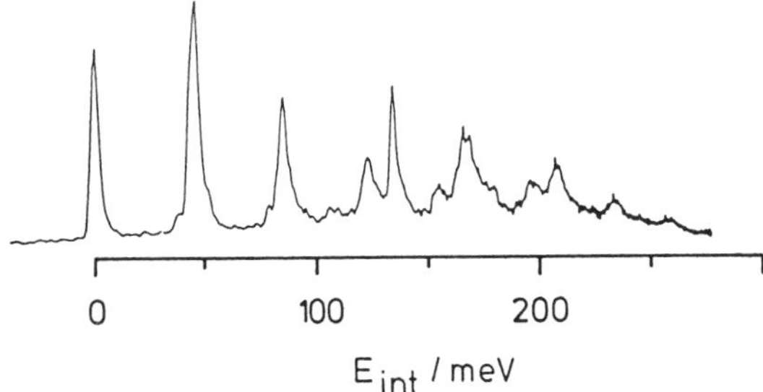

Figure 14. Time-of-flight PE spectrum of benzene with the $S_1\,6^1$ state as intermediate resonance (Redrawn from [76].)

Jahn–Teller splitting. No resolution of the quadratic Jahn–Teller splitting of the $6^1(\frac{3}{2})$ state into the two B_{1g} and B_{2g} components nor any resolved rotational structure was seen in that experiment.

In the ZEKE experiments the C_6H_6 molecules are ionized in a two-photon, two-color $(1+1')$ process via the first vibronically excited state S_16^1 as intermediate resonance (here the symbol 6^1 represents a vibrational state in which the normal vibrational mode ν_6 is excited with one quantum). The first laser selects a rovibronic state in S_16^1 and the second laser ionizes the molecule from this state. Depending on the rovibronic symmetry of the intermediate state, different rovibronic states of the ion are populated. The conservation of the symmetry of the nuclear spin wave functions on transition and the symmetry selection rules and rules for coupling of angular momenta severely restrict the number of possible transitions. The symmetry of the cation can be deduced from the assignment of the transitions and the rotational progressions that occur.

Figure 15 shows the ZEKE spectrum for the transitions into the ion ground state from the intermediate state $S_16^1(J'=1, K'=1, +1)$ (see Fig. 2 for the corresponding ZEKE spectrum from $S_16^1(J'=2, K'=2, +1)$ [8]). The assignment of rotational states in S_16^1 is discussed in Ref. [92]. The most notable feature in the observed rotational structure is the appearance of progressions in the total angular momentum quantum number N^+ (without electron spin) of the ionic core. The selection rules $\Delta J=0, \pm 1$ and $\Delta K=0, \pm 1$ are valid for transitions between bound states. Here, however, for a transition from a bound to an unbound state the ZEKE electron can compensate for the angular momentum by vector coupling depending on its angular momentum $(s, p, d, \ldots)$, thus enabling higher rotational angular momentum changes [6, 12].

The following model was developed to explain the observed transitions

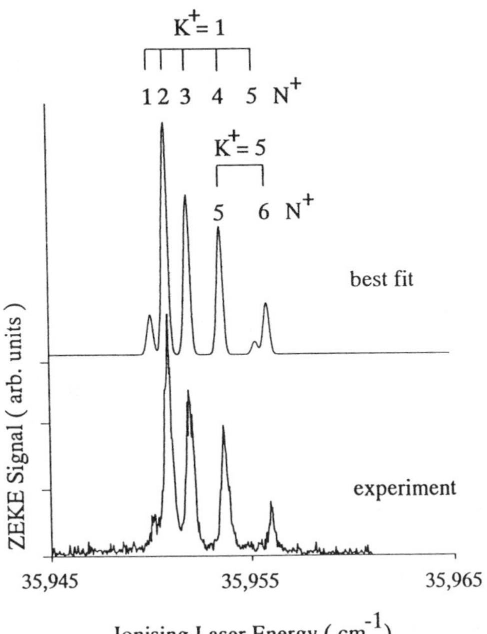

Figure 15. Measured and fitted Zeke spectrum of benzene for the ionizing transition $D_0\,{}^2E_{1g} \leftarrow S_1\,6^1E_{1u}$ ($J' = 1$, $K' = 1 + l$). (Redrawn from [8].)

[6]: By starting with the vibronic symmetry of the intermediate state, the vibronic symmetry species of the final state(s) that can be reached in an electronically allowed one-photon transition can be determined. The vibronic symmetry species of this state is decoupled into the vibronic symmetry species of the ion core and the symmetry species of the ejected electron, according to the molecular symmetry group [96]. From the correlation of the symmetry group of the cation with the $D_{\infty h}$ group the partial wave character of the photoelectron is obtained [53] and hence the possible rotational states of the ion core [6]. Those transitions for which the partial wave functions of the ZEKE electron correlate with symmetries that are permitted in an allowed electronic transition are expected to be the strongest. For the D_{6h} symmetry of the benzene cation, this will be the symmetry species e_{2g} with partial wave functions $|\ell = 2, \lambda = 2\rangle$, $|\ell = 4, \lambda = 2\rangle$, and $|\ell = 4, \lambda = 4\rangle$ where λ is the projection of the orbital angular momentum ℓ of the ZEKE electron on the sixfold principal axis. Analogously, one obtains for D_{3h} the species e' with the additional partial wave function $|\ell = 1, \lambda = 1\rangle$ and for D_{2h} the species $a_g + b_{1g}$ with the additional partial wavefunction $|\ell = 0, \lambda = 0\rangle$. In all three cases only the

lowest electron angular momentum partial waves from s to g are considered. A determination of the symmetry of the cation is thus possible through the assignment of the rotational transitions in the ZEKE spectrum. On the basis of the above-mentioned model, all spectral lines are consistent with the assumption of D_{6h} symmetry. None of the intense lines expected from the additional partial wave functions for D_{3h} or D_{2h} are observed in the spectra, hence D_{3h} or D_{2h} symmetry can be excluded. Hence, the benzene cation in its electronic ground state is completely compatible with D_{6h} symmetry [8]. Therefore, the conclusions are that the stabilization by Jahn–Teller distortion is smaller than the energy of the zero-point vibration, and the reduction in symmetry induced by the distortion has no influence on the observed rotational transitions.

With 10 rotational states in the P and R branches of the intermediate resonance $S_1 6^1$ as initial state, 44 lines could be assigned in the ZEKE spectra [8] and the rotational constants of the cation have been determined. In Table I these constants, as well as the values for the ground state S_0 and the intermediate resonance state $S_1 6^1$ of the neutral molecule, are listed [97]. A comparison of the ground-state constants of the neutral and the ionized molecule reveals that the benzene ring is only marginally influenced by the removal of an electron. In particular, the values of the rotational constant C, corresponding to the moment of inertia about the axis perpendicular to the ring, are almost identical. A slight contraction or expansion of the ring on ionization is compatible with the error limits. The deviation from the planarity condition $C = B/2$ lies within the error limits for the rotational constants of the cation. This result is consistent with the deduced D_{6h} symmetry, which requires a planar structure. The rotational constants were therefore also fitted under the assumption of planarity, and the constants obtained in this way were thus given preference. These values are given in Table II and contrasted with the results of an *ab initio* calculation [56]. The pairs of values of C shown in Table II suggest a stretching of the ring bonds, in agreement with the intuitive expectation of what happens when an electron is removed from a bonding orbital. The experimentally obtained stretch of

TABLE I
Spectroscopic Constants of the Ground State of Neutral Benzene [97], the S_1 State [92, 93] and the D_0 Ground State of the Benzene Cation [8][a]

	S_0	$S_1\,6^1$	D_0
B (cm^{-1})	0.18977389	0.1817904	0.18912 ± 0.00056
C (cm^{-1})	0.0948637	0.0908482	0.09494 ± 0.00022
ζ_{eff}	0	−0.57873	0.4534 ± 0.0065

[a] For the latter constants the constraint $B = C/2$ was not used.

TABLE II
Spectroscopic Constants of the Ground State $D_0\ {}^2E_{1g}$ of the Cation Using the Relation $C = B/2$ for the Fit [8][a]

	D_0	Calculation
B (cm^{-1})	0.18956 ± 0.00034	0.18844
C (cm^{-1})	0.09478 ± 0.00017	0.09422
ζ_{eff}	0.4543 ± 0.0066	

[a]For comparison, the values from an *ab initio* calculation [56] are also reported [5].

approximately 0.1% is, however, about a factor of 5 smaller than that predicted by theory. The strength of the bond in the benzene molecule is apparently affected less by the removal of one of the four e_{1g} electrons than previously assumed.

2. The 6^1 (e_{2g}) Excited Jahn–Teller State

It is interesting to see the effect of excitation of a Jahn–Teller active vibrational mode on the structure of the ion and how vibronic interactions affect the vibrational structure of the ion's ground state. According to theory, excitation of a degenerate vibration in a degenerate electronic state should lift the degeneracy, and splitting into vibronic states of different energies is expected [98, 99]. Because of vibronic coupling, the electronic wave function depends on the vibrational coordinate. The Jahn–Teller effect is thus one of the most significant examples of the breakdown on the Born–Oppenheimer (BO) approximation, which has as its basic tenet the separability of the motion of electrons and nuclei.

In the case of linear vibronic coupling, a good quantum number for the characterization of the vibronic wave function for a doubly degenerate electronic state is simply the sum $j = \rho \pm \lambda$ of the vibrational angular momentum $\rho = \pm\nu$ and one half of the electronic angular momentum $\lambda = \frac{1}{2}\Lambda$ $(\Lambda = \pm 1)$, and not the individual angular momenta alone [98, 100]. For the benzene cation this implies that excitation of one quantum of the degenerate ν_6 (e_{2g}) vibrational mode (6^1) results in a splitting into two states with different vibronic angular momenta $j = \pm\frac{1}{2}$ and $j = \pm\frac{3}{2}$. The level with angular momentum $j = \pm\frac{3}{2}$ should split further into two states due to quadratic JT coupling. This differing behavior of the two components with $j = \pm\frac{1}{2}$ and $j = \pm\frac{3}{2}$ can also be explained by group theory. The direct product of the irreducible representations of the electronic state E_{1g} and the ν_6 vibrational mode e_{2g} generates one doubly degenerate state with the symmetry species E_{1g} and two states with the symmetry species $B(\Gamma_e \otimes \Gamma_v = E_{1g} \otimes e_{2g} = E_{1g} \oplus B_{1g} \oplus B_{2g})$. Higher order couplings can only split the state $\Gamma_{ev} =$

$B_{1g} \oplus B_{2g}(j = \pm\frac{3}{2})$; the state $\Gamma_{ev} = E_{1g}(j = \pm\frac{1}{2})$ remains degenerate. An unambiguous assignment of the individual, split vibronic states and an exact determination of their energies is necessary for a precise determination of the strengths of the vibronic coupling. Rotationally resolved ZEKE spectroscopy is the method for the unambiguous assignment of the symmetry species of the vibronic states and thus the vibrational modes of the benzene cation, thereby providing reliable values for the vibronic coupling strengths.

The rotational state of an oblate symmetric rotor is characterized by two quantum numbers. For the ground state (S_0) and the intermediate state (S_1) of the neutral benzene molecule, these are the quantum numbers J of the total angular momentum and the quantum number K for its projection along the molecule-fixed principal axes. For the benzene cation with the electronic spin $S = \frac{1}{2}$, the rotational states are characterized (with the assumption of only a weak spin–orbit interaction) by the quantum number N^+, which describes the total angular momentum without electronic spin, and the projection quantum number K^+. In degenerate vibronic states the rotational states split by Coriolis interactions must be differentiated by an additional quantum number [53] ± 1.

The determination of the rotational structure of the benzene cation and the assignment of the rotational transitions observed in the ZEKE spectra allow the determination of the vibronic symmetry species of the corresponding vibrational state. This determination is done in a similar way for ammonia [5, 6] and for the vibrationless ground state of the benzene cation (see above). If we start with the known symmetry of the intermediate state, in which only one rovibronic state is populated, the possible final states are determined based on group theoretical considerations. The conservation of symmetry of the nuclear spin wave function and the rules for electronic symmetry selection and the coupling of angular momenta greatly restrict the possible transitions; depending on the vibronic symmetry of the final state, only a few rotational progressions can be observed [6, 9].

For the benzene cation the $\nu_6(e_{2g})$ mode is the Jahn–Teller active mode with the lowest frequency [56, 101]. Photoelectron spectra of benzene [95] indicate a splitting of the $j = \frac{3}{2}$ and $j = \frac{1}{2}$ components of the 6^1 band by roughly 300 cm^{-1}. Measurements with a variant of the ZEKE method, in which the ions remaining after the pulsed-field ionization are detected (referred to as mass analyzed threshold ionization, MATI [41, 42]) show vibrational structure in the region up to 800 cm^{-1} above the first ionization potential [43]. In this vibrational structure the splitting of the two components with $j = \pm\frac{3}{2}$ and $j = \pm\frac{1}{2}$, and to some extent the separation of the quadratic JT components, can be seen. However, rotational resolution that would ensure the unambiguous assignment of the bands, was not achieved in these measurements [43]. This resolution is especially necessary for the $j = \pm\frac{1}{2}$ band, as it lies close to high-intensity combination modes of other vibrations.

Figure 16 shows the completely rotationally resolved ZEKE spectrum in the region of the $6^1(j = \pm\frac{1}{2})$ band of the benzene cation [9, 85]. The first laser is tuned to the $S_1 6^1 \leftarrow S_0 R_0(0)$ transition, and thus a single rovibronic level $(J' = 1, K' = 1, +l)$ is populated in the intermediate state [92]. The second laser is scanned over the range $\nu_2 = 36{,}598$–$36{,}638\ \text{cm}^{-1}$, and the ZEKE signal is recorded versus ν_2. Figure 16 shows two vibrational bands in this region with almost the same intensity and both with resolved rotational structure. Without rotational resolution it is not possible to decide with certainty which of the two states should be

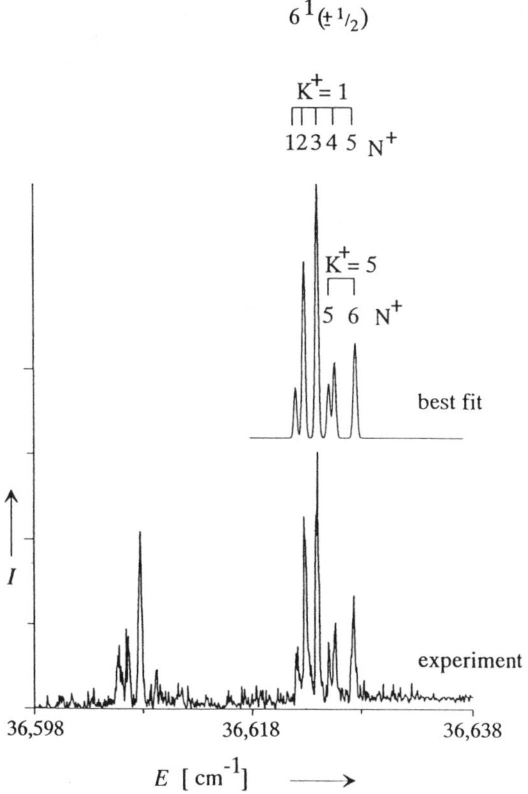

Figure 16. Measured and fitted ZEKE spectrum of the transition $6^1(\pm\frac{1}{2})$ $^2E_{1g} \leftarrow S_1\, 6^1\, {}^1E_{1u}$ $(J' = 1,\ K' = 1,\ +l)$. (Redrawn from [9].)

assigned to the $6^1(j=\pm\frac{1}{2})E_{1g}$ band. Starting from the rotational transitions characteristic of a given vibronic symmetry, two rotational progressions with $K^+=1$ and $K^+=5$ can be assigned to the state higher in energy (Fig. 16, right). The rotational structure indicates that E_{1g} is the vibronic symmetry species of this band, which can thus be unambiguously assigned to the $6^1(j=\pm\frac{1}{2})E_{1g}$ state. This assignment is supported by comparison with the spectrum of the vibrationless ground state [8] (see above), which also has E_{1g} symmetry. The two spectra are very similar with the exception of the Coriolis splittings, which result from Coriolis coupling constants ζ of different magnitude and sign. In contrast, the rotational structure of the lower energy state (Fig. 16, left) is not compatible with an E_{1g} symmetry species. This result probably corresponds to a component of the combination mode $16^1 6^1(j=\pm\frac{3}{2})$.

The second component of the split ν_6 band, the state with vibronic angular momentum $j=\pm\frac{3}{2}$, is split into two states with B_{1g} and B_{2g} symmetry due to the quadratic, dynamic Jahn–Teller effect [98]. Very similar rotational structures are expected for the two states since both vibronic states have the same symmetry species B and their rotational constants should have almost the same values. Figure 17 shows the ZEKE spectrum via the same intermediate state $S_1 6^1$ ($J'=1, K'=1, +l$) [9], in which the second laser is scanned over the range $\nu_2=36{,}288$–$36{,}328\ \mathrm{cm}^{-1}$. Two clearly separated vibronic states with almost identical rotational structures can be seen, which can both be assigned to a progression with $K^+=3$. This result leads to the vibronic symmetry species B_{1g} and B_{2g}, which is in agreement with the symmetry selection rules for ZEKE spectroscopy; these components must be assigned to the two JT states 6^1 $(j=\pm\frac{3}{2})$. So far it cannot be determined which state has the vibronic symmetry species B_{1g} and which has B_{2g}. The separations of each set of two transitions with the same quantum numbers N^+ and K^+ agree to within $\pm 0.1\ \mathrm{cm}^{-1}$. The mean splitting between the states is $20\ \mathrm{cm}^{-1}$.

The differences in band origin energy between the 6^1 $(j=\pm\frac{1}{2})$ bands and the 6^1 $(j=-\frac{3}{2})$ and 6^1 $(j=+\frac{3}{2})$ bands is 331 and $311\ \mathrm{cm}^{-1}$, respectively. This result corresponds to roughly one half of the vibrational energy of the ν_6 mode. Thus the benzene cation is a good example of intermediate coupling, in which the vibrational frequency of the Jahn–Teller active mode and the accompanying coupling strengths have almost the same values [101]. With the precise values given here for the energies of the components of the ν_6 band that are affected by vibronic interaction, the results of quantum chemical *ab initio* calculations can be examined very closely. Comparison with values published recently shows how well these latest dynamic Jahn–Teller calculations can take into

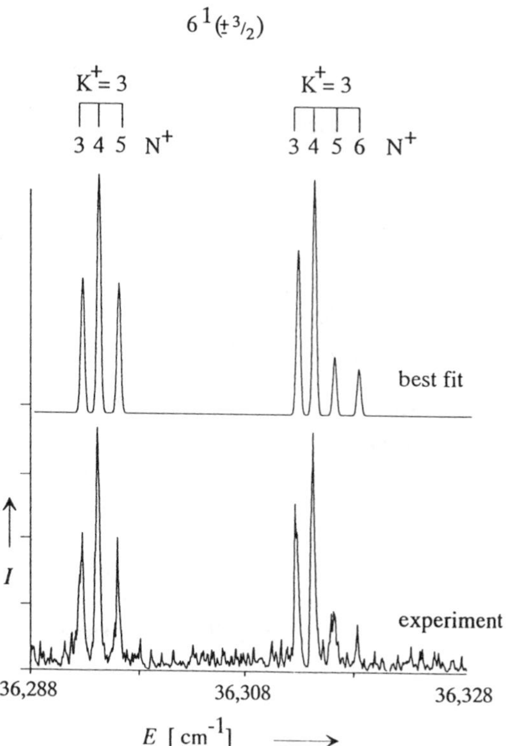

Figure 17. Measured and fitted ZEKE spectrum of the transition $6^1(\pm\frac{3}{2})$ $^2B_{1g} \oplus {}^2B_{2g} \leftarrow S_1\, 6^1\, {}^1E_{1u}$ ($J' = 1$, $K' = 1 + l$). The two components originating from the quadratic Jahn–Teller splitting show clearly resolved rotational structure (Redrawn from [9].)

account several active modes [102]. The difference between the calculated value for the 6^1 ($j = \pm\frac{1}{2}$) band of 694 cm^{-1} and the measured value of 676.4 cm^{-1} is only 3%.

The determination of the symmetry of a vibronic state by rotationally resolved ZEKE spectroscopy, demonstrated here for the ν_6 mode, would allow the elucidation of the full vibrational structure of the benzene cation, which is predicted to be very irregular due to multiple-mode Jahn–Teller coupling. This would be a significant contribution to the understanding of the Jahn–Teller effect and vibronic interactions. High-resolution experimental analysis of the cation of the benzene molecule, the prototypical aromatic molecule, which is frequently the subject of theoretical studies, was not possible previously [102].

III. THEORY AND FORMULATION

Section II has given some examples of rotationally resolved ZEKE spectra; in this section we will focus on *ab initio* theoretical concepts for calculations of intensity for such spectra.

Rotationally resolved PE spectra and their associated PE angular distributions can clearly provide significant insight into the dynamics of molecular photoionization, one of the simplest of molecular fragmentation processes. Such state-resolved spectra are an obvious signature of the exchanges of energy and angular momentum between the photoelectron and molecular ion. The recent development of ZEKE photoelectron spectroscopy now makes it possible to exploit the narrow bandwidth of laser radiation to routinely achieve subwavenumber resolution in ion rovibronic state distributions [3, 4, 11, 13, 20]. These and other rotationally resolved measurements [148] are providing useful insight into significant spectral features, such as the dependence of photoelectron angular distributions on rotational levels [77, 103], rotational propensity rules, parity selectivity in transitions involving electronically degenerate states [75, 105–107, 148], autoionization [108, 109, 140], effects of alignment [110], and the influence of Cooper minima [111–113] and shape resonances [114] on ion distributions. The underlying dynamics of such rotationally resolved PE spectra can clearly be expected to present new theoretical and computational challenges.

In this section, we present an *ab initio* theoretical framework and computational approach [115, 116] for studying vibrational and rotational distributions of ions resulting from the REMPI of single rotational levels of excited electronic states of molecules and from single-photon ionization of ground electronic states of molecules. This formulation, in which an intermediate coupling scheme between Hund's cases (a) and (b) is used to represent the initial and ionic states of linear molecules and where multiplet-specific final state wave functions are assumed, has been used to carry out quantitative theoretical studies of rotationally resolved PE spectra for a wide range of diatomic molecules. A similar formulation for a pure Hund's case (b) coupling scheme is also given for specific $\Sigma - \Sigma$ ionization processes. Several of these examples and the dynamical insight they provide will be reviewed here. An extension of this formulation to symmetric top [117] and asymmetric top [118] nonlinear polyatomic molecules as well as several applications of this extension are also discussed.

Numerous examples illustrating the widespread implications of parity selection rules, effects arising from Cooper minima, and strong ℓ mixing in molecular PE wave functions will be used to highlight the underlying dynamics of rotationally resolved molecular photoionization. These include spectra at low PE energies resulting from the (2 + 1) REMPI of a single rotational level of HBr and OH and ZEKE–PFI photoelectron spectra for single-photon ionization of the ground states of OH, NO, CO, N_2, H_2O, H_2CO, and CH_3. These studies reveal the rich underlying dynamics of quantum state specific studies of molecular photoionization and provide a robust prediction of key spectral features of interest in related experimental studies.

A. $(n + 1')$ REMPI for Linear Molecules

1. *Rotationally Resolved Photoelectron Spectra*

In an $(n + 1')$ REMPI process of interest here, the resonant intermediate state is created by absorption of n photons from an initially unaligned (all M_{J_0} levels equally populated) state. This aligned state (levels with different $|M_{J_i}|$ values have different populations) is subsequently ionized by absorption of another photon. The photon energies of the excitation and ionization steps of REMPI may differ from each other in a two-color experiment. Under collision-free conditions, ionization out of each M_{J_0} magnetic sublevel of the initial state forms an independent channel for linearly polarized light. The differential cross-section for photoionization of the resonant state by a single photon can then be written as

$$\frac{d\sigma}{d\Omega} = \frac{\sigma}{4\pi}\left[1 + \sum_{L=1}^{n+1} \beta_{2L} P_{2L}(\cos\theta)\right] \tag{1.1}$$

where σ is the total cross-section, β_{2L} are the asymmetry parameters, $P_{2L}(\cos\theta)$ are the Legendre polynomials, and θ is the angle between the direction of the PE and the electric vector of the polarized light. The total cross-section σ and asymmetry parameters β for ionization of a J_i level of the resonant or initial ($n = 0$) state can be written as [116]

$$\sigma \propto \sum_{\substack{\ell m \\ M_{J_i}, M_{J_+}}} \rho_{M_{J_i} M_{J_i}} |C_{\ell m}(M_{J_i} M_{J_+})|^2 \tag{1.2}$$

and

$$\beta_{2L} = \frac{4L+1}{\sigma} \sum_{\substack{\ell \ell' m \\ M_{J_i}, M_{J_+}}} (-1)^m (2\ell + 1)(2\ell' + 1) \rho_{M_{J_i} M_{J_i}}$$
$$\times C_{\ell m}(M_{J_i} M_{J_+}) C^*_{\ell' m}(M_{J_i} M_{J_+}) \begin{pmatrix} \ell & \ell' & 2L \\ m & -m & 0 \end{pmatrix} \begin{pmatrix} \ell & \ell' & 2L \\ 0 & 0 & 0 \end{pmatrix} \tag{1.3}$$

where ℓ is an angular momentum component of the PE and m is its projection in the laboratory frame. In Eq. (1.2), $|C_{\ell m}(M_{J_i}M_{J_+})|^2$ is the probability for photoionization of the M_{J_i} level of the resonant or initial state leading to the M_{J_+} level of the ion. The expression for $C_{\ell m}(M_{J_i}M_{J_+})$ will be derived later. For the branching ratios of interest here, the constant implied in Eq. (1.2) is unimportant and will be suppressed.

In Eqs. (1.2) and (1.3), $\rho_{M_{J_i}M_{J_i}}$ is the population of a specific M_{J_i} level in the optically aligned resonant state. For single-photon ionization processes $(n=0)\rho_{M_{J_i}M_{J_i}}$ is the population of the unaligned M_{J_0} levels of the initial state. For the REMPI examples of interest here, we will limit our discussion to only the $n=2$ case, that is, $(2+1')$ REMPI processes. For the low-intensity experiments of interest here, $\rho_{M_{J_i}M_{J_i}}$ for two-photon excitation has the general form

$$\rho_{M_{J_i}M_{J_i}} = \mathcal{N}\left|\sum_k \begin{pmatrix} J_i & 1 & J_k \\ -M_{J_i} & 0 & M_{J_i} \end{pmatrix}\begin{pmatrix} J_k & 1 & J_0 \\ -M_{J_i} & 0 & M_{J_i} \end{pmatrix} B_k\right|^2 \qquad (1.4)$$

where $\mathcal{N}$ is a normalization constant, J_k is the total angular momentum of a dipole-allowed virtual state $|k\rangle$, and B_k is related to the rotational line strength [119]. Evaluation of Eq. (1.4) generally requires summation over all dipole-allowed virtual states $|k\rangle$. For rotational branches other than the Q branch, Eq. (1.4) can be further simplified into a product of a 3-j symbol and rotational line strength B [120],

$$\rho_{M_{J_i}M_{J_i}} = \mathcal{N}\begin{pmatrix} J_i & 2 & J_0 \\ -M_{J_i} & 0 & M_{J_i} \end{pmatrix}^2 B \qquad (1.5)$$

However, in the case of a Q branch, evaluation of $\rho_{M_{J_i}M_{J_i}}$ requires two factors B_0 and B_2 [119],

$$\rho_{M_{J_i}M_{J_i}} = \mathcal{N}\left|\begin{pmatrix} J_i & 2 & J_0 \\ -M_{J_i} & 0 & M_{J_i} \end{pmatrix} B_2 + B_0\right|^2 \qquad (1.6)$$

Note that B_0 contains no polarization information, but is crucial in determining the alignment $\rho_{M_{J_i}M_{J_i}}$ of the intermediate state for $(2+1')$ REMPI via the Q rotational branch. An expression for $\rho_{M_{J_i}M_{J_i}}$ for the $n\neq 2$ case is given in [116].

In Eqs. (1.2) and (1.3), $C_{\ell m}(M_{J_i}M_{J_+})$ are just the expansion coefficients of the bound-free matrix element $\langle f|D_{\mu_0}|i\rangle$, that is

$$\langle f|D_{\mu_0}|i\rangle = \sum_{\ell m} C_{\ell m}(M_{J_i}M_{J_+})Y_{\ell m}(\hat{\mathbf{k}}) \qquad (1.7)$$

for photoionization of an M_{J_i} sublevel of a parity level of the resonant state leading to an M_{J_+} level of a parity component of the ion. In Eq. (1.7), D_{μ_0} is the dipole moment operator with μ_0 as the light polarization index in the laboratory frame, and the total final state continuum wave function $|f\rangle$ is defined as

$$|f\rangle = \sum_{\Lambda_f \Sigma_f} |\Lambda_f \Sigma_f\rangle \langle \Lambda_f \Sigma_f | f\rangle \tag{1.8}$$

where $|\Lambda_f \Sigma_f\rangle$ are the multiplet-specific final state wave functions including spin. The $|\Lambda_f \Sigma_f\rangle$ are antisymmetrized products of the photoelectron $|\phi_e\rangle$ and ion wave functions $|f^+\rangle$, which may consist of several spin components

$$|\Lambda_f \Sigma_f\rangle = \sum_s C_s \,|\, \Lambda_+ S_+ \Sigma_+ J_+ \Omega_+ M_{J_+};\, \ell m S_e \Sigma_e\rangle \tag{1.9}$$

with $S_e = \frac{1}{2}$ and $\Sigma_e = \pm\frac{1}{2}$ for the PE and C_s is an expansion coefficient. Therefore, the bound-free transition moment of Eq. (1.7) can be rewritten as

$$\langle f|D_{\mu_0}|i\rangle = \sum_{\Lambda_f \Sigma_f} \langle \phi_e | \langle f^+ | \Lambda_f \Sigma_f\rangle \langle \Lambda_f \Sigma_f | D_{\mu_0} | i\rangle \tag{1.10}$$

In Eq. (1.10) we use an intermediate coupling scheme between Hund's cases (a) and (b) to represent the intermediate (or initial) state $|i\rangle$ and the ionic state wave functions $|f^+\rangle$ [116]. The dipole moment operator D_{μ_0} and the PE wave function $|\phi_e\rangle$ are given, respectively, as

$$D_{\mu_0} = (4\pi/3)^{1/2} r \sum_{\mu} (-1)^{\mu-\mu_0} \mathscr{D}^{1}_{\mu_0 \mu} Y_{1\mu}(\hat{\mathbf{r}}') \tag{1.11}$$

and

$$\begin{aligned} |\phi_e\rangle = & \sum_{\Sigma_e} (-1)^{\Sigma_e - m_e} \mathscr{D}^{S_e}_{m_e \Sigma_e} |S_e \Sigma_e\rangle \\ & \times \sum_{\ell m \lambda} i^{\ell} e^{-i\eta_\ell} (-1)^{m-\lambda} Y^{*}_{\ell m}(\hat{k}) \mathscr{D}^{\ell}_{m\lambda} \psi_{k\ell\lambda} \end{aligned} \tag{1.12}$$

where μ is the light polarization index in the molecular frame, $\mathscr{D}^{\ell}_{m\lambda}$ are the rotational matrices in Edmonds' notation [121], η_ℓ is the Coulomb phase shift, λ is the projection of ℓ in the molecular frame, and $\psi_{k\ell\lambda}$ is a partial wave component of the PE orbital [122] with momentum $\mathbf{k}$.

With the following basis for both the intermediate and ionic state wave functions

$$|\gamma_i\Lambda_i S_i \Sigma_i J_i \Omega_i M_{J_i}\rangle = \left(\frac{2J_i+1}{8\pi^2}\right)^{1/2} \psi^e_{\gamma_i\Lambda_i}(\mathbf{r};R)\chi_{v_i}(R)(-1)^{M_{J_i}-\Omega_i}\mathscr{D}^{J_i}_{M_{J_i}\Omega_i}|S_i\Sigma_i\rangle \tag{1.13}$$

$$|\gamma_+\Lambda_+S_+\Sigma_+J_+\Omega_+M_{J_+}\rangle = \left(\frac{2J_++1}{8\pi^2}\right)^{1/2} \psi^e_{\gamma_+\Lambda_+}(\mathbf{r};R)\chi_{v_+}(R)(-1)^{M_{J_+}-\Omega_+}\mathscr{D}^{J_+}_{M_{J_+}\Omega_+}|S_+\Sigma_+\rangle \tag{1.14}$$

and making use of properties of the 3-j symbol, we obtain

$$\begin{aligned} C_{\ell m}(M_{J_i}M_{J_+}) &= (\pi/3)^{1/2}[(2J_++1)(2J_i+1)(2S_i+1)]^{1/2} \\ &\sum (-1)^{M_{J_i}+S_+-\Omega_i+\Sigma_i+\frac{1}{2}}(2J_t+1)(2J_r+1)C_{n'_+n_+}C_{n'n} \\ &[1+(-1)^P]\begin{pmatrix} J_+ & J_i & J_t \\ -M_{J_+} & M_{J_i} & m_t \end{pmatrix}\begin{pmatrix} S_e & \ell & J_r \\ -m_e & -m & m_r \end{pmatrix} \\ &\begin{pmatrix} J_r & 1 & J_t \\ -m_r & \mu_0 & -m_t \end{pmatrix}\Big[\sum \tilde{I}_{\ell\lambda\mu;\Sigma_e}\begin{pmatrix} J_r & 1 & J_t \\ -\lambda_r & \mu & -\lambda_t \end{pmatrix} \\ &\begin{pmatrix} S_e & \ell & J_r \\ -\Sigma_e & -\lambda & \lambda_r \end{pmatrix}\begin{pmatrix} J_+ & J_i & J_t \\ -\Omega_+ & \Omega_i & \lambda_t \end{pmatrix}\begin{pmatrix} S_+ & S_e & S_i \\ \Sigma_+ & \Sigma_e & -\Sigma_i \end{pmatrix} \\ &+\sum \tilde{I}_{\ell\lambda\mu;\Sigma_e}(-1)^{p_i+q_i}\begin{pmatrix} J_r & 1 & J_t \\ -\lambda_r & \mu & -\lambda_t \end{pmatrix}\begin{pmatrix} S_e & \ell & J_r \\ -\Sigma_e & -\lambda & \lambda_r \end{pmatrix} \\ &\begin{pmatrix} J_+ & J_i & J_t \\ -\Omega_+ & -\Omega_i & \lambda_t \end{pmatrix}\begin{pmatrix} S_+ & S_e & S_i \\ \Sigma_+ & \Sigma_e & \Sigma_i \end{pmatrix}\Big] \end{aligned} \tag{1.15}$$

with

$$\tilde{I}_{\ell\lambda\mu;\Sigma_e} = \sum_{\Lambda_f\Sigma_f} \langle\Lambda_+\lambda|\Lambda_f\rangle\langle\Sigma_+\Sigma_e|\Sigma_f\rangle I_{\ell\lambda\mu(\Lambda_f\Sigma_f)} \tag{1.16}$$

$$I_{\ell\lambda\mu} = (-i)^\ell e^{i\eta_\ell}\int dR\chi^*_{v_+}(R)r^{\ell\lambda\mu}_{fi}(R)\chi_{v_i}(R) \tag{1.17}$$

and

$$P = \Delta J + \Delta S + 2S_e + \Delta p + \Delta q + \ell + 1 \tag{1.18}$$

where $\Delta J = J_+ - J_i$, $\Delta S = S_+ - S_i$, $\Delta p = p_+ - p_i$, $\Delta q = q_+ - q_i$, p and q

are parity indexes [123], J_r and J_t are angular momentum transfers, $C_{n_+'n_+}$ and $C_{n'n}$ are expansion coefficients for the wave functions of the rovibronic states of the ion and resonant states [116], respectively, and the summations in Eq. (1.15) go over all possible indexes. In Eq. (1.17) $r_{fi}^{\ell\lambda\mu}(R)$ is the PE matrix element and $I_{\ell\lambda\mu}$ is the vibrationally averaged PE matrix element between the resonant state and the PE continuum wave function.

If pure Hund's case (b) bases are used for the ionic and intermediate state wave functions, which are appropriate for $\Sigma - \Sigma$ type transitions, the coefficient $C_{\ell m}(M_{J_i}M_{J_+})$ becomes

$$C_{\ell m}(M_{J_i}M_{J_+}) = (\pi/3)^{1/2}[(2N_+ + 1)(2N_i + 1)(2J_+ + 1)(2J_i + 1)(2S_i + 1)]^{1/2}$$

$$\sum (-1)^{P'}(2N_t + 1)\begin{pmatrix} N_i & S_i & J_i \\ M_{N_i} & M_{S_i} & -M_{J_i} \end{pmatrix}\begin{pmatrix} N_+ & S_+ & J_+ \\ M_{N_+} & M_{S_+} & -M_{J_+} \end{pmatrix}$$

$$\begin{pmatrix} S_+ & S_e & S_i \\ M_{S_+} & M_{S_e} & -M_{S_i} \end{pmatrix}\begin{pmatrix} N_+ & N_i & N_t \\ -M_{N_+} & M_{N_i} & m_t \end{pmatrix}\begin{pmatrix} N_t & 1 & \ell \\ -m_t & \mu_0 & -m \end{pmatrix}$$

$$[1 + (-1)^{P''}]\Big[\sum \tilde{I}_{\ell\lambda\mu;M_{S_e}}\begin{pmatrix} N_+ & N_i & N_t \\ -\Lambda_+ & \Lambda_i & \lambda_t \end{pmatrix}\begin{pmatrix} N_t & 1 & \ell \\ -\lambda_t & \mu & -\lambda \end{pmatrix}$$

$$+ \sum \tilde{I}_{\ell\lambda\mu;M_{S_e}}(-1)^{p}\begin{pmatrix} N_+ & N_i & N_t \\ -\Lambda_+ & -\Lambda_i & \lambda_t \end{pmatrix}\begin{pmatrix} N_t & 1 & \ell \\ -\lambda_t & \mu & -\lambda \end{pmatrix}\Big] \qquad (1.19)$$

with

$$P' = M_{N_+} + M_{J_+} - M_{N_i} + \mu_0 - N_i - N_+ + S_i - \Lambda_+ - \mu + \tfrac{1}{2} \qquad (1.20)$$

and

$$P'' = N_+ - N_i + \ell + p_+ - p_i + 1 \qquad (1.21)$$

where N_i and N_+ are total angular momenta apart from spin of the resonant and ionic states, respectively, M_{N_i} and M_{N_+} are their projections along the internuclear axis in the laboratory frame, N_t is the angular momentum transfer, and M_{S_e} is the spin projection along the internuclear axis in the laboratory frame.

2. *Photoelectron Matrix Element*

The matrix element for photoejection of an electron from a bound molecular orbital ϕ_i (or $|i\rangle$) into a PE continuum orbital $\Psi_{f,\mathbf{k}}^{(-)}(\mathbf{r})$ (or ϕ_e) is a central quantity in these studies. Here $\mathbf{k}$ is the momentum of the PE and the $(-)$ denotes incoming-wave boundary conditions. For linear molecules, the partial wave components $\psi_{k\ell m}^{(-)}$ of $\Psi_{f,\mathbf{k}}^{(-)}(\mathbf{r})$ are defined by an expansion in spherical harmonics about $\hat{\mathbf{k}}$ of the PE

$$\Psi_{f,\mathbf{k}}^{(-)}(\mathbf{r}) = \left(\frac{2}{\pi}\right)^{1/2} \sum_{\ell m} i^{\ell} \psi_{k\ell m}^{(-)}(\mathbf{r}) Y_{\ell m}^{*}(\hat{\mathbf{k}}) \tag{1.22}$$

Single-center expansions of $\psi_{k\ell m}^{(-)}(\mathbf{r})$ and $\phi_i(\mathbf{r}')$, for example,

$$\psi_{k\ell m}^{(-)} = \sum_{\ell'\lambda} g_{\ell,\ell'\lambda}(k, r) \mathscr{D}_{m\lambda}^{\ell} Y_{\ell'\lambda}(\hat{\mathbf{r}}') \tag{1.23}$$

define partial wave PE matrix elements $r_{fi}^{\ell\lambda\mu}(R)$ in the molecular frame for ionization of orbital $\phi_i(\mathbf{r}')$,

$$r_{fi}^{\ell\lambda\mu}(R) = \sum_{\ell',\ell_0} \langle g_{\ell,\ell'\lambda}(k, r, R) Y_{\ell'\lambda}(\hat{\mathbf{r}}') | r Y_{1\mu}(\hat{\mathbf{r}}') | \phi_{i\ell_0}(r, R) Y_{\ell_0,\lambda_0}(\hat{\mathbf{r}}') \rangle \tag{1.24}$$

where R denotes a dependence on internuclear distance.

Equation (1.24) reveals an important underlying dynamical aspect of molecular PE wave functions. Whereas only $\ell = \ell'$ terms are allowed in Eq. (1.24) for atomic systems with their central fields, where the angular momentum of the PE must be conserved, $\ell \neq \ell'$ terms arise in Eq. (1.24) due to the nonspherical potential fields of molecular ions. This angular momentum coupling between partial waves ℓ and ℓ' is brought about by the torques associated with the molecular ion potential and makes a molecular PE orbital an admixture of angular momentum components. Viewed very simply, this means that whatever angular momentum ℓ a PE may have at a given instant can be changed as the PE scatters off the molecular ion. The use of molecular PE orbitals that correctly incorporate such angular momentum coupling is essential at the low PE energies of interest here.

3. *Parity Selection Rules*

Parity selection rules [115, 116, 124, 125] govern changes of rotational angular momentum upon ionization. Equations (1.15) and (1.18) yield the parity selection rule

$$\Delta J + \Delta S + \Delta p + \Delta q + \ell = even \tag{1.25}$$

for a coupling scheme intermediate between Hund's cases (a) and (b). Note that Eq. (1.25) is also suitable for pure Hund's case (a) coupling since the intermediate coupling scheme assumes a Hund's case (a) basis. This parity selection rule is the same as that of Xie and Zare [124] and of

Dixit and McKoy [115]. The Δq term does not arise in the results of Dixit and McKoy [115] since Σ^- states were not explicitly considered there. For these states the spin–orbit splittings are always negligible and pure Hund's case (b) applies. In Hund's case (b) limit, Eqs. (1.19) and (1.21) lead to [115, 116, 124, 125]

$$\Delta N + \Delta p + \ell = odd \tag{1.26}$$

with $\Delta N = N_+ - N_i$. For a $\Sigma - \Sigma$ transition, Eq. (1.26) reduces to $\Delta N + \ell = odd$ [51].

B. $(n + 1')$ REMPI for Asymmetric Tops

1. *Rotationally Resolved Photoelectron Spectra*

The formulation for rotationally resolved PE spectra resulting from $(n + 1')$ REMPI of asymmetric tops is basically the same as that for linear molecules. The defining equations for the differential cross section [Eq. (1.1)], total cross section [Eq. (1.2)], and asymmetry parameters [Eq. (1.3)] for linear molecules still hold for nonlinear polyatomic molecules except that the coefficients $C_{\ell m}(M_{J_i} M_{J_+})$ must be redefined. In this section, we will derive an expression for these $C_{\ell m}(M_{J_i} M_{J_+})$ coefficients primarily for asymmetric tops with C_{2v} molecular symmetry even though it can also be applied to other molecular symmetries.

To evaluate $C_{\ell m}(M_{J_i} M_{J_+})$, explicit forms for the wave functions of the intermediate state $|i\rangle$, the ionic state $|f^+\rangle$, and the PE $|\phi_e\rangle$ are needed

$$|i\rangle = \psi_i \chi_{v_i} (-1)^{N_i - S_i + M_{J_i}} (2J_i + 1)^{1/2} \begin{pmatrix} N_i & S_i & J_i \\ M_{N_i} & M_{S_i} & -M_{J_i} \end{pmatrix} \times |S_i M_{S_i}\rangle |N_i M_{N_i} K_{a_i} K_{c_i}\rangle \tag{1.27}$$

$$|f^+\rangle = \psi_+ \chi_{v_+} (-1)^{N_+ - S_+ + M_{J_+}} (2J_+ + 1)^{1/2} \begin{pmatrix} N_+ & S_+ & J_+ \\ M_{N_+} & M_{S_+} & -M_{J_+} \end{pmatrix} \times |S_+ M_{S_+}\rangle |N_+ M_{N_+} K_{a_+} K_{c_+}\rangle \tag{1.28}$$

and

$$|\phi_e\rangle = \left(\frac{2}{\pi}\right)^{1/2} \left|\frac{1}{2} m_\sigma\right\rangle \sum_{\gamma\tau h\ell\lambda} \frac{i^\ell}{k} \psi_{h\ell\lambda}^{(-)\gamma\tau}(\mathbf{r}) X_{h\ell\lambda}^{*\gamma\tau}(\hat{\mathbf{k}}) \tag{1.29}$$

where $|\frac{1}{2} m_\sigma\rangle$ is the spin eigenfunction of the PE, k defines the PE kinetic energy, γ is one of the irreducible representations (IR) of the molecular point group, τ is a component of this representation, h distinguishes between different bases for the same IR corresponding to the same value of ℓ, $\psi_{h\ell\lambda}^{(-)\gamma\tau}(\mathbf{r})$ is a partial wave component of the PE orbital, $X_{h\ell\lambda}^{\gamma\tau}(\hat{\mathbf{k}})$ is a generalized harmonic, and ψ_i and χ_{v_i} are the electronic and vibrational wave functions of the state $|i\rangle$, respectively. Note that a Hund's case (b) coupling scheme is chosen to represent both the intermediate and ionic states [see Eqs. (1.27) and (1.28)]. The rotational levels of both the intermediate and ionic states are given in terms of asymmetric top wave functions

$$|NM_N K_a K_c\rangle = \sum_K a_{N\zeta K} S(NM_N Kp) \tag{1.30}$$

where $\zeta = K_a - K_c$, K_a, and K_c are the projections of the total angular momentum along the a and c axes, respectively, and the coefficients $a_{N\zeta K}$ are determined by diagonalizing the rigid rotor Hamiltonian

$$\mathcal{H}_\tau = \frac{J_x^2}{2I_x} + \frac{J_y^2}{2I_y} + \frac{J_z^2}{2I_z} \tag{1.31}$$

in a basis of symmetric top eigenfunctions. In Eq. (1.30) $S(NM_N Kp)$ is a linear combination of symmetric top functions and has the form [126]

$$S(NM_N Kp) = \frac{1}{\sqrt{2}}(|NM_N K\rangle + (-1)^p|NM_N - K\rangle) \tag{1.32}$$

for $K > 0$ and

$$S(NM_N 00) = |NM_N 0\rangle \tag{1.33}$$

for $K = 0$. The parity index p in Eq. (1.32) takes a value of 0 or 1. The symmetric top wave functions $|NM_N K\rangle$ in Eqs. (1.32) and (1.33) are given by

$$|NM_N K\rangle = \left(\frac{2N+1}{8\pi^2}\right)^{1/2} (-1)^{K-M_n} \mathcal{D}_{M_N K}^N \tag{1.34}$$

Note that the $S(NM_N Kp)$ functions adopted in Eq. (1.30), instead of symmetric top eigenfunctions $|NM_N K\rangle$, fulfill the symmetry requirements of the V_4 group [126]. In Eq. (1.29) we use the generalized harmonics $X_{h\ell\lambda}^{\gamma\tau}(\hat{\mathbf{k}}')$ as bases for the IR of the molecular point group. These symmetry-adapted angular functions satisfy well-known orthonormality relations [127] and can be expressed in terms of the usual spherical harmonics

$$X^{\gamma\tau}_{h\ell\lambda}(\hat{\mathbf{k}}') = b^{\gamma\tau}_{h\ell\lambda} Y_{\ell\lambda}(\hat{\mathbf{k}}') + b^{\gamma\tau}_{h\ell-\lambda} Y_{\ell-\lambda}(\hat{\mathbf{k}}') \tag{1.35}$$

in the body-fixed frame. The coefficients $b^{\gamma\tau}_{h\ell\lambda}$ for C_{2v} and O_h symmetries have been given by Burke et al. [127]. These $Y_{\ell\lambda}(\hat{\mathbf{k}}')$ can be further written in terms of the spherical harmonics $Y_{\ell m}(\hat{\mathbf{k}})$ in the laboratory frame as

$$Y_{\ell\lambda}(\hat{\mathbf{k}}') = \sum_m (-1)^{\lambda-m} \mathscr{D}^{\ell}_{m\lambda} Y_{\ell m}(\hat{\mathbf{k}}) \tag{1.36}$$

With steps similar to those for the case of linear molecules, we obtain

$$\begin{aligned} C_{\ell m}(M_{J_i} M_{J_+}) = {} & \sqrt{\frac{4\pi}{3}}\frac{1}{2}[(2J_i+1)(2J_+ +1)(2N_i+1)(2N_+ +1)(2S_i+1)]^{1/2} \\ & \sum (-1)^Q (2N_t+1) \begin{pmatrix} S_+ & \frac{1}{2} & S_i \\ M_{S_+} & m_\sigma & -M_{S_i} \end{pmatrix} \begin{pmatrix} N_+ & S_+ & J_+ \\ M_{N_+} & M_{S_+} & -M_{J_+} \end{pmatrix} \\ & \begin{pmatrix} N_i & S_i & J_i \\ M_{N_i} & M_{S_i} & -M_{J_i} \end{pmatrix} \begin{pmatrix} N_+ & N_i & N_t \\ -M_{N_+} & M_{N_i} & m_t \end{pmatrix} \begin{pmatrix} N_t & 1 & \ell \\ -m_t & \mu_0 & m \end{pmatrix} \\ & a_{N_i\zeta_i K_i} a_{N_+\zeta_+ K_+} \tilde{I}^{\gamma\tau}_{h\ell\lambda\mu}(\Lambda_f\Sigma_f) b^{\gamma\tau}_{h\ell\lambda}(\Lambda_f\Sigma_f) \\ & [1+(-1)^{\Delta p+\Delta N+\ell+1}] \left[\begin{pmatrix} N_+ & N_i & N_t \\ -K_+ & K_i & K_t \end{pmatrix} \begin{pmatrix} N_t & 1 & \ell \\ -K_t & \mu & \lambda \end{pmatrix} \right. \\ & \left. + (-1)^{p+} \begin{pmatrix} N_+ & N_i & N_t \\ K_+ & K_i & K_t \end{pmatrix} \begin{pmatrix} N_t & 1 & \ell \\ -K_t & \mu & \lambda \end{pmatrix} \right] \end{aligned} \tag{1.37}$$

$$Q = \Delta N + \Delta M_J - S_i + M_{S_i} - \mu_0 - m + M_{N_+} + K_i - \tfrac{1}{2} \tag{1.38}$$

$$\tilde{I}^{\gamma\tau}_{h\ell\lambda\mu}(\Lambda_f\Sigma_f) = \langle \gamma_+\tau_+\gamma_e\tau_e | \Lambda_f \rangle \langle M_{S_+} m_\sigma | \Sigma_f \rangle I^{\gamma\tau}_{h\ell\lambda\mu}(\Lambda_f\Sigma_f) \tag{1.39}$$

and

$$\begin{aligned} I^{\gamma\tau}_{h\ell\lambda\mu}(\Lambda_f\Sigma_f) = {} & \sqrt{\frac{2}{\pi}}\frac{(-i)^\ell}{k}\int dRdr\psi^*_+(\mathbf{r},R)\chi^*_{v_+}(R) \\ & \times \psi^{*(-)\gamma\tau}_{h\ell\lambda}(\mathbf{r},R) r Y_{1\mu}\psi_i(\mathbf{r},R)\chi_{v_i}(R) \end{aligned} \tag{1.40}$$

2. *Photoelectron Matrix Element*

For asymmetric top molecules, the partial wave components $\psi^{(-)\gamma\tau}_{kh\ell m}$ of $\Psi^{(-)}_{f,k}(\mathbf{r})$ are similarly defined by an expansion in generalized harmonics about $\hat{\mathbf{k}}$ of the PE

$$\Psi^{(-)}_{f,\mathbf{k}}(\mathbf{r}) = \left(\frac{2}{\pi}\right)^{1/2} \sum_{\ell m} i^{\ell} \psi^{(-)\gamma\tau}_{kh\ell m}(\mathbf{r}) X^{*\gamma\tau}_{h\ell m}(\hat{\mathbf{k}}) \tag{1.41}$$

With single-center expansions for $\psi^{(-)\gamma\tau}_{kh\ell m}(\mathbf{r})$ and $\phi^{\gamma'\tau'}_{i}(\mathbf{r}')$, partial wave PE matrix elements $r^{\gamma\tau}_{h\ell\lambda\mu}(R)$ in the molecular frame for ionization out of orbital $\phi^{\gamma'\tau'}_{i}(\mathbf{r}')$ are given by [118]

$$r^{\gamma\tau}_{h\ell\lambda\mu}(R) = \sum_{\ell',\ell_0} \langle g^{\gamma\tau}_{h\ell,\ell'\lambda}(k, r, R) X^{\gamma\tau}_{h\ell'\lambda}(\mathbf{r}') | r Y_{1\mu}(\hat{\mathbf{r}}') | \times \phi^{\gamma'\tau'}_{ih'\ell_0}(r, R) X^{\gamma'\tau'}_{h'\ell_0,\lambda_0}(\hat{\mathbf{r}}') \rangle \,. \tag{1.42}$$

Note that while $\ell \neq \ell'$ terms arise in both Eqs. (1.24) and (1.42) due to the nonspherical potential fields of molecular ions. For nonlinear molecules, coupling of the projections of angular momenta along appropriate axes can become pervasive.

3. *Parity Selection Rules*

Equation (1.37) yields the general parity selection rule

$$\Delta N + \Delta p + \ell = odd \tag{1.43}$$

which has the same form as diatomic molecules with Hund's case (b) coupling [115, 116, 124, 125]. However, this selection rule is not particularly transparent since the parity indexes p_i and p_+ are defined with respect to the $S(NM_N K_p)$ functions used to expand the asymmetric top wave functions [see Eq. (1.32)]. Equation (1.43) must be further developed to explicitly reveal its dependence on the angular momentum changes ΔK_a and ΔK_c.

Here we choose a left-handed coordinate system for the molecular internal x, y and z axes with the molecular z axis as the symmetry axis. With the symmetry properties of the asymmetric top [126], it can be shown that ΔK_η is *even* (*odd*) when $\Delta N + \Delta p$ is *even* (*odd*). Here η is the principal axis lying along the molecular x axis and the K_η is the projection of the total angular momentum along the axis. The parity selection rule of Eq. (1.43) then reduces to

$$\Delta K_\eta + \ell = odd \tag{1.44}$$

The selection rules also depend on which principal axis coincides with the molecular z axis. With the properties of 3-j symbols, Eq. (1.37) provides additional selection rules [126]

$$\mu + \lambda = \begin{cases} \Delta K_a & \text{if } a//\text{z}//\text{the symmetry axis} \\ \Delta K_b & \text{if } b//\text{z}//\text{the symmetry axis} \\ \Delta K_c & \text{if } c//\text{z}//\text{the symmetry axis} \end{cases} \tag{1.45}$$

To relate the selection rules of Eq. (1.44) and (1.45), a relationship among ΔK_a, ΔK_b, and ΔK_c is essential. This relationship, which can be obtained from the symmetry properties of an asymmetric top [126], is

$$\Delta K_a + \Delta K_c = even(odd) \leftrightarrow \Delta K_b = even(odd) \tag{1.46}$$

Determination of $\mu + \lambda$ now becomes a critical step in the application of these selection rules. Whereas Eqs. (1.44–1.46) are suitable for any nonlinear polyatomic molecule, $\mu + \lambda$ must be determined specifically for a given molecular symmetry. These selection rules will be further illustrated in later sections.

C. $(n + 1')$ REMPI for Symmetric Tops

In view of the applications to CH_3, to be discussed later, it is convenient to give explicit expressions for molecules of D_{3h} symmetry. To obtain the $C_{\ell m}(M_{J_i} M_{J_+})$ of Eqs. (1.2) and (1.3) for $(n + 1')$ REMPI of symmetric tops, the wave functions of the intermediate state $|i\rangle$ [Eq. (1.27)] and ionic state $|f^+\rangle$ [Eq. (1.28)] are replaced by

$$|i\rangle = \psi_i \chi_{v_i} (-1)^{N_i - S_i + M_{J_i}} (2J_i + 1)^{1/2} \times \begin{pmatrix} N_i & S_i & J_i \\ M_{N_i} & M_{S_i} & -M_{J_i} \end{pmatrix} |S_i M_{S_i}\rangle |N_i M_{N_i} K_i p_i\rangle \tag{1.47}$$

$$|f^+\rangle = \psi_+ \chi_{v_+} (-1)^{N_+ - S_+ + M_{J_+}} (2J_+ + 1)^{1/2} \times \begin{pmatrix} N_+ & S_+ & J_+ \\ M_{N_+} & M_{S_+} & -M_{J_+} \end{pmatrix} |S_+ M_{S_+}\rangle |N_+ M_{N_+} K_+ p_+\rangle \tag{1.48}$$

where

$$|NM_N Kp\rangle = \tfrac{1}{2}[|NM_N K\rangle + (-1)^p |NM_N - K\rangle] \tag{1.49}$$

With manipulations similar to those for asymmetric tops, we obtain

$$C_{lm}(M_{J_i}M_{J_+}) = \sqrt{\frac{4\pi}{3}}\,[(2J_i+1)(2J_++1)(2N_i+1)(2N_++1)(2S_i+1)]^{1/2}$$
$$\times \sum (-1)^Q (2N_t+1) \begin{pmatrix} S_+ & \frac{1}{2} & S_i \\ M_{S_+} & m_\sigma & -M_{S_i} \end{pmatrix} \begin{pmatrix} N_+ & S_+ & J_+ \\ M_{N_+} & M_{S_+} & -M_{J_+} \end{pmatrix}$$
$$\times \begin{pmatrix} N_i & S_i & J_i \\ M_{N_i} & M_{S_i} & -M_{J_i} \end{pmatrix} \begin{pmatrix} N_+ & N_i & N_t \\ -M_{N_+} & M_{N_i} & m_t \end{pmatrix} \begin{pmatrix} N_t & 1 & \ell \\ -m_t & \mu_0 & m \end{pmatrix}$$
$$\times \tilde{I}^{\gamma\tau}_{h\ell\lambda\mu}(\Lambda_f\Sigma_f) b^{\gamma\tau}_{h\ell\lambda}(\Lambda_f\Sigma_f) \left[\begin{pmatrix} N_+ & N_i & N_t \\ -K_+ & K_i & K_t \end{pmatrix} \begin{pmatrix} N_t & 1 & \ell \\ -K_t & \mu & \lambda \end{pmatrix} \right.$$
$$+ (-1)^{p_i} \begin{pmatrix} N_+ & N_i & N_t \\ -K_+ & -K_i & K_t \end{pmatrix} \begin{pmatrix} N_t & 1 & \ell \\ -K_t & \mu & \lambda \end{pmatrix}$$
$$+ (-1)^{p_+} \begin{pmatrix} N_+ & N_i & N_t \\ K_+ & K_i & K_t \end{pmatrix} \begin{pmatrix} N_t & 1 & \ell \\ -K_t & \mu & \lambda \end{pmatrix}$$
$$\left. + (-1)^{p_i+p_+} \begin{pmatrix} N_+ & N_i & N_t \\ K_+ & -K_i & K_t \end{pmatrix} \begin{pmatrix} N_t & 1 & \ell \\ -K_t & \mu & \lambda \end{pmatrix} \right] \tag{1.50}$$

Note that no parity selection rules relating Δp, ΔN, and ℓ are obtained for photoionization of symmetric tops since the partial wave components of each continuum have different phase factors [117]. However, Eq. (1.50) does reveal that $\mu + \lambda = \pm\Delta K$ and $\mu + \lambda = \pm(K_+ + K_i)$.

D. Computational Procedures

In applications of our procedure to REMPI of Rydberg states we use the improved virtual orbital (IVO) method [128] to obtain the wave function of the resonant state. The core orbitals of these Rydberg states are taken to be those of the fully relaxed ion. For single-photon ionization, we assume a self-consistent-field (SCF) wave function for the ground state of the molecule. For the final state we assume a frozen-core Hartree–Fock model in which the PE orbital is obtained as a solution of a one-electron Schrödinger equation containing the Hartree–Fock potential of the molecular ion $V_{\text{ion}}(\mathbf{r}, R)$, that is

$$\left(-\frac{1}{2}\nabla^2 + V_{ion}(\mathbf{r}, R) - \frac{k^2}{2} \right) \phi_k^{(-)}(\mathbf{r}, R) = 0 \tag{1.51}$$

with $\phi_k^{(-)} = \psi_{k\ell m}^{(-)}$ for linear molecules and $\phi_k^{(-)} = \psi_{kh\ell m}^{(-)\gamma\tau}$ for nonlinear molecules.

To obtain the PE orbital $\phi_k^{(-)}$, we have used an iterative procedure, based on the Schwinger variational principle [122], to solve the Lippmann–Schwinger equation associated with Eq. (1.51). This procedure begins by approximating the static-exchange potential of the ionic core by a separable potential of the form

$$U(\mathbf{r}, \mathbf{r}') = \sum_{ij} \langle \mathbf{r}|U|\alpha_i\rangle (U^{-1})_{ij} \langle \alpha_j|U|\mathbf{r}'\rangle \tag{1.52}$$

where the matrix U^{-1} is the inverse of the matrix with the elements $(U)_{ij} = \langle \alpha_i|U|\alpha_j\rangle$ and the α values are discrete basis functions, such as Cartesian or spherical Gaussian functions. The parameter U is twice the static-exchange potential in Eq. (1.51) with the long-term Coulomb potential removed. The Lippmann–Schwinger equation with this separable potential $U(\mathbf{r}, \mathbf{r}')$ can be readily solved and provides an approximate PE orbital $\phi_k^{(0)}$. These $\phi_k^{(0)}$ orbitals can be iteratively improved to yield converged solutions to the Lippmann–Schwinger equation containing the full static-exchange potential. A few iterations of this iterative variational method provide highly converged solutions for the molecular PE orbitals.

IV. RESULTS AND DISCUSSION

A. Linear Molecules

1. *Single-Photon Ionization of NO*

Figure 18 shows the (*a*) measured and (*b*) calculated ZEKE–PFI photoelectron spectra for single-photon ionization of rotationally cold NO ($X\,^2\Pi_{1/2}, v'' = 0$) molecules leading to $NO^+(X\,^1\Sigma^+, v_+ = 1)$ by coherent VUV radiation [129]. The calculated ion rotational distributions assume a temperature of 5 K. These spectra were calculated for a PE energy of 50 meV and convoluted with a Gaussian detection function with a full-width at half-maximum (fwhm) of 2 cm^{-1}. In Fig. 18(*b*) each branch is associated with a letter designation that refers only to the change in angular momentum apart from spin, that is, $\Delta N = N_+ - N''$. The label 1$'$ denotes the $N'' = 1$, $J'' = \frac{1}{2}$ level. The agreement between these measured and calculated ZEKE spectra is excellent. To provide further insight into the underlying dynamics of these ZEKE photoelectron spectra, Fig. 19 shows the (*a*) measured and (*b*) calculated ion rotational branching ratios for photoionization of the $N'' = 1$, $J'' = \frac{3}{2}$ rotational level of the $X\,^2\Pi_{1/2}$ ground state of NO. The data of Fig. 19(*a*) are extracted from the measured PE spectrum of Fig. 18(*a*) for rotationally cold NO where, in addition to the $N'' = 1$ level, a few other rotational levels are also

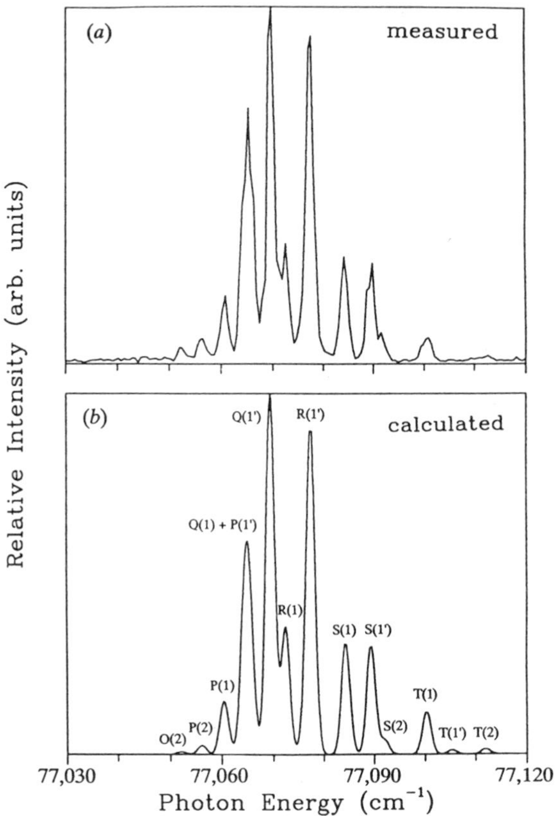

Figure 18. (*a*) measured and (*b*) calculated ZEKE–PFI photoelectron spectra for single-photon ionization of rotationally cold NO ($X\,^2\Pi_{1/2}$) by coherent VUV radiation. The calculated spectrum is for 5 K. (Reproduced from [129] with permission.)

populated. These rotational branching ratios for ionization of this valence 2π orbital differ somewhat from those generally seen in ZEKE spectra of Rydberg states.

In Fig. 18, only small changes in total angular momentum are observed ($|\Delta J| \leq \frac{5}{2}$) with branch intensities that fall-off rapidly as $|\Delta J|$ increases. The observation of branches with $|\Delta J| \leq \frac{5}{2}$ suggests that the dominant contributions to the PE matrix element at threshold are from $\ell \leq 1$. This result differs from expectations based on the calculated magnitudes of these partial wave dipole elements $|D_\ell^{(-)}|$, which are shown in Fig. 20, where the $d(\ell = 2)$ and $f(\ell = 3)$ components are seen to be substantial [129]. Unusually strong s and d waves are also predicted in addition to

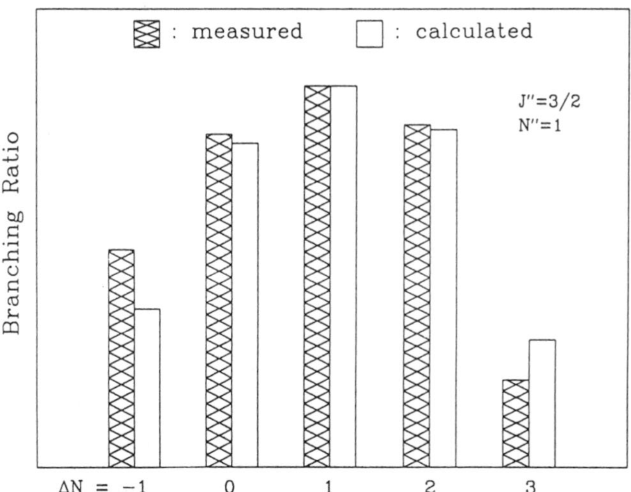

Figure 19. (*a*) measured ▩ and (*b*) calculated □ ZEKE photoelectron spectra for single-photon ionization of the $N' = 1$, $J'' = \frac{3}{2}$ level of the $X\,^2\Pi_{1/2}$ state of NO by coherent VUV radiation. (Reproduced from [129] with permission.)

the p and f partial waves expected in photoionization of the 2π orbital with its 84% $d(\ell_0 = 2)$ character. This behavior reflects the significant coupling of angular momentum in the PE continuum due to torques associated with the molecular ion potential. Given that these calculated PE matrix elements yield an accurate representation of the measured threshold PE spectra (see Figs. 18 and 19), one may conclude that photoexcitation into high ℓ partial waves is not accompanied by observable changes in ion-core angular momentum of comparable magnitude. That such large changes in angular momentum $\Delta J(\Delta N)$ are not seen in these spectra in spite of the significant magnitudes of the PE matrix element for higher $\ell(=3)$ is probably due to interference between these partial waves in the PE continua. The effect of such interference on ion rotational distributions has also been seen in $(2+1)$ REMPI via the $f\,^1\Pi\,(3p\sigma)$ and the $g\,^1\Delta(3p\pi)$ Rydberg states of NH [112].

2. *Single-Photon Ionization of OH*

Figure 21 shows (*a*) measured and (*b*) calculated ZEKE–PFI photoelectron spectra for single-photon ionization of OH ($X\,^2\Pi_i, v'' = 0$) leading to OH^+ ($X\,^3\Sigma^-, v_+ = 0$) by coherent VUV radiation [130]. Agreement between the calculated and measured spectra is again quite encouraging. The PE spectrum [Fig. 21(*b*)] was calculated assuming a kinetic energy of

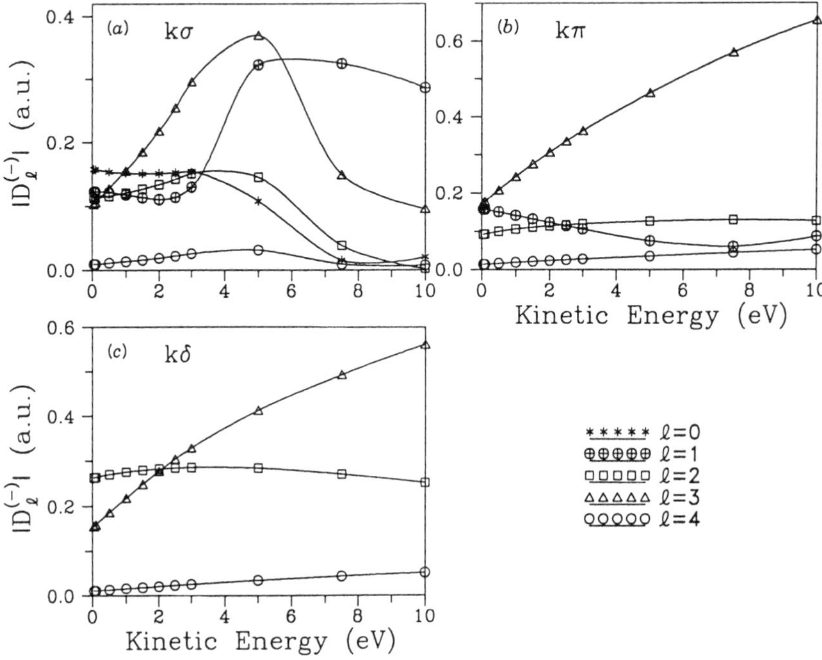

Figure 20. Magnitude of the partial wave components of the PE matrix element $|D_\ell^{(-)}|$ for photoionization of the $X\,^2\Pi$ ground state of NO. (*a*) $2\pi \rightarrow k\sigma$, (*b*) $2\pi \rightarrow k\pi$, and (*c*) $2\pi \rightarrow k\delta$ ionization channels. (Reproduced from [129] with permission.)

25 meV and a temperature of 220 K. This spectrum was convoluted with a Gaussian detection function with an fwhm of 3 cm^{-1}. Furthermore, the ground state was assumed unaligned and the *e*/*f* components of rotational levels of the ground state were equally populated. In Fig. 21(*a*) each branch is associated with a letter designation referring only to the change in angular momentum apart from spin, that is, $\Delta N = N^+ - N''$, a subscript denoting the spin (*F* level) of the originating state, and a double prime indicating the quantum numbers associated with the ground state of the neutral molecule.

The most prominent branches are the P_1, Q_1, and R_1, which are associated with the normal one-photon allowed electronic transitions since $|\Delta N| \leq 1$. The corresponding branches originating from the F_2 components of the ground state are nearly as strong since the spin–orbit splitting is small relative to the rotational temperature. The observation of one-photon branches with $|\Delta N| > 1$ (*N*, *O*, and *S*) has no counterpart in non-ionizing electronic transitions and is a consequence of the parti-

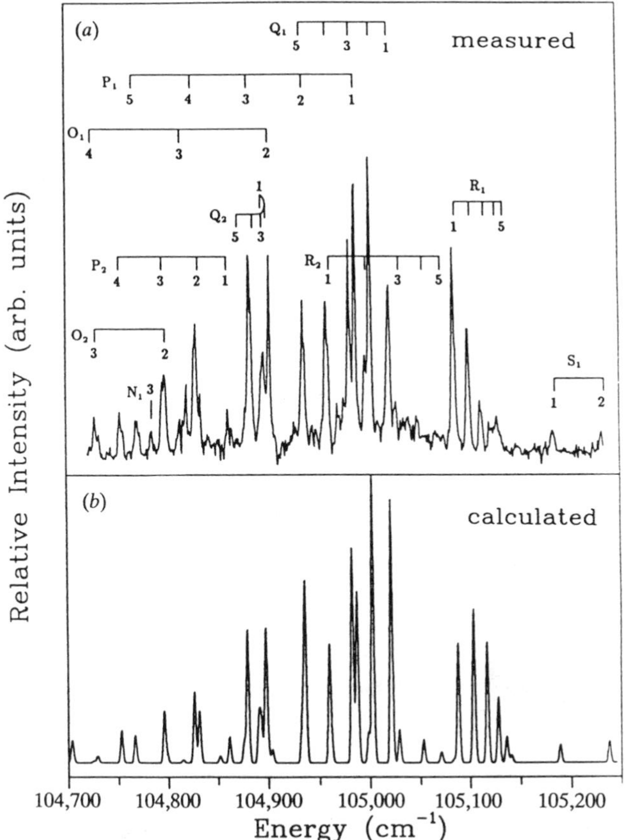

Figure 21. (*a*) measured and (*b*) calculated pulsed-field ionization spectra of OH^+ ($X\,^3\Sigma^-, v_+ = 0$) for the $X\,^2\Pi$ ground state of OH. The branch designations refer to ΔN and the subscript designates whether the transition originates from the F_1 or the F_2 level. The triplet splitting in the ion is not resolved and numbers on each ladder designate the value of N''. (Reproduced from [130] with permission.)

tioning of angular momentum between the ion core and PE in the photoionization process. Of particular interest here is the low intensity of the *S* branches compared to the *O* branches, both of which involve the same absolute change in angular momentum, that is, $|\Delta N| = 2$. This behavior is also observed in the threshold photoionization spectra of N_2O [140] and HCl [30, 145]. It is also evident in several $(1 + 1')$ REMPI–ZEKE photoelectron spectra for a single rotational level of the $A\,^2\Sigma^+$ ($3s\sigma$) Rydberg state of NO [7, 11, 131]. A field-induced autoionization mechanism involving the interaction of high *n* Rydberg states ($n > 150$)

with lower n levels $(20 < n < 80)$ converging to thresholds for higher rotational levels of the cation has successfully accounted for this behavior [140].

3. *Single-Photon Ionization of CO and N_2*

Figure 22 shows the ZEKE–PFI photoelectron spectra for single-photon ionization of rotationally cold CO ($X\,^1\Sigma^+$) molecules leading to the $v_+ = 0$ level of CO^+ ($X\,^2\Sigma^+$) by coherent extreme ultraviolet (XUV) radiation [132]. The calculated spectrum assumes a temperature of 8 K, a PE energy of 50 meV, and is convoluted with a Gaussian detection

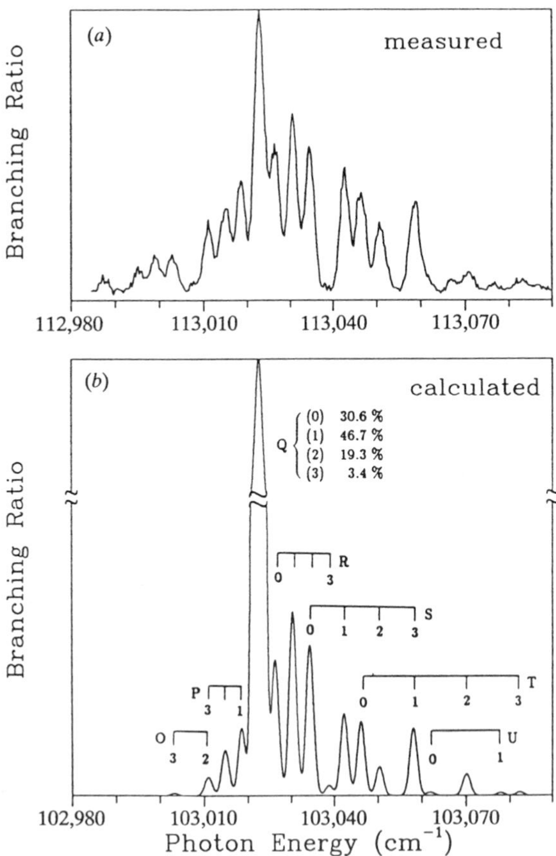

Figure 22. (*a*) measured and (*b*) calculated ZEKE photoelectron spectra for single-photon ionization of rotationally cold CO ($X\,^1\Sigma^+$) by coherent XUV radiation. The calculated spectrum is for 8 K. (Reproduced from [132] with permission.)

function with an fwhm of 2 cm^{-1}. The letter designation on each branch again refers only to the change of angular momentum apart from spin. The agreement between these calculated and measured rotational branching ratios is excellent for all branches except the $Q(\Delta N = 0)$ branches. Although the $Q(N'' = 0, 1, 2)$ branches are the strongest in both spectra, its calculated spectral intensity is about 10 times larger than that of the $R(1)$ branch. This behavior is indicated by the broken scale in Fig. 22(*b*). The strength of the Q branches in these spectra is consistent with the parity selection rule, $\Delta N + \ell = odd$ [115, 116, 124, 125], and atomic-like behavior for photoionization of the 5σ orbital (45.5% s, 24.1% p, 25.2% d, and 2.7% f character at $R_e = 2.1322\, a_0$). Another factor that significantly enhances the Q branches is that these branches are much denser than the spectral resolution due to the small difference in the rotational constants for the ground states of CO and CO^+. The origin of this disagreement between these calculated and measured spectra is not yet clear.

The calculated spectral intensity of the $O(2)(\Delta N = -2)$ branch is about one half that of the $S(2)(\Delta N = 2)$ branch, while these branches have almost the same intensities in the measured spectrum of Fig. 22. Similar behavior favoring negative ΔN transitions is also seen in the $N(4)$, $O(4)$, $N(3)$, and $O(3)$ branches in the measured spectrum. Such enhancement of negative ΔN branches over their positive counterparts is due to field-induced rotational autoionization [140].

Figure 23 shows the (*a*) measured and (*b*) calculated ZEKE–PFI photoelectron spectra for single-photon ionization of rotationally cold N_2 $(X\,^1\Sigma_g^+, v'' = 0)$ molecules leading to the $v_+ = 0$ level of N_2^+ $(X\,^2\Sigma_g^+)$ by coherent XUV radiation [132]. Again, the calculated spectrum assumes a temperature of 8 K and a PE energy of 50 meV. This spectrum is convoluted with a Gaussian detection function with an fwhm of 2.7 cm^{-1}. Nuclear spin statistics are also taken into account in determining the populations of the rotational levels of N_2. Agreement between the calculated and measured spectra is very encouraging, except for the O branches. It is again apparent that the abnormal intensities of these O branches with negative $\Delta N(\Delta N = -2)$ seen in the measured spectrum must arise from rotational autoionization [140]. Note that no $\Delta N = odd$ (N, P, R, or T branch) transitions occur in single-photon ionization of the $3\sigma_g$ orbital of the $X\,^1\Sigma_g^+$ state of N_2 in both the measured and calculated spectra. However, these $\Delta N = odd$ branches are quite prominent in the PE spectrum for photoionization of the 5σ orbital of the $X\,^1\Sigma^+$ state of CO (Fig. 22). On the basis of the selection rule of Eq. (1.26), these $\Delta N = odd$ transitions must arise from even partial wave contributions to the PE matrix elements. Since the $3\sigma_g$ orbital of N_2 has only even partial

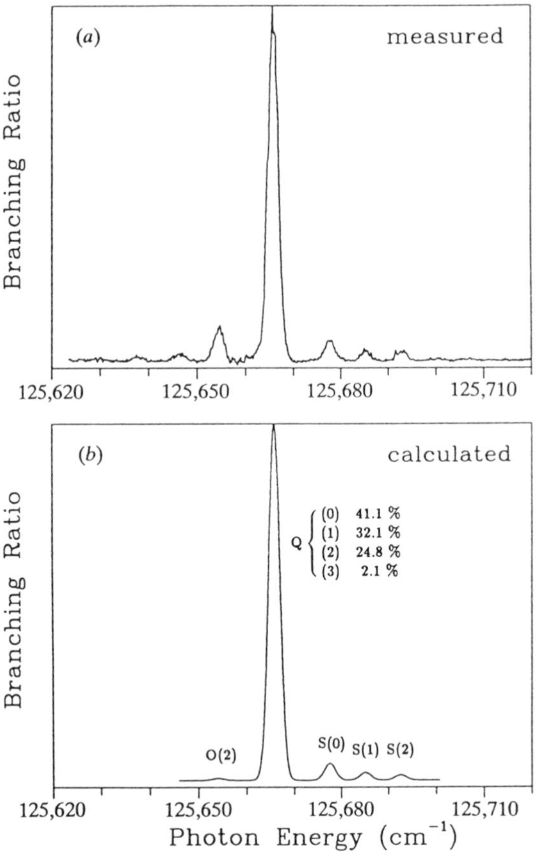

Figure 23. (*a*) measured and (*b*) calculated ZEKE photoelectron spectra for single-photon ionization of rotationally cold $N_2(X\,^1\Sigma_g^+)$ by coherent XUV radiation. The calculated spectrum is for 8 *K*. (Reproduced from [132] with permission.)

wave character, the PE has only odd partial wave components for single-photon ionization of this orbital, and, hence, no $\Delta N = odd$ transitions can occur.

Another significant difference between the calculated and measured ZEKE–PFI spectra of Figs. 22 and 23 is the intensities of the Q branches relative to other branches, such as the $S(0)$ branch for N_2 and CO. The measured N_2 ZEKE–PFI spectrum shows a much larger Q-to-S branch intensity ratio than for CO. This ratio is in contrast to the calculated spectra where these ratios are comparable for N_2 and CO. Since the *even* partial wave character of the 5σ (45.5% s and 25.2% d) orbital of CO and the $3\sigma_g$ (55.5% s and 41.0% d) orbital of N_2 are quite similar, we do not expect the underlying photoionization dynamics of these orbitals to differ significantly. This is the case in the calculated spectra. The reason for this disagreement between the calculated and measured intensities for CO is not clear.

4. (2+1) REMPI of HBr

In Fig. 24 we compare our calculated ion rotational distributions of HBr^+ in their X $^2\Pi_{1/2}$ ground state for (2+1) REMPI of HBr via the $S(2)$ branch of the $F\ ^1\Delta_2(5p\pi)$ Rydberg state with the experimental data of Xie and Zare [133, 148]. Our studies of these spectra were motivated by the recent measurements of Xie and Zare in which the populations of the individual parity components of each ion rotational level were obtained using laser-induced fluorescence [148]. These spectra show a strongly (−) parity-favored ion rotational distribution for photoionization of the (+) parity component of the $J_i = 4$ level of the $5p\pi$ Rydberg orbital of the $F\ ^1\Delta_2$ state, which has about 97% $p(\ell_0 = 1)$ character. These (−) parity-favored ion distributions can be readily understood on the basis of the parity selection rules [115, 116, 124, 125, 133] of Eq. (1.25) and the dominance of the PE matrix element by its $\ell = 0$ and $\ell = 2$ angular momentum components for photoionization of the $5p\pi$ orbital with 96% p character. However, the most striking result is that, in addition to the dominant population seen in the (−) parity components of the Λ doublet of the ion rotational levels, about 20% population is observed in the (+) parity components that cannot be accounted for on the basis of atomic-like behavior. Earlier theoretical studies of these ion rotational distributions had predicted extremely small populations for the (+) parity levels [134]. Autoionization was subsequently proposed as a possible underlying mechanism responsible for the population observed in the (+) parity components of these ion rotational levels [134].

Figure 24 shows our calculated spectra for photoionization of an optically aligned $J_i = 4$ level, and an unaligned $J_i = 4$ level, along with the measured spectra of Xie and Zare [133, 148]. The agreement between the calculated and measured ion distributions is excellent. Note that, on the basis of selection rules [115, 116, 124, 125, 133], this 20% population seen in the (+) parity component (cross-hatched bars) of the Λ doublet is due to odd partial wave components of the PE matrix elements that arise from angular momentum coupling in the molecular PE orbitals. The present results clearly demonstrate that the 20% population seen in the (+) levels arises quite naturally in a correct quantitative description of the direct photoionization and an autoionization mechanism need not be

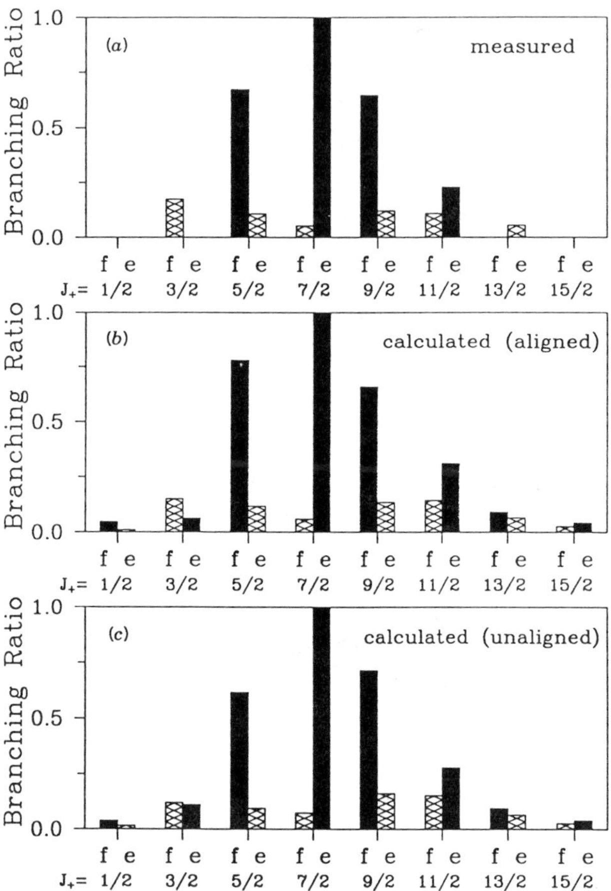

Figure 24. The measured and calculated rotational distributions of HBr^+ for (2 + 1) REMPI via the $S(2)$ branch of the $F\,^1\Delta_2$ state. The calculated distributions in (*b*) and (*c*) are for an aligned and unaligned resonant state, respectively, and for a PE energy of 2.33 eV. (Reproduced from [133] with permission.)

invoked. Examination of the PE matrix element $r_{fi}^{\ell\lambda\mu}$ of Eq. (1.24) reveals that the f wave of the $5p\pi \rightarrow k\delta$ ionization channel makes the dominant contribution to the population of these (+) parity levels of the ion. Such behavior is entirely nonatomic-like [133]. Comparison of the ion distributions for an unaligned $J_i = 4$ level with those for an optically aligned level and with the measured spectra serves to illustrate that, although not large, the effect of alignment can be important. Furthermore, the ZEKE photoelectron spectra for photoionization of this $F\,^1\Delta_2$

Rydberg state of HBr is predicted to display the same nonatomic-like behavior seen in the spectra of Fig. 24.

5. (2 + 1) REMPI of OH

Figure 25 shows the measured and calculated rotationally resolved PE spectra for (2 + 1) REMPI of OH via the $O_{11}(11)$ branch of the

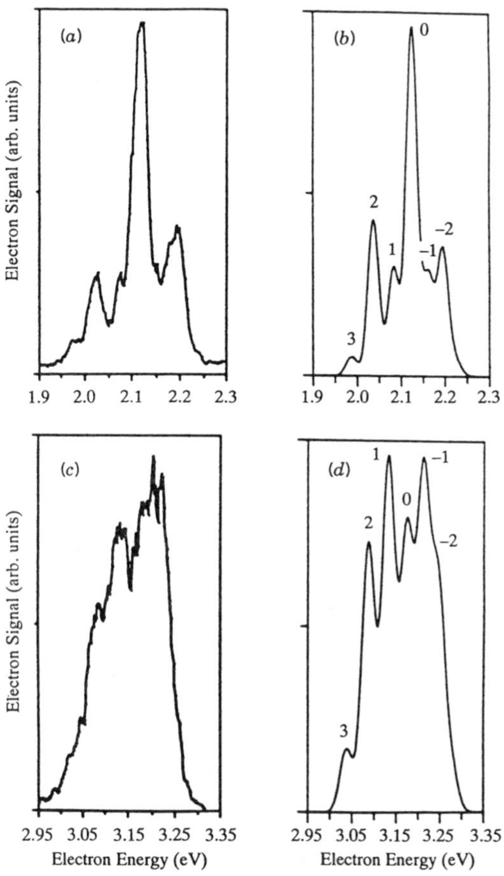

Figure 25. Measured and calculated rotationally resolved photoelectron spectra for (2 + 1) REMPI of OH: (*a*) measured spectrum for the $D\,^2\Sigma^-$ ($3p\sigma$) state, $v_i = 0 \rightarrow v_+ = 0$, $O_{11}(11)$ rotational branch; (*b*) calculated $D\,^2\Sigma^-$ photoelectron spectrum, assuming a Gaussian line shape with an fwhm of 30 meV; (*c*) measured spectrum for the $3\,^2\Sigma^-$ ($4s\sigma$) state, $v_i = 0 \rightarrow v_+ = 0$, O_{11} (11) rotational branch; (*d*) calculated $3\,^2\Sigma^-$ photoelectron spectrum, assuming a Gaussian line shape with an fwhm of 35 meV. The labeling of peaks in the calculated spectra indicates the change of rotational quantum number $\Delta N = N^+ - N_i$. (Reproduced from [113] with permission.)

$D\ ^2\Sigma^-$ ($3p\sigma$) [Fig. 25(a) and (b)] and the $3\ ^2\Sigma^-$ ($4s\sigma$) [Fig. 25(c) and (d)] Rydberg states [113]. For ionization via the $D\ ^2\Sigma^-$ state, we observe strong $\Delta N = even$ signals in the PE spectrum, in contrast to the $\Delta N = odd$ distribution expected for ionization of a $3p\sigma$ (34.9% s and 63.3% p character at the R_e of 2.043 a_0) Rydberg orbital with its dominant $3p\sigma \rightarrow ks, kd$ components. On the other hand, the rotationally resolved PE spectra for the $3\ ^2\Sigma^-$ Rydberg state of Figs. 25(c) and (d) reveal a qualitatively different and much broader distribution with prominent $\Delta N = even$ and $\Delta N = odd$ transitions. The appearance of these spectra arises from greater ℓ mixing in this higher Rydberg orbital (54.3% s and 42.7% p character at $R = 2.043\ a_0$).

To illustrate the rich photoionization dynamics of these strong *even* ΔN peaks in the spectra, Fig. 26 shows the magnitude of the (incoming-wave normalized) partial wave amplitude $|D_\ell^{(-)}|$ as a function of PE kinetic energy for the (a) $3p\sigma \rightarrow k\sigma$ and (b) $3p\sigma \rightarrow k\pi$ channels for photoionization of the $3p\sigma$ orbital of the $D\ ^2\Sigma^-$ Rydberg state of OH. Two Cooper minima, which are due to sign changes in the $d(\ell = 2)$ components around the minimum in $|D_\ell^{(-)}|$, are clearly seen at a kinetic energy of about 3.0 eV for the $k\sigma$ and $k\pi$ channels. The actual sign changes in the dipole matrix elements occur in the principal value (standing-wave normalized) dipole amplitude D_ℓ^P, as shown in the insets of Fig. 26 for the $\ell = 2$ wave. The energy positions of both minima in $|D_\ell^{(-)}|$ differ somewhat from those of Cooper zeros in D_ℓ^P. These shifts reflect the influence of angular momentum coupling in the electronic continua on the transformation relating D_ℓ^P to $D_\ell^{(-)}$. The strong *even* $\Delta N = 0$ peaks of Figs. 26(a) and (b) can, in fact, be shown to be due to the presence of Cooper minima in the $\ell = 2$ wave of the PE channels. Depletion of the d wave ($\ell = 2$) contribution to the PE matrix element in the vicinity of both Cooper minima subsequently enhances the relative importance of the odd ℓ waves and, hence, that of the even ΔN rotational peaks. Similar effects of Cooper minima on rotationally resolved PE spectra have also been observed for REMPI of the $f\ ^1\Pi$ ($3p\sigma$) state of NH [112] and the $D\ ^2\Sigma^+$ ($3p\sigma$) state of NO [135].

B. Nonlinear Molecules

1. *Single-Photon Ionization of H_2O*

Figure 27(a) shows the rotationally resolved ZEKE–PFI spectrum of H_2O recently reported by Tonkyn et al. [31] for single-photon ionization of the $1b_1$ orbital by coherent VUV radiation. This spectrum can be assigned to two types of rotational transitions, corresponding to specific changes in K_a and K_c. Most of the strong spectral lines could be classified

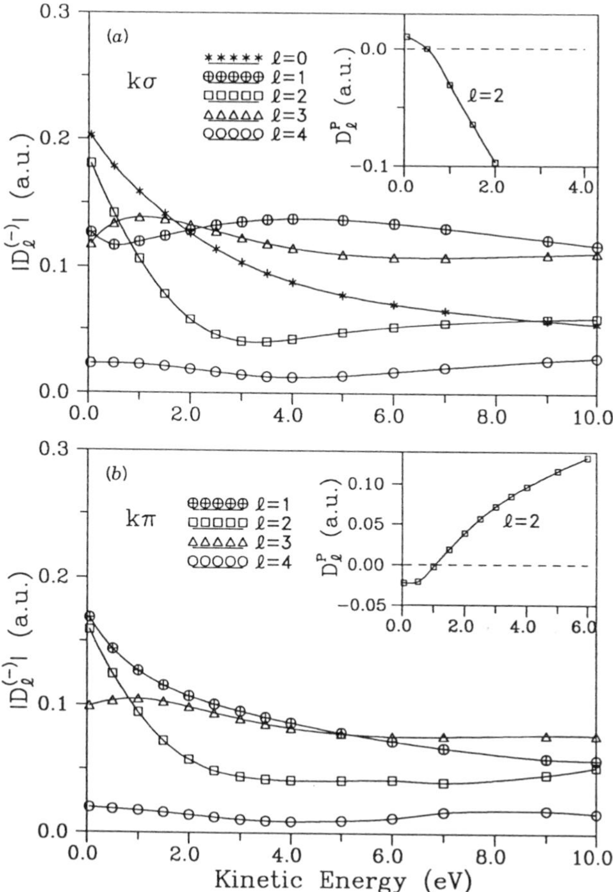

Figure 26. Magnitude of the partial wave photoionization matrix elements as a function of kinetic energy for the (*a*) $3p\sigma \rightarrow k\sigma$ and (*b*) $3p\sigma \rightarrow k\pi$ channels for the $D\ ^2\Sigma^-$ ($3p\sigma$) state of OH. The inset shows the principal-value dipole amplitude D_ℓ^P for the $\ell = 2$ component. (Reproduced from [44] with permission.)

as type *c* rotational transitions ($\Delta K_a = odd$, $\Delta K_c = even$), but type *a* transitions ($\Delta K_a = even$, $\Delta K_c = odd$) are also clearly evident. Figure 27(*b*) shows the calculated ion rotational distributions of Lee et al. [32, 136] for photoionization of the $1b_1$ orbital of the $\tilde{X}\ ^1A_1$ (000) ground state of jet-cooled H_2O leading to the $\tilde{X}\ ^2B_1$ (000) ground state of the ion. A PE kinetic energy of 50 meV and a rotational temperature of 15 K are assumed in these calculations. Furthermore, we assume that there is no spin exchange taking place during the jet-cooled expansion of room

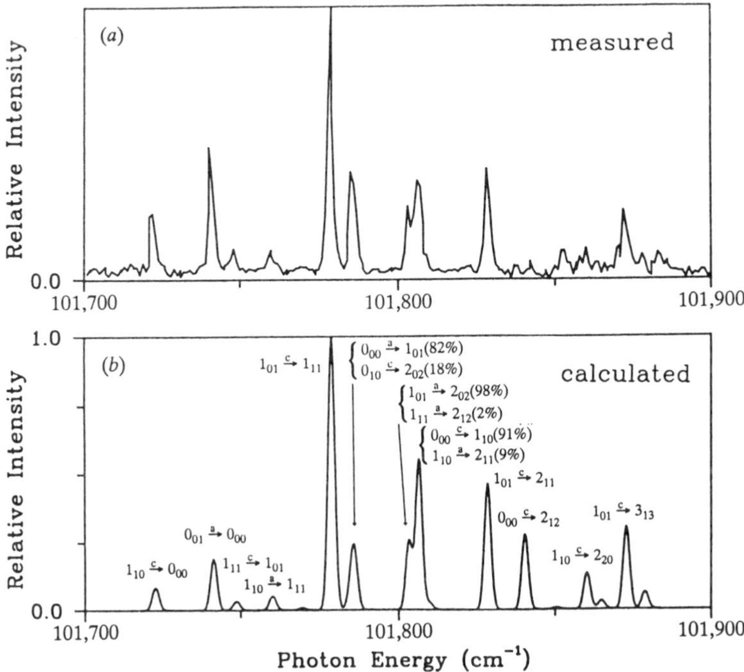

Figure 27. (*a*) measured and (*b*) calculated ion rotational distributions for single-photon ionization of the $1b_1$ orbital of the $\tilde{X}\,^1A_1$ ground state of jet-cooled H_2O. The calculated spectrum is for 15 K and has an fwhm of 1.5 cm^{-1}. The *a* and *c* labels indicate type *a* and type *c* transitions, respectively. (Reproduced from [32] with permission.)

temperature water. The calculated spectrum is convoluted with a Gaussian detection function with an fwhm of 1.5 cm^{-1}. The agreement between the calculated and measured spectra is clearly encouraging except for the $0_{00} \rightarrow 2_{12}$ transition for which the calculated intensity is somewhat stronger than that of the measured value.

The underlying dynamics of these PE spectra is quite rich. From the parity selection rules of Eqs. (1.44) and (1.45), we obtain [32, 136]

$$\Delta K_a + \ell = odd \tag{1.53}$$

and

$$\mu + \lambda = \Delta K_b \tag{1.54}$$

Here we assume that the molecular z axis coincides with the C_2 symmetry axis and the x axis lies in the plane of the molecule (ion). The molecular x, y, and z axes hence coincide with the a, c, and b axes, respectively. For photoionization of the $1b_1$ orbital of H_2O, $\mu + \lambda$ is always odd and, hence, from Eq. (1.46) we have

$$\Delta K_a + \Delta K_c = odd \tag{1.55}$$

Clearly, both types *a* and *c* transitions are allowed and type *b* [$\Delta K_a = even(odd)$ and $\Delta K_c = even(odd)$] transitions are forbidden. Furthermore, Eq. (1.53) shows that type *a* transitions ($1_{01} \rightarrow 0_{00}$, $0_{00} \rightarrow 1_{01}$, $1_{01} \rightarrow 2_{02}$, and $1_{11} \rightarrow 2_{12}$) arise from odd (almost pure *p*) wave contributions to the PE matrix element of Eq. (1.42). These *p* waves of the ka_1 and kb_1 continua are entirely molecular in origin, since the almost pure *p* (99.7%) character of the $1b_1$ orbital would lead only to *s* and *d* (*even*) PE continua in an atomic-like picture. The strong type *c* transitions in these spectra arise from *s* and *d* (even) wave components of the PE matrix elements. The ka_2 continuum makes almost no contribution to these type *a* transitions due to the negligible *f* (*odd*) wave. In a recent study using MQDT, Child and Jungen [137] predicted that only type *c* transitions are allowed for photoionization of the $1b_1$ orbital of ground-state water. Gilbert and Child [138] further proposed a rotational autoionization mechanism, based on polarization-induced quasiautoionizing state-mixing between the $1b_1 \rightarrow nd$ and $1b_1 \rightarrow np$ Rydberg series, in an effort to account for these type *a* transitions. In contrast, the above analysis [32, 136] shows that these type *a* transitions arise quite naturally in a correct quantitative description of the direct molecular photoionization process itself and a polarization-induced autoionization mechanism need not be invoked.

2. *Single-Photon Ionization of H_2CO*

Figure 28 shows the (*a*) measured [139] and (*b*) calculated rotationally resolved ZEKE–PFI spectra of H_2CO for single-photon ionization of the nonbonding $2b_2$ orbital by coherent VUV radiation. The calculated spectrum assumes a rotational temperature of 7 K, a PE kinetic energy of 50 meV, and is convoluted with a Gaussian detection function with an fwhm of 1.8 cm^{-1}. Furthermore, we assume that no spin exchange takes place during the jet-cooled expansion of room temperature formaldehyde, that is, a population ratio of 2:1 is kept for the ortho (*B* symmetry) to para (*A* symmetry) species. The agreement between calculated and measured spectra is very encouraging.

In our calculations we assume that the molecular z axis coincides with the C_2 symmetry axis and the x axis lies in the plane of the formaldehyde molecule (ion). The molecular x, y, and z axes hence coincide with the b,

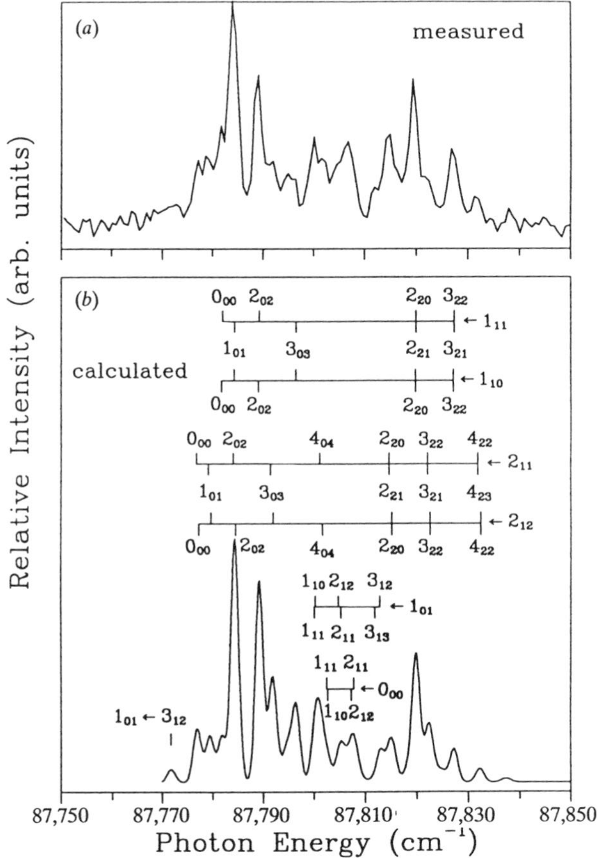

Figure 28. (*a*) measured and (*b*) calculated ion rotational distributions for single-photon ionization of the $2b_2$ orbital of the $\tilde{X}\,^1A_1$ ground state of jet-cooled H_2CO. The calculated spectrum is for 7 K and has an fwhm of $1.8\,\text{cm}^{-1}$. (Reproduced from [139] with permission.)

c, and a axes, respectively. From the parity selection rules of Eqs. (1.44) and (1.45), we obtain

$$\Delta K_b + \ell = odd \tag{1.56}$$

and

$$\mu + \lambda = \Delta K_a \tag{1.57}$$

For photoionization of the $2b_2$ orbital of H_2CO, $\mu + \lambda$ is always odd and, hence, from Eq. (1.46) we have

$$\Delta K_b + \Delta K_c = odd \tag{1.58}$$

Clearly, both type *c* ($\Delta K_a = odd$ and $\Delta K_c = even$) and type *b*($\Delta K_a = odd$ and $\Delta K_c = odd$) transitions are allowed, and type *a*($\Delta K_a = even$ and $\Delta K_c = odd$) and other type *b* ($\Delta K_a = even$ and $\Delta K_c = even$) transitions are forbidden. Furthermore, Eqs. (1.56–1.58) show that the allowed type *b* transitions arise from *odd* partial wave contributions to the PE matrix elements of Eq. (1.42), whereas the type *c* transitions arise from *even* angular momentum components of the PE.

In Fig. 28(*b*), we label several of the more important transitions out of rotational levels of the ground state of the neutral species leading to different rotational levels of the ion. The quantum numbers used as labels are $N_{K_aK_c}$. Note that in this figure we use the same set of quantum numbers to designate type *c* transitions out of the 1_{11} level and type *b* transitions out of the 1_{10} level, since the energies of these transitions are essentially equal. A similar labeling is also adopted for transitions out of the 2_{11} and 2_{12} rotational levels. Since the moment of inertias I_b and I_c are almost the same for H_2CO (H_2CO^+), the rotationally resolved ZEKE–PFI spectra are very congested. Unlike the spectra for H_2O, type *b* transitions, which arise from *odd* waves, cannot be distinguished from type *c* transitions, which arise from *even* waves.

Figure 29 shows the calculated threshold PE spectra for the individual type *c* [Fig. 29(*a*)] and type *b* [Fig. 29(*b*)] transitions of Fig. 28 for single-photon ionization of the $2b_2$ orbital of the ground state of H_2CO. Furthermore, our calculated spectrum shows a total intensity for type *b* transitions about twice as large as that for type *c* transitions. This large intensity of the type *c* transitions cannot be accounted for on the basis of atomic-like propensity rules since the $2b_2$ orbital of the H_2CO has 4.8% *p*, 80.0% *d*, 3.7% *f*, 9.1% *g* ($\ell_0 = 4$), and 1.5% *h* ($\ell_0 = 5$) character. The strong even-wave character of the $2b_2$ orbital would be expected to lead to dominant *odd* partial wave contributions to the PE matrix elements and, hence, to dominant type *b* transitions. This unexpected contribution from *even* partial waves of the PE matrix element is quite molecular in origin and arises from strong ℓ-mixing due to the nonspherical molecular ion potentials. Such comparisons between measured and calculated spectra can be helpful in unraveling severely congested PE spectra. Note that a shape resonance is predicted in the ka_1 photoelectron continuum at kinetic energies of about 5 eV [139]. However, this shape resonance

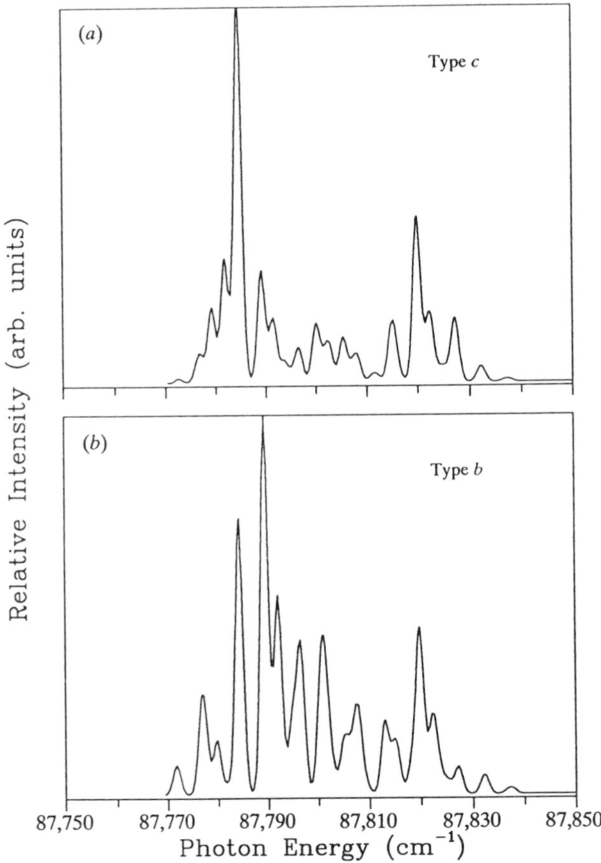

Figure 29. Calculated PE spectra for the (*a*) type *c* and (*b*) type *b* transitions of Fig. 28(*b*) for single-photon ionization of the $2b_2$ orbital of the $\tilde{X}\,^1A_1$ ground state of jet-cooled H_2CO. The type *b* to type *c* intensity ratio is about 1.9. (Reproduced from [139] with permission.)

should have very little effect on the threshold photoionization ZEKE–PFI spectrum.

3. Single-Photon Ionization of CH_3

Figure 30 shows the (*a*) measured [89] and (*b*) calculated rotationally resolved ZEKE–PFI spectra of CH_3 ($\tilde{X}\,^2A_2''$) for single-photon ionization of the $1a_2''$ orbital by coherent VUV radiation leading to the $\tilde{X}\,^1A_1'$ ground state of the ion. The calculated spectrum assumes a rotational temperature of 250 K and a PE kinetic energy of 50 meV. This spectrum is

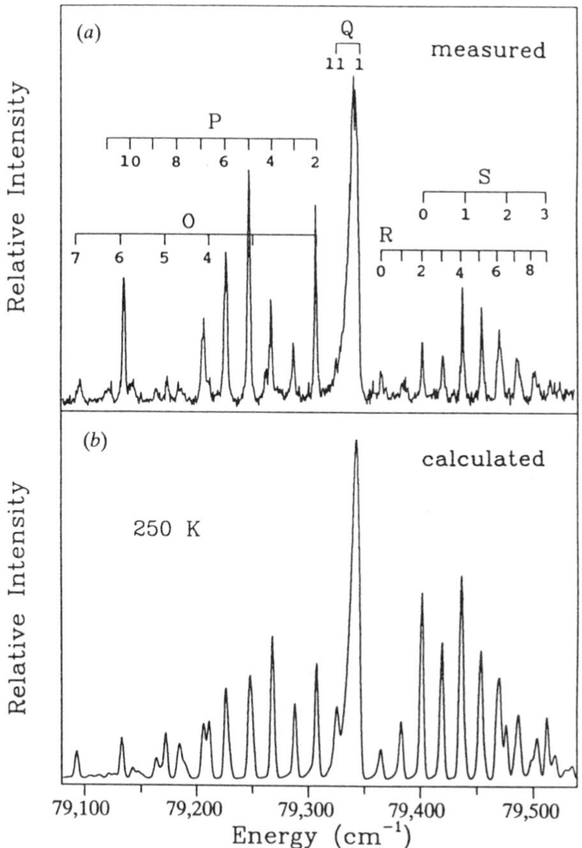

Figure 30. (*a*) measured and (*b*) calculated ion rotational distributions for single-photon ionization of the $1a_2''$ orbital of the $\tilde{X}\,^2A_2''$ ground state of jet-cooled CH_3 leading to the $\tilde{X}\,^1A_1'$ ground state of the ion. The calculated spectrum is for 250 K and has an fwhm of 2.5 cm^{-1}. (Reproduced from [139] with permission.)

convoluted with a Gaussian detection function that has a fwhm of 2.5 cm^{-1}. The CH_3 and CH_3^+ are oblate symmetric tops, belonging to the D_{3h} point group. For the D_{3h} group, the *total* wave function of CH_3 should have either A_2' or A_2'' symmetry to obey the Pauli principle. The wave functions for ortho and para species have A_1' $(I=\frac{3}{2})$ and E' $(I=\frac{1}{2})$ symmetries, respectively. Since $\Gamma_{\text{spin}} \otimes \Gamma_{\text{rot}} \supset \Gamma_{\text{total}}$, the rotational wave function must have A_2', A_2'', E', or E'' symmetry. Therefore, the rotational levels belonging to A_1' and A_1'' symmetries are not initially populated, that is, levels with $J = even$, $K = 0$ and the A_1 $(K = 3n)$ level of the $A_1 - A_2$ pair are forbidden. Due to the spin statistical weight of

4:2 between ortho and para species, the $K = 3n$ levels (A_2' or A_2'') have a statistical weight twice that of the $K = 3n + 1$ levels (E' or E'').

Agreement between the calculated and measured spectra is very encouraging in spite of some discrepancies resulting from rotational autoionization at negative ΔN transitions [140]. Our calculations reveal that $\Delta K = 0$ are the dominant transitions (up to 95%) and the $\Delta K = \pm 2$ (not labelled) transitions are much weaker. Even though the $\Delta K = odd$ transitions are also dipole allowed, their intensities are very weak and they do not appear in the spectrum. Note that $\Delta K = odd$ transitions are associated with $\mu + \lambda = odd$ partial waves of the PE, which, in turn, must originate from the $f(\ell_0 = 3)$ component of the $1a_2''$ orbital of the ground state. Since the $1a_2''$ orbital is essentially an out-of-plane $2p$ orbital localized on the central carbon atom, its f wave is negligible. This result leads to a $\Delta K = even$ selection rule.

V. DEVIATIONS OF ZEKE INTENSITIES FROM THE PREDICTIONS OF DIRECT IONIZATION

The understanding of ZEKE intensities has been an issue of keen interest since the first rotational structure was resolved for NO. A fundamental distinction between ZEKE band strengths and those derived from conventional PES can be clearly made and firmly grounded in differences long recognized between PES and TPES. The relation between ZEKE spectroscopy and TPES is more subtle. The determination of transition moments in both cases is a multichannel problem. Threshold scans invariably encounter intervals in which discrete–discrete interactions (autoionization in the case of TPES) distribute oscillator strength. The ZEKE methods, however, operate on bound states and incorporate delay as an essential element of threshold discrimination. This introduces the possibility for radiationless decay and raises the issue of intramolecular dynamics and the effect of relaxation processes on ZEKE observables.

Intramolecular coupling effects can be divided into two categories based on the nature of the interaction and its likely effect on intensities observed in high-resolution threshold photoionization. Discrete–discrete coupling distributes oscillator strength and lifetime over a band of isoenergetic ZEKE Rydberg states and lower principal quantum number higher core energy Rydberg states. The result generally leads to increased intensity, as lower thresholds, often ones forbidden by simple rotational angular momentum selection rules, borrow intensity from allowed transitions to interloping Rydberg states converging to higher limits. Alternatively and equivalently, the ZEKE spectrometer records allowed transitions to lower n Rydberg states that, because of their location with respect to an ionization threshold, acquire sufficient lifetime to survive the discrimination delay. Examples of such effects include discrete–discrete intensity-sharing interactions assignable to specific interlopers in the sparse threshold photoionization spectra of H_2 and N_2 [19, 25], as well as evidence for multichannel coupling identified in the ZEKE spectrum of NO [2, 10]. Generalization of such interactions explain asymmetries in rotational branch intensities found in early VUV studies of N_2O and HCl [30, 140], as well as anomalous transitions with seemingly forbidden large negative changes in rotational angular momentum in such systems as NO_2 [29], and specific rotational branch anomalies in (HCl) [141]. Results for the latter two systems will be discussed below.

Discrete-continuum interactions form a second category that until recently has not been fully appreciated. Such coupling confers continuum character on ZEKE Rydbergs. Regardless of its effect on the transition moment, such coupling reduces ZEKE intensity because of the delay required for threshold-electron (high-Rydberg) discrimination. In this respect, observable ZEKE intensities can differ substantially from conventional TPES band strengths. The most definitive example identified so far is that of HCl, where spin–orbit branching ratios in conventional PE spectroscopy are in accord with theoretical expectation, but are reversed in the ZEKE spectrum, as signal from the higher cross-section upper $^2\Pi_{1/2}$ threshold is depleted by spin–orbit autoionization [141, 142]. As detailed below, vibrational autoionizing continua have similar but mode-specific effect on threshold intensities for the production of vibrationally excited NO_2.

The study of intramolecular coupling on ZEKE intensities is complicated by the ability of external factors to influence both discrete–discrete and discrete–continuum coupling. External fields can assist coupling that facilitates intensity transfer by Stark mixing within ℓ manifolds converging to various core-rotational limits [143]. Stark mixing can also open continuum channels. Collisions have been implicated both in lifetime-lengthening ℓ-scrambling interactions [36], as well as detachment-type lifetime-shortening processes [35]. Complete understanding will come only with careful systematic study of suitable systems.

Hydrogen chloride is instructive, but because of its open-shell cation core, rotational effects are bound up with spin–orbit coupling. Nitrogen dioxide is in many respects more ideal for detailing intramolecular effects in threshold photoionization spectroscopy. Like NO, this molecule presents an electronic structure that is well suited for an examination of the intramolecular dynamics of threshold photoionization. It forms a closed-shell cation, leaving only vibrational–rotational interactions to play a primary role in threshold intramolecular coupling. Also like NO,

NO_2 is a system for which a great deal of spectroscopic information has been accumulated. It is less characterized theoretically, but efforts along those lines are in progress [144]. Among its advantages over NO for studies of coupling patterns near threshold is a smaller rotational constant, which places rotational thresholds nearer one another energetically and thus able to sample interactions with greater regularity from a denser manifold of interloping series. In addition, triatomic NO_2 has more than one vibrational mode, so discrete continuum effects that involve different pathways for vibrational relaxation can be compared in the same molecule. The following sections compare theory with experiment for HCl. Suggestive results are then taken as a guide of effects to expect in NO_2.

A. Effects of Spin–Orbit Relaxation and Rotational Coupling in the (2 + 1) Threshold Photoionization of HCl and DCl

1. *Rotational Line Intensities*

Figure 31 shows threshold photoionization scans over $^2\Pi_{3/2}$ and $^2\Pi_{1/2}$ thresholds taken from the $J' = 2(+)$ intermediate level in the $F\ ^1\Delta$ state of deuterium chloride (DCl). Observed rotational intensities are for the most part comparable. Overall, transitions to the lower spin–orbit state of the cation show somewhat greater signal than those to the upper component, and in most cases the lowest accessible rotational state for each component displays the largest delayed pulsed-field ionization signal. However, resonances to the higher rotational levels ($J^+ = \frac{7}{2}$ and $\frac{9}{2}$) of the $^2\Pi_{3/2}$ component are noticeably enhanced for DCl compared with their intensity in HCl [145].

The lowest frames of Fig. 31 compare these experimental threshold photoionization spectra to theoretical ones, calculated using the methods discussed above. Spin–orbit branching ratios implied in the calculated spectra shown here are arbitrary, a point to which we will return below. In other respects, experimental and theoretical spectra correspond reasonably, with the exception of higher rotational levels of the $^2\Pi_{3/2}$ substrate in DCl, which are noticeably enhanced compared to theoretical predictions, and to corresponding intensities in HCl. We suggest that these particular intensity anomalies for DCl arise from discrete–discrete intensity-sharing interactions with underlying Rydberg states converging to the upper spin–orbit component. When they are isoenergetic with threshold high Rydbergs, such states can form long-lived complex resonances with absorption cross sections that are enhanced by the low principal quantum number of the interloper. With a larger rotational constant, and thus sparse distribution of interlopers, such effects are

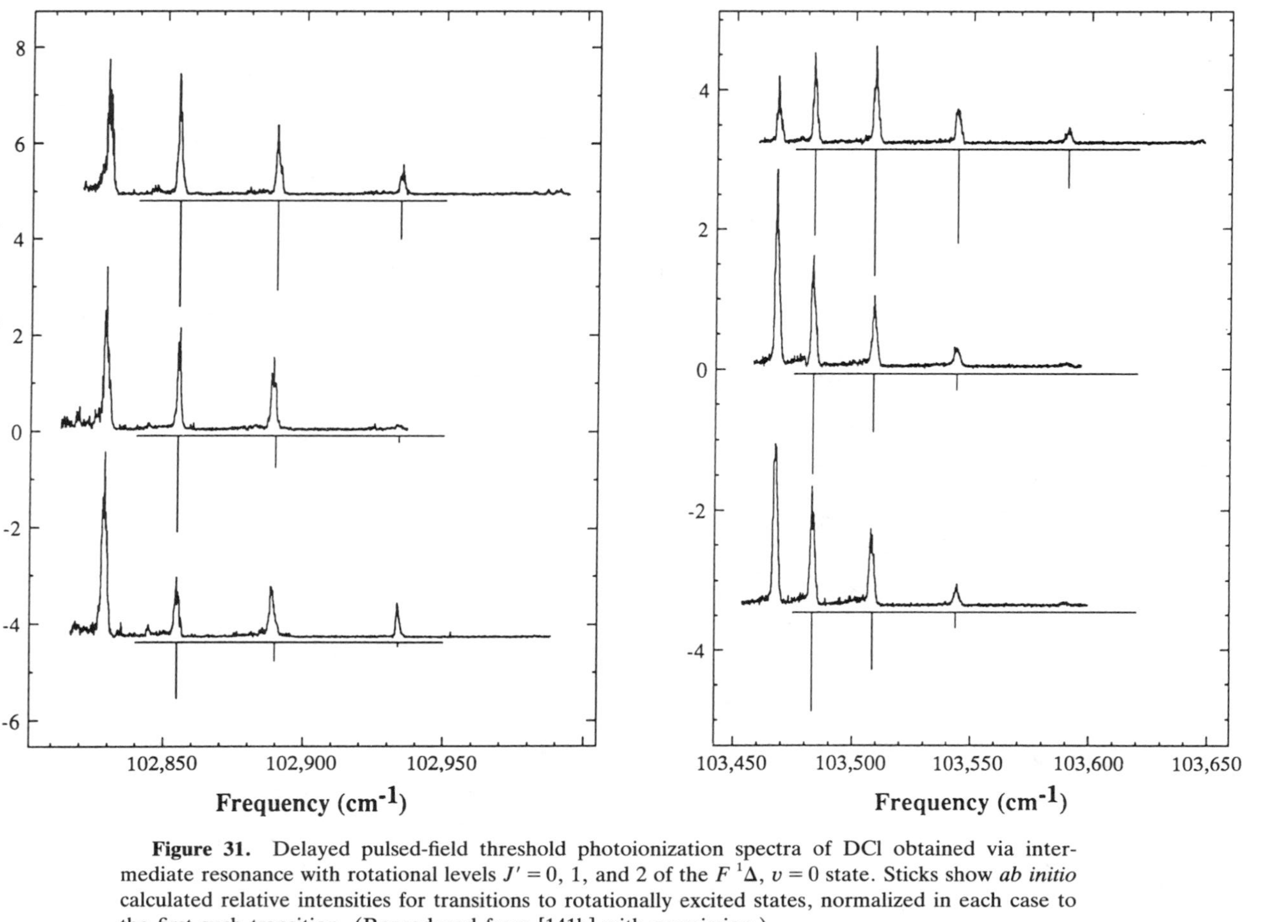

Figure 31. Delayed pulsed-field threshold photoionization spectra of DCl obtained via intermediate resonance with rotational levels $J' = 0$, 1, and 2 of the $F\,^1\Delta$, $v = 0$ state. Sticks show *ab initio* calculated relative intensities for transitions to rotationally excited states, normalized in each case to the first such transition. (Reproduced from [141b] with permission.)

expected to be less important for HCl. Discrete–discrete intensity sharing interactions across spin–orbit manifolds are obviously not possible for transitions to the upper $^2\Pi_{1/2}$ substrate. In accounting for such complex resonance effects just mentioned in transitions to the $^2\Pi_{3/2}$ threshold in HCl, and in DCl, we see a slight but systematic disagreement between experiment and calculation for the lowest, $J^+ = \frac{3}{2}$, rotational state. This level is special in that the high-Rydberg states sampled there by the ZEKE experiment have no accessible ionization continuum channels, and thus perhaps it is not surprising that delayed threshold-field ionization records a higher signal. The extent of this effect appears to depend on the distribution of Rydberg orbital angular momentum sampled by absorption, as evidenced by an apparent variation with the initial state selected for the photoionizing transition.

Figure 32 presents a set of spectra showing structure associated with $^2\Pi_{3/2}$ and $^2\Pi_{1/2}$ states of HCl^+ obtained by delayed pulsed-field threshold photoionization from the $J' = 0$, 1, and 2 levels of the $E\ ^1\Sigma^+$ intermediate state. For ionization at the $^2\Pi_{3/2}$ threshold from $J' = 2$, the strongest transition terminates on the lowest rotational state, $J^+ = \frac{3}{2}$. Much weaker structure appears in transitions to cation levels $J^+ = \frac{5}{2}$ and $\frac{7}{2}$. Weak resonances to $J^+ = \frac{5}{2}$ and $\frac{7}{2}$ are also seen from $J' = 1$ and to $\frac{5}{2}$ from $J^+ = 0$, but in both cases $J^+ = \frac{3}{2}$ still dominates. The lowest rotational state also yields the strongest resonance for threshold photoionization in the $^2\Pi_{1/2}$ manifold, but transitions terminating on higher rotational states are much more evident. Trends in rotational intensities from the $J' = 0$, 1, and 2 levels of the $g\ ^3\Sigma^-$ state parallel those found for the E state. As seen in Fig. 33, $J^+ = \frac{3}{2}$ dominates in the spectrum of the lower spin–orbit component of the ion, while rotational intensities are more broadly distributed for transitions to $^2\Pi_{1/2}$.

Comparison of the calculated intensities with measured ones for the $E\ ^1\Sigma^+$ and $g\ ^3\Sigma^-$ states, shows that the signal measured for the lowest rotational state in each spin–orbit component is disproportionally large. This finding is particularly evident for the lower $^2\Pi_{3/2}$ state of HCl^+. The magnitude of this effect compared with the comparatively small departures from predictions based on direct ionization out of the F state is particularly striking. Perturbative interactions of these $E\ ^1\Sigma^+$ and $g\ ^3\Sigma^-$ states with the valence $V\ ^1\Sigma^+$ state certainly contribute to this behavior.

Thus, we have a pattern of intensities in which first rotational levels appear enhanced, particularly for the lower spin–orbit substate, while higher thresholds appear to agree at least qualitatively with simple theoretical line strength expectations (apart from specific perturbations). This finding fits with a general picture of the effect of rapid decay channels on intensities in delayed pulsed-field threshold photoionization

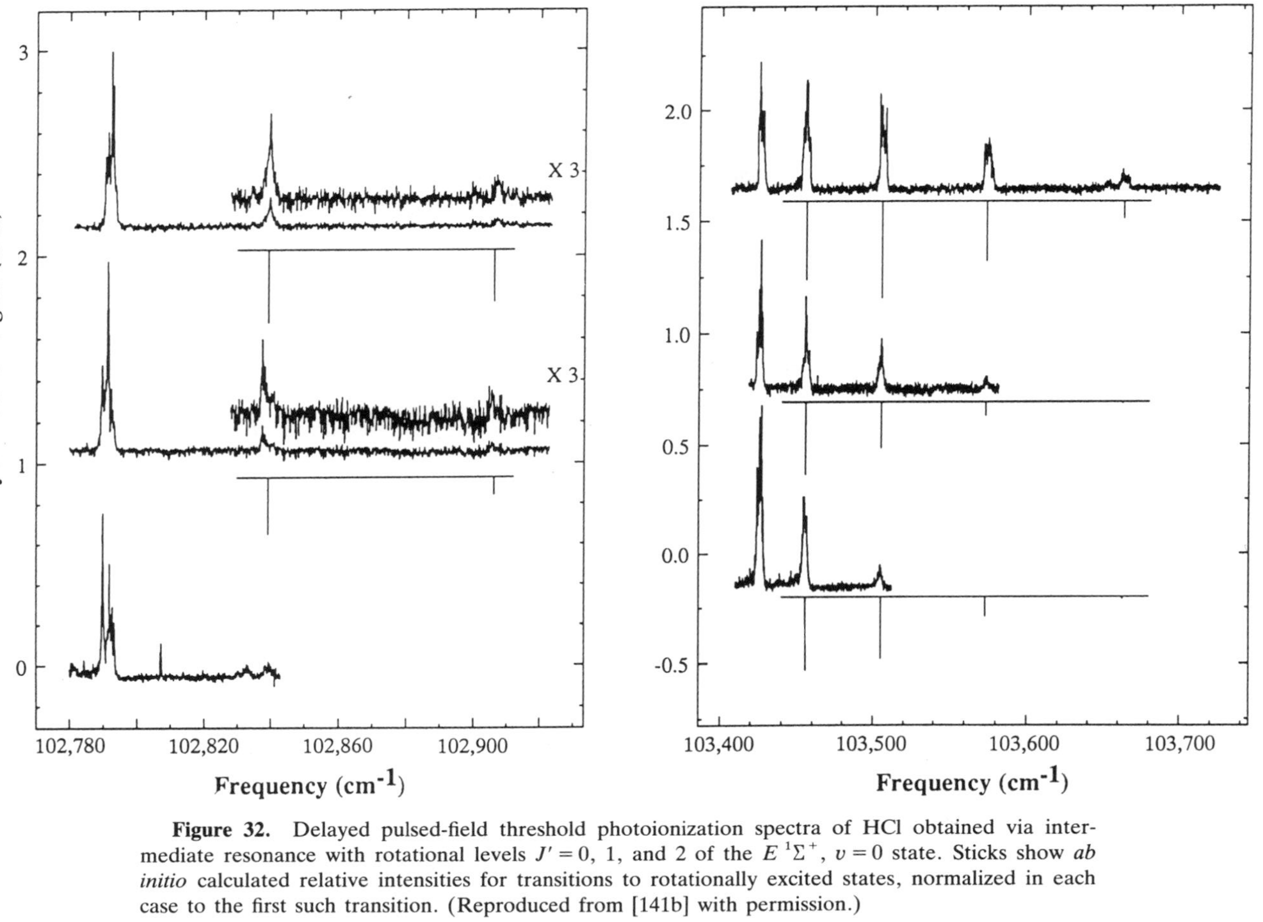

Figure 32. Delayed pulsed-field threshold photoionization spectra of HCl obtained via intermediate resonance with rotational levels $J' = 0$, 1, and 2 of the $E\,^1\Sigma^+$, $v = 0$ state. Sticks show *ab initio* calculated relative intensities for transitions to rotationally excited states, normalized in each case to the first such transition. (Reproduced from [141b] with permission.)

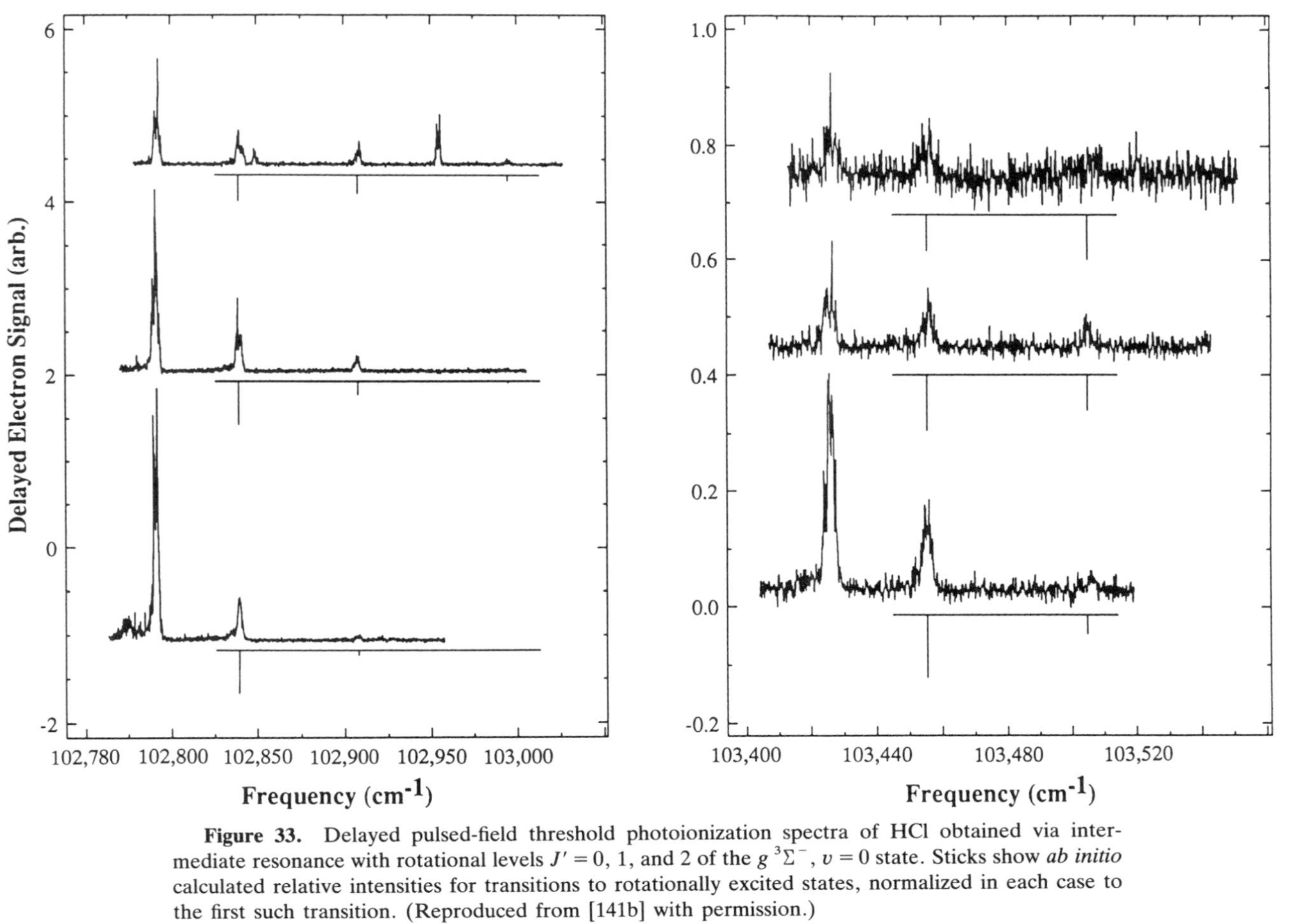

Figure 33. Delayed pulsed-field threshold photoionization spectra of HCl obtained via intermediate resonance with rotational levels $J' = 0$, 1, and 2 of the $g\,^3\Sigma^-$, $v = 0$ state. Sticks show *ab initio* calculated relative intensities for transitions to rotationally excited states, normalized in each case to the first such transition. (Reproduced from [141b] with permission.)

spectroscopy, which can best be seen by looking at the comparative intensities of spin–orbit components.

2. *Intensities of Transitions to the Spin–Orbit Substates of HCl^+*

The intensities of threshold photoionizing resonances to the two cation substates convey information on the electronic character of the intermediate state. Theoretically, if the intermediate state is of pure singlet (or triplet) character and the ion core is treated in pure Hund's case (a), the total intensities for the $^2\Pi_{1/2}$ and $^2\Pi_{3/2}$ thresholds will be equal. Conventional photoelectron spectra out of the $X\,^1\Sigma^+$ ground states of HCl and HF show a slightly larger intensity for the $^2\Pi_{3/2}$ component [142, 146]. Recent calculations [141], assuming an intermediate coupling scheme between Hund's case (a) and (b) for the $^2\Pi$ ion, reproduce well the measured intensity ratio of 1.20 for the spin-orbit components of the ion observed in photoionization of HF. The calculated intensity ratio of 1.06 is also in excellent agreement with the measured value of 1.06 ± 0.05 for HCl. This is the pattern observed, for example, in conventional PE spectra starting from the $X\,^1\Sigma^+$ ground state of HCl [142, 146]. The initial states for the present photoionization spectra are Rydberg states, for which it is well known that singlet and triplet states with the same Rydberg orbital configuration are strongly mixed by the spin–orbit interaction (see, e.g., [147]). If this mixing is complete, only one substate is accessible by photoionization. This is the case for the above described experiments on HBr [148], where the intermediate state, $F\,^1\Delta_2$ can be described purely in terms of Hund's case (c) as a complete mixture of $^1\Delta_2$ and $^3\Delta_2$ with the $5p\pi$ Rydberg configuration. In the state assigned $F\,^1\Delta_2$, the core has the spin–orbit configuration $^2\Pi_{1/2}$. Photoionization from this state populates only the $^2\Pi_{1/2}$ component of the cation. In HCl, the spin–orbit singlet–triplet mixing of the $4p\pi$ configuration is not complete. Starting from $F\,^1\Delta_2$, the conventional PE spectra of de Beer et al. [142] show a ratio for the two components $^2\Pi_{1/2}/^2\Pi_{3/2}$ of about 6.5 [9, 15], which corresponds well to a singlet–triplet mixing of 22% as estimated recently [149]. Further experiments in the study by de Beer et al. [142] also confirm that this ratio is insensitive to the energy of ionization.

This value of spin–orbit branching ratio is not found in rotationally resolved threshold photoionization spectra of HCl recorded in transitions from the $F\,^1\Delta$ state by delayed pulsed-field threshold photoionization. For example, in the case of $F\,^1\Delta J' = 2$ in HCl, which is typical, the experimental ratio summed over rotational lines is $I(^2\Pi_{1/2})/I(^2\Pi_{3/2}) = 0.56$. This result is opposite to the trend observed in the conventional PE spectrum. This discrepancy, together with anomalous rotational intensities, can be explained when we consider effects of rotational coupling and spin–orbit autoionization.

Delayed pulsed-field threshold photoionization spectroscopy relies on the long lifetimes of high Rydberg states that lie just below the convergence limit corresponding to each internal state of the ion. Any factor that shortens the lifetime of such high-Rydberg states will affect the spectrum by reducing the population of neutrals available to field ionize after a given delay, and thereby diminish intensities. All high Rydbergs converging to the $^2\Pi_{1/2}$ limit of HCl^+ can autoionize by spin–orbit coupling to the $^2\Pi_{3/2}$ continuum. Series of different ℓ's can be expected to have different rates of autoionization, and, indeed, we have found evidence for very long-lived series of near-zero quantum defect [145]. Nevertheless, the accessibility of channels for decay by autoionization can do nothing but shorten the lifetimes of these near threshold states, and apparently do so by enough to account for a systematically diminished intensity for threshold photoionization from the $F\ ^1\Delta$ state to the upper spin–orbit component.

Rydberg states converging to rotational limits of the lower spin–orbit component can rotationally autoionize, except for those built on the lowest such state $J^+ = \frac{3}{2}$. Indeed, as noted above, in every case this lowest ionization limit gives rise to the most intense threshold transition in the $^2\Pi_{3/2}$ substate, often despite theoretical predictions of rotational propensities favoring higher rotational levels of the cation. Interestingly, for the upper spin–orbit component, where all levels can autoionize by spin–orbit coupling, the lowest rotational level is not so favored in delayed pulsed-field threshold photoionization, and experimental intensities approach theoretically predicted values. This finding suggests a zero-order model that recognizes the special character of the lowest rotational state of the cation, and then it groups levels with open autoionization channels as having diminished intensity in proportion to their absorption cross-sections. Refinements to this picture will distinguish between rotational and spin–orbit autoionization, recognize possible J^+ dependences, and include neutral continua (predissociation).

VI. THREE-COLOR TRIPLE-RESONANT THRESHOLD PHOTOIONIZATION OF NO_2

Threshold photoionization of NO_2 shows distinctively how relaxation pathways that couple Rydberg electronic motion with core vibrational and rotational degrees of freedom can effect ZEKE intensities. This molecule is ideal for probing such issues. Its electronic structure becomes separable at higher energies and offers convenient pathways for state-specific excitation. Hierarchies in Rydberg core-internuclear coupling give rise to vibrational mode selectivity in ZEKE intensity envelopes. Factors governing these effects also determine rotational details in threshold photoionization spectra.

A. Electronic Structure and Photoselection

The electronic structure of NO_2 naturally focuses attention on pure Rydberg rovibrational pathways for state mixing and intramolecular decay. Its 17-electron $\ldots(1a_2)^2(3b_2)^2(4a_1)^1$ orbital configuration supports a bent $^2A_1$ ground state and low-lying Renner and Jahn–Teller coupled $^2B_1$, $^2B_2$, and $^2A_2$ excited states.[150] These give way with increasing excitation to an extensive system of Rydberg states that are well described in terms of a 16-electron closed-shell cation core and single extravalent electron [151–155]. Resonant transitions through the vibronically coupled and distorted valence intermediate states then conveniently bridge the Franck–Condon gap between the ground state and linear low-lying Rydberg states [151, 156]. Such optically prepared Rydberg states in turn provide a convenient gateway for final vertical steps of threshold photoionization.

Unfavorable Franck–Condon factors for direct transitions from the bent ground state of NO_2 to low-lying vibrational levels of the cation obscure the adiabatic ionization threshold and prevent the determination of cation fundamental frequencies by conventional VUV photoelectron spectroscopy [157]. Photoionization from intermediate Rydberg states solves this problem, and, with rotational selection provided by an initial step of double resonance, three-color stepwise excitation resolves individual rotational thresholds in the photoionization of NO_2 to successive final vibrational states beginning with the cation ground state [29, 158]. The results establish the ionization potential of this important molecule and provide a first view of its cation vibrational frequencies.

Figure 34 shows an energy level diagram. First-photon absorption in the visible selects a rotational state in the vibronically mixed $^2B_1/^2B_2$ valence system of NO_2. A second step of UV excitation elevates population to a chosen rovibrational level of the $3p\sigma\ ^2\Sigma_u^+$ state. Broad Franck–Condon envelopes for transitions from the vibronically mixed intermediate state to the linear excited state make a large number of vibrational levels in this long-lived Rydberg state accessible. Double resonance provides baseline separation of rotational levels. Figure 35 shows a double resonance spectrum obtained for an intermediate (valence) level that bridges ground-state $N''=2$ with $3p\sigma\ ^2\Sigma_u^+(100)$ $N'=1,3$. Experiments confirm such assignments by scanning the first-photon frequency while fixing the second on the transition from the valence

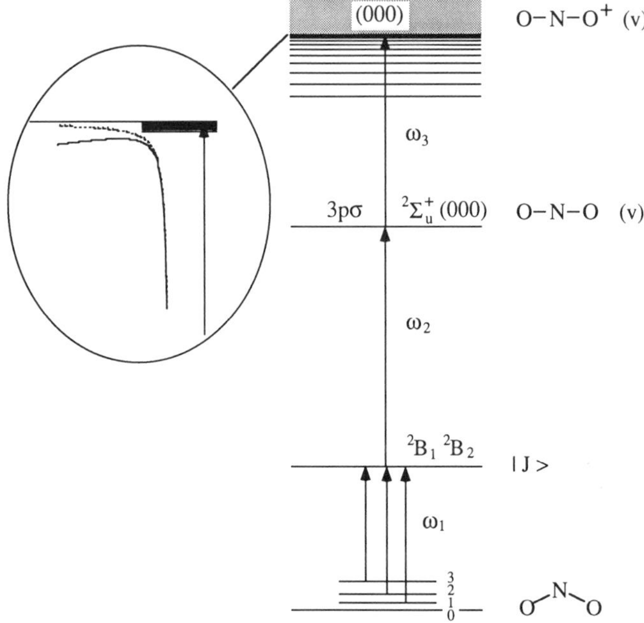

Figure 34. Level diagram showing intermediate states and excitation schemes for multiresonant threshold photoionization of NO_2. Two-color preparation of the $3p\sigma\ ^2\Sigma_u^+$ Rydberg state of NO_2 proceeds through bound states of the $^2B_1 - {}^2B_2$ visible absorption system. Excitation to selected vibrational thresholds from two-color prepared $3p\sigma$ levels is achieved by tunable third-color absorption.

intermediate to the Rydberg final rotational state. The ionization detected absorption spectrum shows resonance enhancement whenever the first photon frequency connects a ground rotational level to that valence intermediate. Thus, the first-photon scan recovers ground-state spacings unambiguously establishing the identity of the double-resonantly selected Rydberg rotational level.

B. State-to-State Threshold Photoionization of NO_2: Vibrational Structure of NO_2^+

From individual rotational levels selected in this way, tunable third-photon absorption promotes transitions to state-resolved thresholds for photoionization. The linear cation-like structure of the $^2\Sigma_u^+$ state together with its long lifetime suggests a well-developed Rydberg character, which in turn implies highly vertical photoionizing transitions. Such expectations are confirmed by experiment. Figure 36 shows a sequence of threshold photoionization bands observed in transitions from single rotational levels

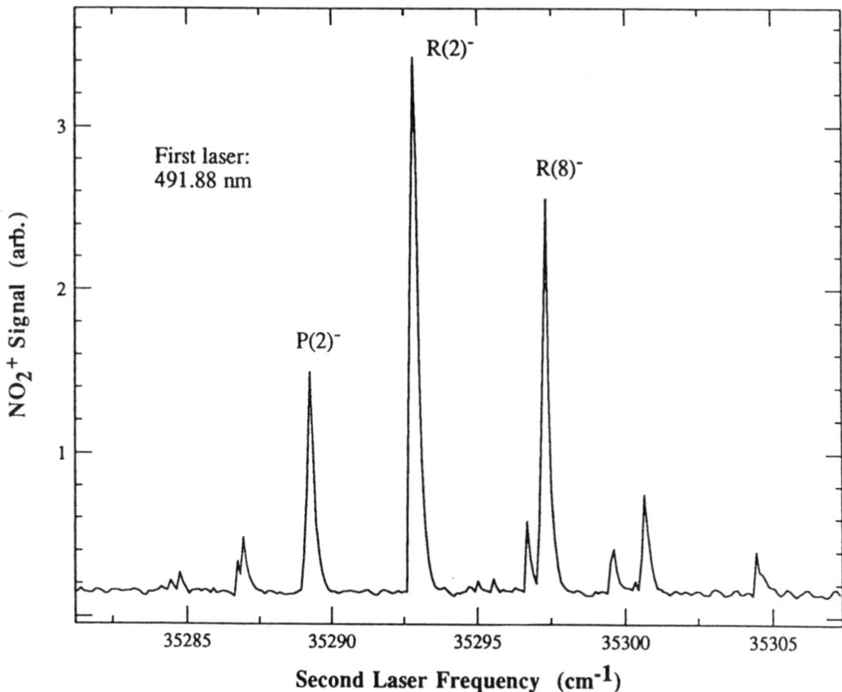

Figure 35. Ionization-detected double-resonance spectrum of the (100) level of $3p\sigma$ $^2\Sigma_u^+$ Rydberg state of NO_2 obtained by setting the first photon frequency to resonance with an intermediate level that optically connects the $N'' = 2$ ground state with $N' = 1$ and 3 in the excited state.

prepared by double resonance in the corresponding sequence of $3p\sigma$ $^2\Sigma_u^+$ vibrational states.

Note that the rotational structure in each case confirms the vibrational assignment. Nuclear spin statistics for NO_2^+ with terminal ^{16}O nuclei allow only levels that are totally symmetric. For (000) and (100) levels in the $^1\Sigma_g^+$ electronic ground state, this restriction excludes odd rotational levels. Totally symmetric components exist for each rotational level of the vibronic Π_u (010) state, and transitions are observed to all quantum numbers starting with $N^+ = 1$. Odd vibronic symmetry of the Σ_u (001) vibrational state confers *even* symmetry only on levels of *odd* rotational quantum number, and the sequence observed is $N^+ = 1, 3, \ldots$.

The structure pictured for each of these bands can be extrapolated to corresponding rotational origins. The relative energies of these origins determine the positions of the vibrationally excited states of NO_2^+. Results are summarized in Table III. Frequencies obtained in recent

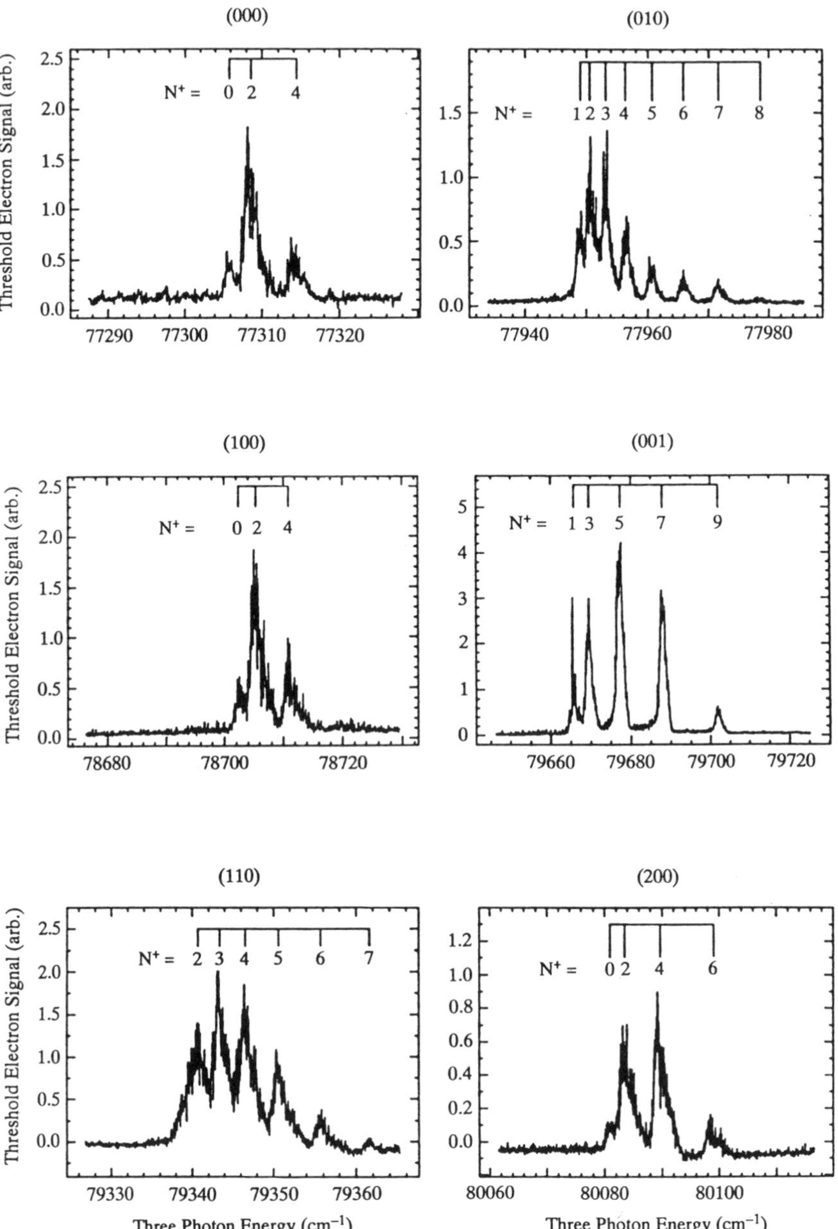

Figure 36. Rotationally resolved threshold photoionizing scans from selected rovibrational levels in the $3p\sigma\ ^2\Sigma_u^+$ state to corresponding vibrational states of NO_2^+. (Reproduced from [158a] with permission.)

TABLE III
Fundamental Frequencies of the Cation and $3p\sigma\ ^2\Sigma_u^+$ Rydberg State of NO_2 Compared with Those of CO_2[a]

	ν_1 Symmetric Stretch	ν_2 Bend	ν_3 Asymmetric Stretch
$3p\sigma\ ^2\Sigma_u^+$ Rydberg State[b]	1410	636	2361
$NO_2^+ X\ ^1\Sigma_g^+$ Experiment[c]	1397	639	2362
$NO_2^+ X\ ^1\Sigma_g^+$ Theory[d]	1384	625	2344
$CO_2 X\ ^1\Sigma_g^+$	1388	667	2349

[a]All values in cm^{-1}.
[b]Ref. 151.
[c]Ref. 158a.
[d]Ref. 159.

high-level *ab initio* calculations [159], which show good agreement, are listed for comparison. The frequencies of the cation are also quite close in magnitude to the vibrational frequencies observed in the $3p\sigma$ Rydberg state. This match confirms that potential surfaces in the Rydberg state and the cation are substantially parallel, reflecting the fact that the presence of a Rydberg electron bound by 20,000 cm^{-1} has little effect on the vibrational frequencies within the NO_2^+ core. The one mode that shows a measurable difference in frequency is that of symmetric stretch, which separate experiments have characterized as a motion in the core that strongly promotes the vibrational autoionization of higher Rydberg states.

1. Rotational Structure

Application of photoselection methods to reach higher rotational states in the $3p\sigma$ (000) band extends the reach of threshold photoionization to higher rotational states of the cation. Figure 37 shows a series of threshold photoionization spectra originating from NO_2 $3p\sigma\ ^2\Sigma_u^+$ (000) rotational states $N' = 1$, 9, 11, and 13. Cation rotational positions obtained from such spectra can be readily fit to the rotational Hamiltonian for a vibrational ground state linear triatomic with centrifugal distortion:

$$E_{N^+} = B_0 N^+(N^+ + 1) - D_0[N^+(N^+ + 1)]^2$$

Figure 38 gives a Fortrat parabola of observed positions. Constants are summarized in Table IV. The zero-point bond distance obtained theoret-

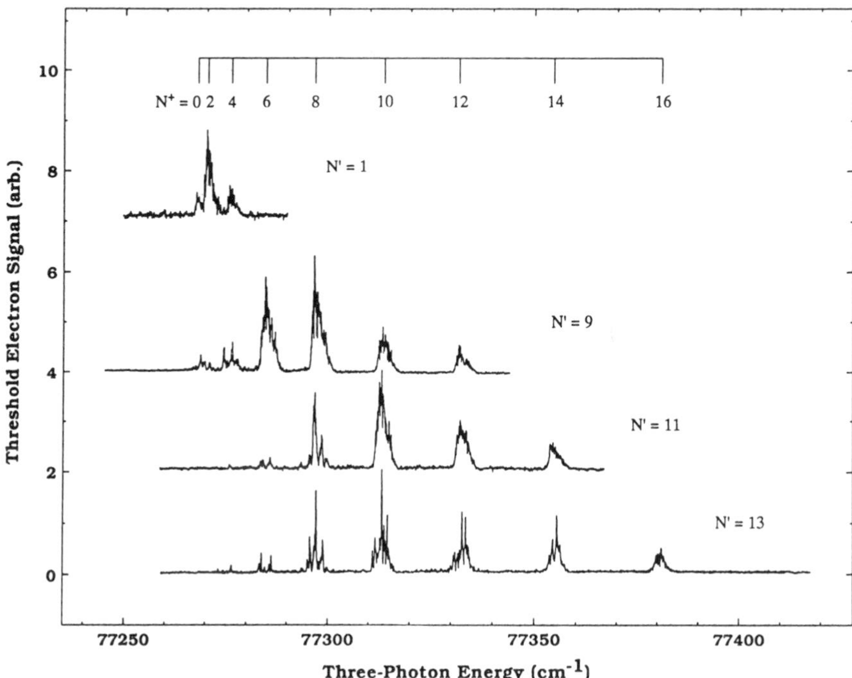

Figure 37. Threshold photoionization spectra originating from NO_2 $3p\sigma\ ^2\Sigma_u^+$ (000) rotational states: $N' = 1$, 9, 11, and 13. (Reproduced from [158b] with permission.)

ically is 1.123 Å, which is in good agreement with that determined from the experimental rotational constant.

2. *Rotational Intensities: Complex Resonances and the Effects of Intramolecular Relaxation*

Rotational intensities for (000) threshold photoionization form a consistent pattern for the highest accessible states. In all cases intensities tail off to a highest observed level at $N^+ = N' + 3$. Transitions to lower N^+, however, do not appear to be bound by a rule limiting the change from $3p\sigma$ N' to cation N^+. The profile of rotational transitions instead exhibits an asymmetry about $\Delta N' = 0$ that favors increasingly negative values of $\Delta N'$ with higher initial rotational state. As discussed above, selection rules for good initial and final rotational and Rydberg orbital angular momentum quantum numbers require $N^+ - N' = \pm(\ell + 2)$. For an initial p orbital, the upper limit observed for N^+ agrees with this prediction. To the extent of its asymmetry, the lower one half of the rotational envelope does not.

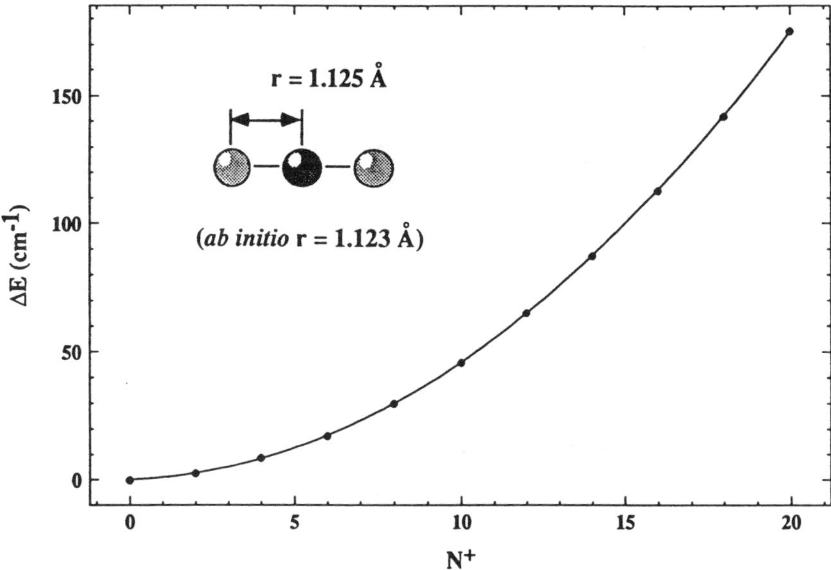

Figure 38. Fortrat parabola of rotational positions observed in transitions from selected levels of $3p\sigma\ ^2\Sigma_u^+$ (000) to the ground vibrational state of NO_2^+.

TABLE IV
Rotational Constants Derived from the High-Resolution Threshold Photoionization of NO_2[a,b]

	B_0 (cm^{-1})	D_0 (cm^{-1})	r_0 (Å)
$NO_2^+ X\ ^1\Sigma_g^+$	0.4167	6.7×10^{-7}	1.125
$CO_2 X\ ^1\Sigma_g^+$	0.3902	1.4×10^{-7}	1.162

[a]From Ref. [158b].
[b]Rotational constants and bond length of CO_2 shown for comparison.

Upon closer examination of the threshold photoionization spectrum one observes that these lower rotational components show reproducible fine structure. This structure holds the key to understanding the source of anomalous rotational intensity. Figure 39 compares a section of the ZEKE spectrum obtained for $N' = 17$ with the accompanying prompt-electron photoionization spectrum. The prompt-electron spectrum shows the structure of rotationally autoionizing Rydberg series converging to accessible higher rotational limits. Over most of this spectrum, discrete states populated by resonant absorption are short lived and decay by

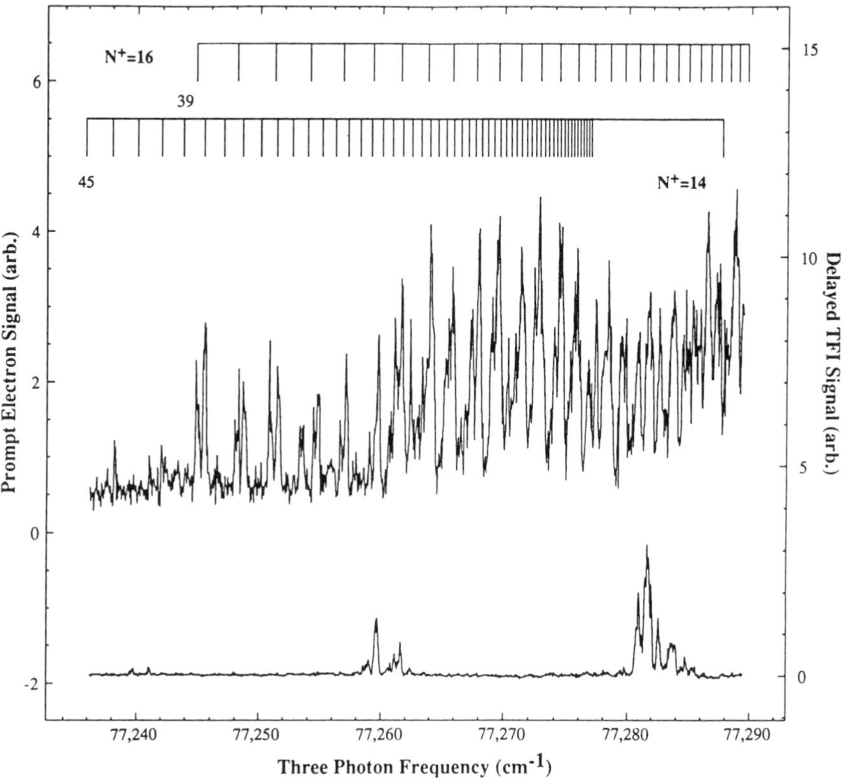

Figure 39. Scans comparing a section of the delayed pulsed-field threshold photoionization spectrum observed for $N' = 17$ with the spectrum of the prompt-electron signal obtained at the same time.

rotational autoionization to form free electrons and lower rotational state cations. Thus, in general, these Rydberg states to not survive to supply electrons for delayed pulsed extraction and the spectrum appears in the prompt-electron channel. However, the same interaction that drives autoionization couples these discrete states to the manifold of very high ZEKE Rydberg states lying just below each ionization threshold. These high Rydberg states have long lifetimes and, when they mix with the strongly absorbing lower principal quantum number (higher core rotation) Rydberg states, they form a long-lived complex resonance. Evidence for this mixing can be seen in the interference profiles present in the prompt spectrum just below successive ionization thresholds [159].

When the pulsed field is applied, mixed neutral states of total energy just below the N^+ threshold can autoionize.

The intensity of the discrete components in the delayed pulsed-field threshold photoionization signal varies with the difference between the rotational quantum number of the threshold and that of the interloping discrete state. At the highest forbidden threshold, structure due to interloping Rydberg states converging to the fully allowed next higher rotational limit is relatively weak. The contributions to the threshold complex resonance from discrete states converging one threshold further are stronger. Thus, the ZEKE resonance for photoionizing transitions from $N' = 17$ to the forbidden $N^+ = 12$ state shows only weak structure due to interloping resonances converging to allowed $N^+ = 14$, while contributions from discrete states converging to $N^+ = 16$ appear with large intensity.

This apparent hierarchy in coupling can be understood when one recognizes the delay inherent in ZEKE discrimination of threshold electrons and considers the pathways for intramolecular relaxation. Figure 40 shows a coupling diagram. High Rydberg states in the ZEKE manifold converging to $N^+ = 12$ couple with discrete states having $N^+ = 14$ and the $N^+ = 10$ continuum by $\Delta N^+ = 2$, $\Delta \ell = 2$ interactions, and to $N^+ = 16$ and the $N^+ = 8$ continuum by $\Delta N^+ = 4$, $\Delta \ell = 4$ interactions. The

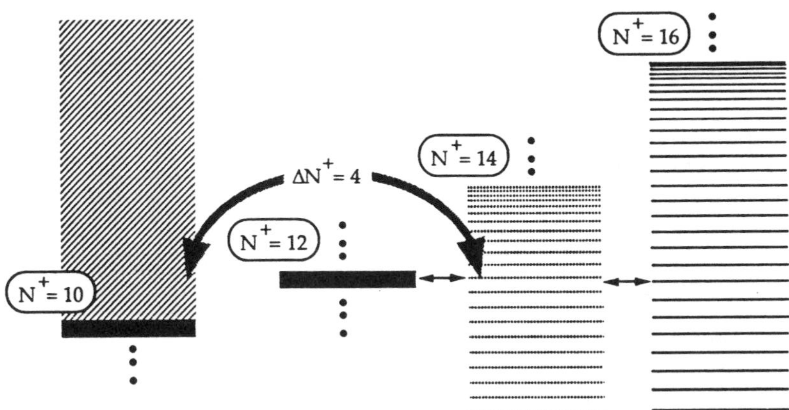

Figure 40. Diagram illustrating the discrete–discrete interactions that appear as structure in the spectrum shown in Fig. 33. Also shown are discrete continuum pathways that, when present, reduce the intensity increasing effect of discrete interlopers on the delayed pulsed-field threshold photoionization spectrum. (Reproduced from [158b] with permission.)

$N^+ = 12$ component of the resulting complex resonance lends lifetime with respect to all pathways for radiationless decay, while discrete states converging to $N^+ = 14$ and 16, which are allowed in transitions from $N' = 17$, lend intensity. While surely dominating the actual composition of the complex resonance, the contributions of $N^+ = 14$ states to the delayed electron signal is evidently moderated by the relative efficiency with which it is coupled by $\Delta N^+ = 4$ matrix elements to the $N^+ = 10$ continuum. Resonance contributions from $N^+ = 16$ states, on the other hand, are more isolated from loss on the timescale of the ZEKE measurement by the weakness of $\Delta N^+ = 6$ coupling to the nearest continuum. Thus, with delay as an essential element of the threshold discrimination process, $N^+ = 16$ features dominate the detectable component of ZEKE absorption. Efficient paths for intramolecular relaxation (in this case, rotational autoionization) diminishes the importance of nearer rotational components of the threshold complex resonance.

C. Intensities at Vibrationally Excited Thresholds: Effects of Vibrational Relaxation

Examination of ZEKE spectra at vibrationally excited thresholds shows how the opening of different vibrationally autoionizing continua can affect rotational intensity patterns. The results are mode selective in a manner that reflects the dynamics of Rydberg core coupling resolved for individual autoionizing resonances observed further below vertical thresholds [2, 30, 140].

Figure 41 shows a spectrum of threshold photoionizing transitions from $N' = 5$ in $3p\sigma$ (010) to the (010) state of NO_2^+. Separate studies of autoionization dynamics show that Rydberg electrons are poorly coupled to the (000) continuum by relaxation in bending. The distribution of rotational intensities in this threshold photoionization spectrum appears much like that for (000) observed from the same rotational level, except for the presence in (010) of *even* and *odd* rotational states, allowed for a state of Π vibronic symmetry.

Autoionization spectra establish that bound Rydberg electrons are also relatively weakly coupled to the cation ground-state continuum by relaxation of the highest frequency fundamental, asymmetric stretch. Figure 42 shows a ZEKE spectrum of transitions from $3p\sigma$ (001) $N' = 6$ to the (001) threshold of NO_2^+. Like the (010) threshold the envelope of rotational intensities resembles that of the cation ground state, favoring transitions to lower values of ΔN with significant intensity in nominally forbidden $\Delta N = -5$.

The spectrum of threshold transitions to rotational states of NO_2^+ (100) differs fundamentally from that of other excited vibrations and the cation

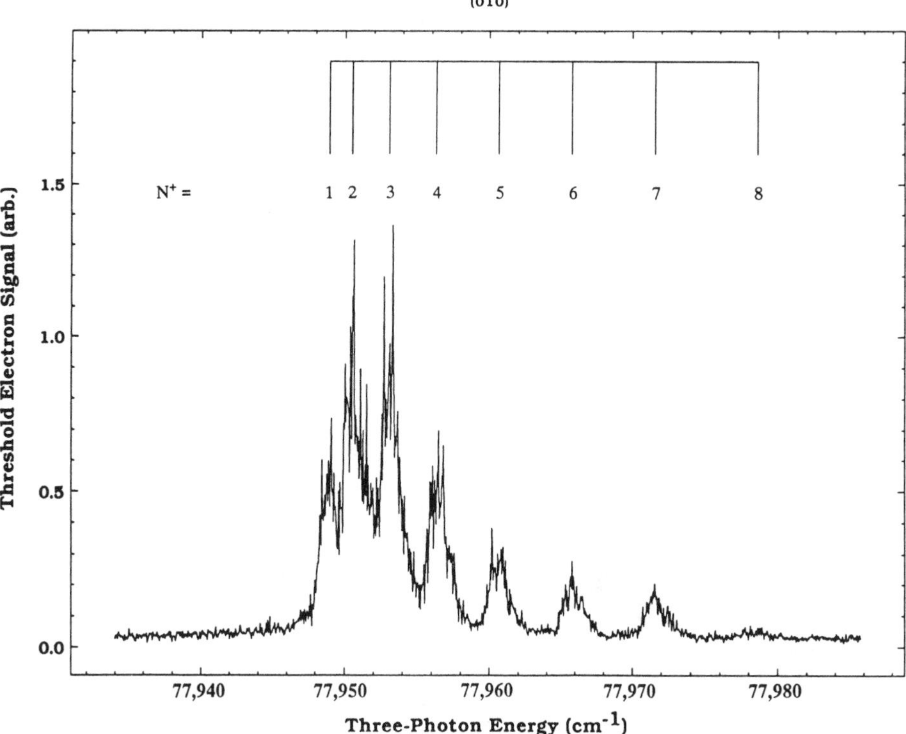

Figure 41. Rotationally resolved threshold photoionization to (010) cation states $N^+ =$ 1, 2, 3, 4, 5, 6, 7, and 8 observed in a one-photon transition from the $N' = 5$ rotational state of $3p\sigma\ ^2\Sigma_u^+$ (010). (Reproduced from [158b] with permission.)

ground state. Though cation-detected vibrational autoionization spectra recorded from $3p\sigma$ (100) are the strongest of the three modes, the ZEKE spectrum of this threshold is by far the weakest. Figure 43 shows a set of (100) threshold scans from $3p\sigma$ (100) $N' = 1, 5, 7$ and 9. The ZEKE intensity is low and rotational asymmetry is barely evident. As mentioned above, symmetric stretch has been characterized as a mode that efficiently promotes vibrational autoionization. The facility of such a pathway for radiationless decay accounts reasonably for the weakness of the spectrum of this state in threshold photoionization. Just as efficient rotational autoionization diminishes the importance of adjacent rotational states in forming long-lived complex resonances that give rise to threshold photoionization structure for (000), fast vibrational autoionization depletes the high-Rydberg states that form the basis for the primary threshold signal

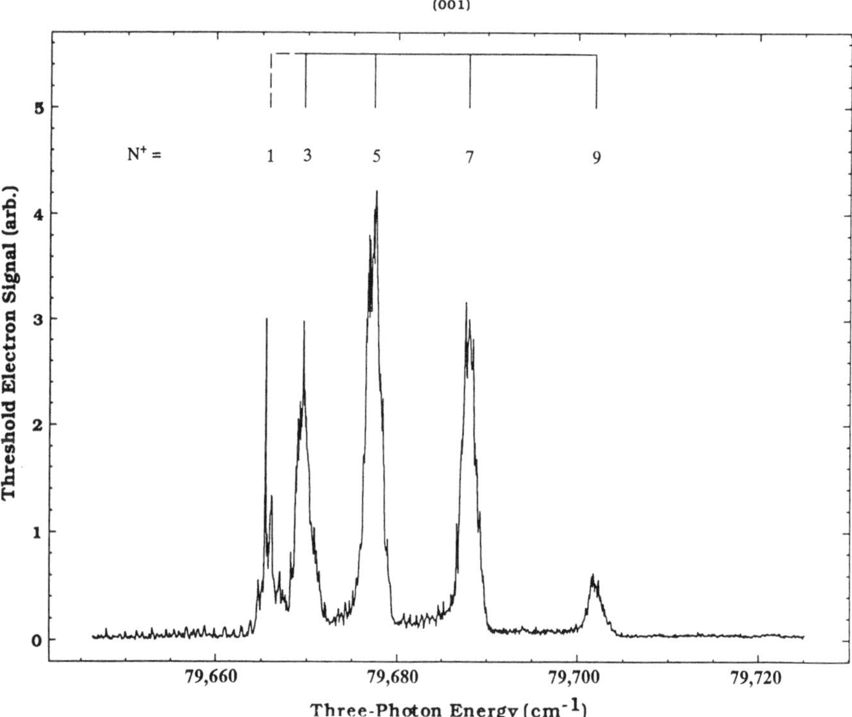

Figure 42. Rotationally resolved threshold photoionization to (001) cation states $N^+ =$ 1, 3, 5, 7, and 9 observed in a one-photon transition from the $N' = 6$ rotational state of $3p\sigma\ ^2\Sigma_u^+$ (001). (Reproduced from [158b] with permission.)

observed in delayed pulsed-field photoionization. Also apparent from the absence of a pronounced asymmetry in the ΔN distribution, vibrational autoionization of lower n Rydberg states responsible for complex resonance formation at nominally forbidden rotational thresholds markedly reduces the effect of these states on the observed rotational intensity profile. In terms of the interactions illustrated in Fig. 40, strong vibrational autoionization opens a loss channel for all coupled levels, which diminishes the delayed signal contribution of every component of the ZEKE threshold complex resonance, but is particularly severe for more distant higher core rotational states with lower principal quantum numbers.

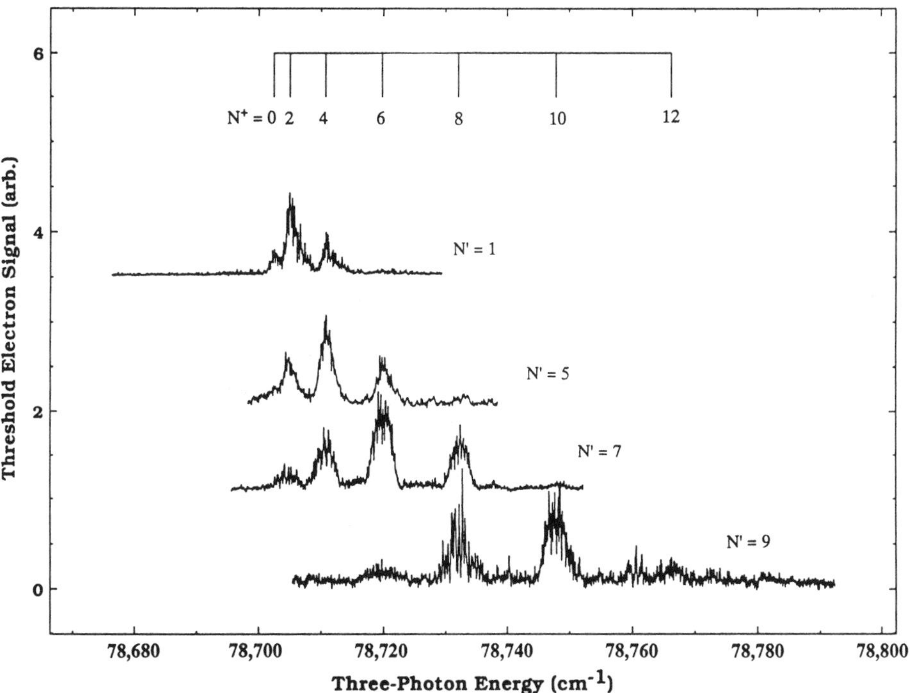

Figure 43. A sequence of spectra of the NO_2^+ (100) threshold obtained from $3p\sigma\ ^2\Sigma_u^+$ (100) initial rotational states: $N' = 1$, 5, 7, and 9. (Reproduced from [158b] with permission.)

VII. VIBRATIONALLY RESOLVED STRUCTURE OF THE *PARA*-DIFLUOROBENZENE CATION

The resolution of the vibrational structure of larger molecular ions is also an important application of ZEKE spectroscopy in chemistry. The considerable gain in resolution makes it possible to measure low-frequency vibrations and to resolve combination bands that are congested in the PE spectrum (see Fig. 4).

The *p*-DFB cation, like most of the larger molecular ions, does not fluoresce, and hence most methods of fluorescence spectroscopy are not applicable. *Ab initio* calculations by von Niessen et al. [160] give an electronic configuration for the outer-molecular orbitals of *p*-DFB of

$\cdots(b_{2u})^2(b_{3g})^2(b_{3u})^2(b_{1g})^2(b_{2g})^2(a_u)^0(b_{3u})^0$ within the D_{2h} symmetry group. From He(I) and He(II) photoelectron spectra [52, 94, 161, 162] it follows that the symmetry species for the electronic ground state for the p-DFB cation is $^2B_{2g}$.

A. ZEKE Spectra

For the ZEKE measurements presented here, p-DFB is prepared in selected vibrational states in S_1 by absorption of a first photon, and then ionized from these vibrationally selected states in S_1 by the second photon. The $S_1 \leftarrow S_0$ excitation spectra and the precise pump energies of the S_1 vibrational intermediate resonances were obtained by measuring the mass-selected ion currents ($m/e = 114$) in one-color (1 + 1) REMPI experiments. In these experiments the first dye laser was tuned through the region of 530–550 nm and the laser output was frequency doubled in a KDP crystal.

ZEKE spectra have been obtained for different $S_1 \leftarrow S_0$ transitions: the 0^0_0, 3^1_0, 6^1_0, 9^2_0, 17^1_1, 27^1_0, and 30^1_1 (the vibrations are denoted according to the Mulliken notation [163]).

B. ZEKE Spectra via Different Vibrational Intermediate S_1 States

1. ZEKE Spectrum from the $S_1 0^0$

The ZEKE spectrum from the vibrationless S_1 state is reproduced in Fig. 44. The scanning region ranges from the ionization energy to approximately 3000 cm^{-1} above. The width of the ZEKE signal peaks is below 6 cm^{-1} (fwhm). The majority of the bands observed in this ZEKE spectrum can be assigned to vibrational states of the p-DFB cation. Besides the vertical transition into the vibrationless state of the cation, the spectrum is dominated by transitions into totally symmetric modes and combination bands of these. The strongest transitions are the 6^1, 5^1, 4^1, and 2^1, and various combinations of two quanta of these. However, the nontotally symmetric modes ν_{30}, ν_{17}, and ν_8 also appear. These modes are also observed in combination with a_g modes. Some transitions observed in the ZEKE spectrum could not be unambiguously identified within the present analysis. The most probable assignments, supported by an *ab initio* calculation, are given in Table IV.

2. ZEKE Spectrum from the $S_1 6^1$

Figure 45 shows the ZEKE spectrum with $S_1 6^1$ as the resonant intermediate state. For the ν_6 mode a progression with four quanta is observed with the maximum intensity for the 6^2 level. This shift of the Franck–Condon (FC) maximum towards higher vibrational excitation in the ν_6

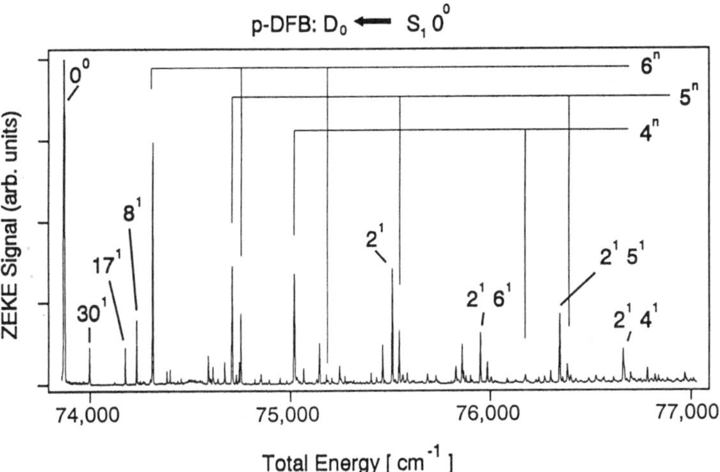

Figure 44. Vibrationally resolved ZEKE spectrum of *p*-difluorobenzene with the vibrationless S_1 as intermediate resonance. The assignments are given above the peaks. (Reproduced from [61] with permission.)

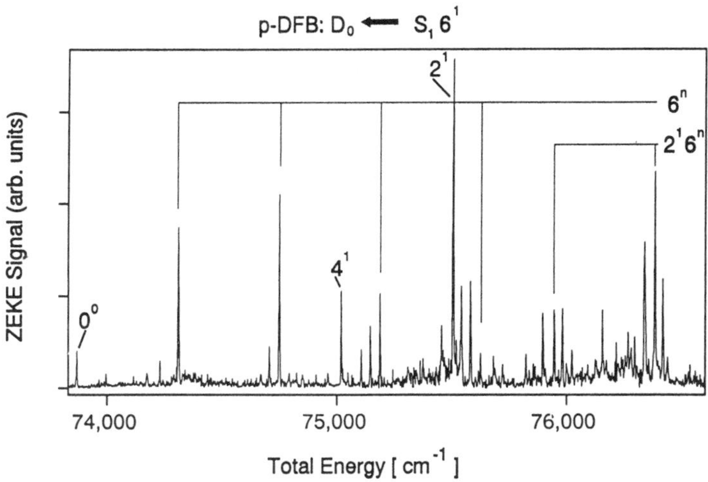

Figure 45. Vibrationally resolved ZEKE spectrum of *p*-difluorobenzene with the S_1 6^1 vibration as intermediate resonance. The assignments are given above the peaks. (Reproduced from [61] with permission.)

was also observed in the corresponding TOF photoelectron spectrum [59]. The spectrum in Fig. 45 also shows mainly transitions into symmetric modes. The most probable assignments, taking into account the combination bands, is also given in Table V.

The most prominent feature of the ZEKE spectra observed is the strong activity of the a_g modes ν_2 to ν_6 (the a_g vibration ν_1 has not been seen in conventional PE spectra and is, with the exception of the ZEKE spectrum from the $S_1 0^0$, outside the scanning range employed in the

TABLE V
Measured and Calculated Frequencies of p-DFB Normal Modes (in cm^{-1})[a]

Mode	Symmetry	$S_0(^1A_g)$		$S_1(^1B_{2u})$	$D_0(^2B_{2g})$	
		Experimental	SCF	Experimental	Experimental	SCF
1	a_g	3088	3044		3098[c]	3047
2		1617	1615		1640	1647
3		1257	1295	1251	1375	1362
4		1140	1152	1116	1148	1167
5		860	833	818	836	794
6		450	452	410	439	434
7	a_u	945	1050	583		1078
8		422	421	179	359	395
9	b_{1g}	800	881	588	726	846
10	b_{1u}	3073	3029			3034
11		1514	1394	1335		1470
12		1228	1155	1015		1292
13		1014	1043	937		943
14		740	752	666		683
15	b_{2g}	928	1007	670		1067
16		692	724	528		750
17		374	392	274	303	326
18	b_{2u}	3073	3042			3045
19		1633	1525			1470
20		1306	1235	1591		1235
21		1085	1078	1100		1096
22		348	318	352		333
23	b_{3g}	3085	3030			3034
24		1617	1578	1516		1371
25		1285	1259	933		1203
26		635	648	558		576
27		446	448	403	430	416
28	b_{3u}	838	910	619	859[b]	924
29		505	538	438	508[c]	526
30		158	164	120	127	132

[a]The SCF values are scaled by a factor of 0.89. (From [61].)
[b]Assignment uncertain.
[c]From [59].

experiments reported here). In particular, the ν_6 vibration is the most prominent mode observed in the ZEKE spectra. On the basis of the FC principle it is expected that the modes excited indicate a geometry change on ionization. (However, Duschinsky rotation [164] of the normal coordinates within the molecular symmetry group D_{2h} may have to be taken into account.)

The restricted validity of the $\Delta v = 0$ propensity rule for *p*-DFB should be noted here. This propensity has previously been observed for ionization of aromatic molecules from their S_1 states [165]. The failure of the rule is not so surprising though, as it is only expected to hold when the geometries of the S_1 and the ionic states are similar (as they are, e.g. for Rydberg states and their corresponding ions). For the different vibrational excitations in the S_1 state there are some bands in combination with a quantum of mode 6 of similar strength as the dominant $\Delta v = 0$ transitions ($D_0 30^1 6^1 \leftarrow S_1 30^1$, $D_0 9^2 6^1 \leftarrow S_1 9^2$, $D_0 17^1 6^1 \leftarrow S_1 17^1$). For the $S_1 6^1$ intermediate state the FC maximum is observed for the 6^2 vibration in the cation. This result is in agreement with the previously published PE spectrum [59] (the stronger appearance of the 2^1 band in the ZEKE spectrum when pumping the $S_1 0^0$ is due to the dye tuning curve). This observation, the shift of the FC maximum towards 6^2, implies that the geometry of the S_1 state and the ionic state differ along this mode so that maximum overlap is achieved for the first overtone in the ion, rather than the fundamental.

Within the FC approximation (and in the absence of any other coupling), no transitions should be observed for which the symmetry species of the vibrational wave function in the S_1 state is different from that of the ion. The most prominent transitions that violate this symmetry restriction correspond to the vibrations ν_8, ν_{17}, and ν_{30}. The reasons for observation of such symmetry forbidden vibrations is still not perfectly clear. Vibronic coupling (to an excited electronic state) in the cation cannot explain the occurrence of these three vibrations of different symmetry species. A possible mechanism could be the final state interaction as described in Section V.

C. Self-Consistent Field *ab initio* Computations of the *p*-DFB Cation

To aid in the assignment of the observed vibrations in the *p*-DFB cation and to check the experimentally observed vibrational frequencies, *ab initio* calculations were performed. The program CADPAC [166] was employed at the restricted (open-shell) Hartree–Fock level, using the standard 3-21G basis set. For the electronic ground states of the neutral and of the cation, geometry optimization leads to D_{2h} symmetry (even with geometry restrictions such that only the planarity of the ring is enforced).

Table V shows the experimentally determined vibrational frequencies for the *p*-DFB cation along with the computed values. The computed values were multiplied by a scaling factor of 0.89, which is typically applied to vibrational frequencies at the SCF level.

The accuracy of the (scaled) computed frequencies is typically of the order of 5%. For modes where data is available, the observed increase or decrease of the frequency of the cationic vibration is predicted correctly by the computed values. (The direct comparison of modes between S_0 and the ion must be treated with some caution though as the ionic vibrational motions may be affected significantly by Duschinsky mixing). Also, a 3-21G/UHF calculation of the vibrational frequencies of the neutral benzyl radical [167] (also scaled by a factor of 0.89) gave good agreement with known experimental values.

In summary, the calculations support the experimental assignments for the ionic frequencies and lead to a D_{2h} structure for the S_0 and the ground ionic state. For the S_1 state deviations from the D_{2h} symmetry, mainly along the ν_8 coordinate are indicated.

VIII. VIBRATIONALLY RESOLVED STRUCTURE OF THE PHENOL–WATER CATION

In recent years the investigation of van der Waals and hydrogen-bonded complexes has attracted considerable attention. Molecular complexes are an important bridge between the isolated molecule and the solid or liquid phase. In particular, it is of interest how chemical and physical properties vary with increasing complex size. For hydrogen-bonded species, the microsolvation process, that is, the successive attachment of solvent molecules to the solute molecule, is of particular interest. Hydrogen-bonded solvents play an important role in all areas of chemistry. Of chemical importance is the proton-transfer reaction which, for instance, is observed in the case of 1:*n* phenol–(ammonia)$_n$ complexes ($n > 3$) [168] and for the solvation of ions in water. Hydrogen bonds similar to those observed in molecular complexes also strongly influence the structure of biomolecules (DNA double helix) as well as the properties of important solvents such as ammonia, water, and ice. The processes leading to solvation or relaxation processes in condensed phases are not yet properly understood on the molecular level. However, it is the properties of the system at the molecular level that define the macroscopic appearance, structure, and dynamics. Information on the intermolecular potential can, in principle, be obtained by measuring the usually low-frequency intermolecular vibrations. Also of interest are dynamical processes such as charge (proton or electron) and energy transfer (e.g., "intracomplex" vibrational redistribution). Experiments on small hydrogen-bonded complexes are well suited for investigating these and similar questions.

Many gas-phase studies on small neutral hydrogen-bonded complexes using various spectroscopic techniques have been carried out in molecular beams to investigate the intermolecular interactions under unperturbed conditions. In contrast to neutral complexes, the amount of work on weakly bound ionic complexes is rather sparse. Ultraviolet photoelectron spectroscopy [169, 170], due to its limited resolution, has mainly provided information on the electronic structure of such complexes. With resonance-enhanced multiphoton ionization PES [171] (REMPI–PES) and molecular beam techniques, somewhat better resolution and mass selectivity could be obtained. However, the low-frequency intermolecular vibrations of ionic complexes (typically 10–400 cm^{-1}) have still not been resolved using this technique.

In order to record two-photon, two-color ZEKE spectra via a resonant intermediate state, an analysis of the intermediate state spectrum is required. In the case of PhOH–$(H_2O)_n$ complexes ($n = 1$–4), where Ph is phenol, the $S_1 \leftarrow S_0(\pi^* \leftarrow \pi)$ transition in the gas phase under supersonic molecular beam conditions is very well understood (for a discussion of the problem and references see [172]). Until very recently, only two out of a possible six intermolecular modes in the S_1 state were identified [173, 174], namely, the stretch (156 cm^{-1}) and in-plane wag vibrations (121 cm^{-1}). Very recently, Leutwyler and co-workers [175] recorded high-quality, two-color REMPI spectra of the h_3 and d_3 complexes. By comparing those spectra with detailed high-level *ab initio* calculations [175], they confirmed the assignment of the stretch and the in-plane wag. However, more peaks in the low-energy region of the S_1 state were identified and assignments for some of these were suggested.

In contrast to the neutral states of the 1:1 phenol–water complex, information about the low-frequency intermolecular vibrations of the *ionic* complex is difficult to obtain, and hence rather sparse. The first experiments resolving intermolecular vibrations were ZEKE experiments [18].

A. ZEKE Spectra

ZEKE spectra of the fully protonated 1:1 phenol–water complex (h_3) (shown in Fig. 46) were obtained via three different intermediate states $S_1 0^0$ (35,996 cm^{-1}), $S_1\sigma^1$ (36,153 cm^{-1}), and $S_1\gamma'^1$ (36,117.5 cm^{-1}). The ZEKE spectrum via the vibrationless S_1 state (36,004,5 cm^{-1}) of the threefold deuterated 1:1 complex (d_3) is shown in Fig. 5. The observed

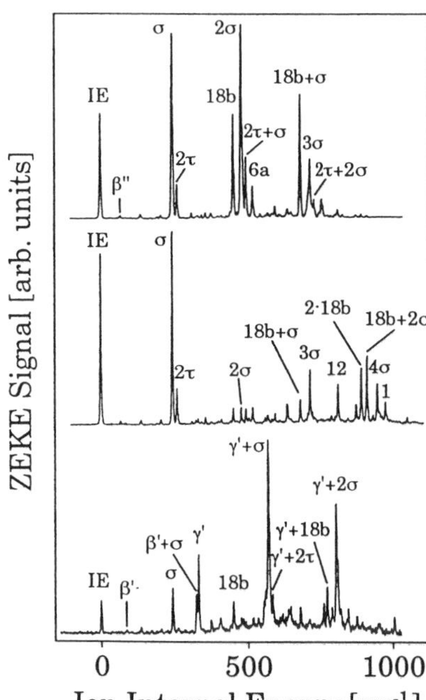

Figure 46. ZEKE spectra of the h_3 1:1 phenol–water complex via different intermediate S_1 levels: $S_1\ 0^0$ (top), $S_1\ \sigma^1$ (middle), and $S_1\ \gamma'^1$ (bottom). The spectra have been corrected for the dye laser output intensity variation of the ionizing laser. (Reproduced from [64] with permission.)

line width is about $5\,\text{cm}^{-1}$ (fwhm) and is mainly due to unresolved rotational structure.

1. *Ionization Energy*

For the phenol–water complex, Fuke et al. [171] deduced an adiabatic ionization energy (IE) of 8.09 eV (~$65250\,\text{cm}^{-1}$) from one-color REMPI–PES spectra via the $S_1 0^0$ state. A much more precise value for the IE was reported by Lipert and Colson [176] who used two-color photoionization efficiency (REMPI–PIE) spectroscopy via the $S_1\ 0^0$ state as the intermediate resonance. In that work, the PIE curves were recorded for variable extraction fields E of $27–500\,\text{V cm}^{-1}$. The field-free ionization energy was calculated by extrapolation to zero field according to the formula for the field shift δ: $\delta[\text{cm}^{-1}] = c \cdot (E[\text{V cm}^{-1}])^{1/2}$. A field-free IE value of $64{,}035 \pm 10\,\text{cm}^{-1}$ was derived from the obtained value of $c = 4.5$. This ionization energy was about $1215\,\text{cm}^{-1}$ lower than the value given by Fuke et al. [171].

From the ZEKE spectra shown in Fig. 46 ionization energies of $64{,}024 \pm 4\,\text{cm}^{-1}$ (via $S_1\ 0^0$), $64{,}023 \pm 4\,\text{cm}^{-1}$ (via $S_1\ \sigma^1$), and $64{,}025 \pm 4\,\text{cm}^{-1}$ (via $S_1\gamma'^1$) are obtained, respectively. These values are not yet corrected for the effect of the pulsed extraction field ($E = 0.7\,\text{V cm}^{-1}$)

employed. The quoted error is due to the width of the IE peaks and the error in the laser calibration. As in the case of the other hydrogen-bonded complexes [177–179] a shift of 3 cm^{-1} is used for the required correction of the ionization energy due to the pulsed extraction field. Thus, a value of 64,027 ± 4 cm^{-1} is derived for the field-free (averaged) adiabatic ionization energy. It should be emphasized that the same value for the ionization energy is obtained (within experimental error) for all three resonant intermediate states. This is a very strong argument for the assumption that the observed ionization energy is indeed the *adiabatic* ionization energy. For the threefold deuterated complex (d_3) an uncorrected ionization energy of 64,017 ± 4 cm^{-1} is measured, and hence a field-corrected value of 64,020 ± 4 cm^{-1} is obtained.

The IE of the 1:1 phenol–water complex (h_3) is lowered by 4601 ± 8 cm^{-1} compared to the IE of PhOH [180] (68,628 ± 4 cm^{-1}). The corresponding decrease of the d_3 complex on complexation is calculated to be 4590 ± 8 cm^{-1} (IE value of PhOD: [180] 68,610 ± 4 cm^{-1}). The decrease of the IE on complexation is much larger than the corresponding decrease of the $S_1 \leftarrow S_0\, 0^0$ transition of 352.5 (h_3) and 341.9 cm^{-1} (d_3), and is mainly due to the additional electrostatic interaction between the charge (localized in the phenol ring) and the dipole moment of the water molecule. This "charge interaction" leads to a very strong increase of the binding energy of the hydrogen bond. In this case, the increase in binding energy (~4600 cm^{-1}), going from the S_0 state to the ground state of the cation, is about twice the *total* binding energy of the neutral complex, which is roughly estimated to be about 2000 cm^{-1}. Hence, for the ionic complex the total stabilization energy of the hydrogen bond is of the order of about 6600 cm^{-1} (800 meV), which is comparable to binding energies of weak chemical bonds (≈1 eV). This increase in binding energy is responsible for the decrease in the hydrogen-bond length in the ionic complex.

2. *Intermolecular Vibrations*

The different intermolecular vibrational fundamentals are now extracted from the ZEKE spectra, while the assignment of those vibrations to the corresponding normal coordinates will be discussed in the following section (together with the *ab initio* results [65]). The dominant feature of the ZEKE spectra of the h_3 and d_3 complexes (Figs. 5 and 46) are progressions of the totally symmetric intermolecular stretch vibration (σ) in combination with different inter- and intramolecular vibrational origins. The stretch frequency is measured as 240 and 221 cm^{-1} for the h_3 and d_3 complex, respectively. The strong stretch progression reflects the change of intermolecular bond length upon ionization: *ab initio* calcula-

TABLE VI
Comparison of Experimental and Calculated (Harmonic) Frequencies (in cm^{-1}) of the Intermolecular Vibrations of the Phenol–Water Cation [a]

Vibration	Experiment from Ref. 64	*ab initio*[b] from Ref. 65
Out-of-plane bend	67/64(?)	87/82
β''	(1.05)	(1.06)
In-plane bend	84/-	124/117
β'	(–)	(1.06)
Stretch	240/221	275/259
σ	(1.09)	(1.06)
Torsion[c]	≈130/100	202/163
τ	(≈1.30)	(1.24)
Out-of-plane wag	-/-	451/330
γ''	(–)	(1.37)
In-plane wag	328/264	358/259[d]
γ'	(1.24)	(1.38)

[a]Shown are the values for the $PhOH–H_2O$ and $PhOD–D_2O$ complex, as well as the corresponding ratio (in brackets).
[b][ROHF/3-21G*(O)].
[c]The torsion is observed as overtone 2τ at 257 (h_3) and 197 cm^{-1} (d_3).
[d]One-dimensional anharmonic approach (see text and [65] for details).

tions [65] predict a value of 0.03 nm (on going from the S_0 to the ion). The full assignment of all vibrational features is discussed in Refs. 64 and 65. The results are given in Table VI and compared to the *ab initio* results.

The ZEKE spectra recorded via the different S_1 levels clearly show remarkably different Franck–Condon factors for the transitions from the intermediate resonances into the ionic vibrational levels (Fig. 46). In summary, there are now five intermolecular and one intramolecular vibrations of the 1:1 phenol–water complex determined from the experiment with frequencies for the $h_3(d_3)$ complex of 240 (221), 257 (197), 328 (264), 84, and 67 (64?) cm^{-1} for the intermolecular fundamentals and 450 cm^{-1} for the strong intramolecular 18*b* vibration (see also Table VI). The d_3 frequency corresponding to the 84-cm^{-1} vibration is not identified.

B. Comparison with *ab initio* Results

The assignment of these intermolecular vibrations to the corresponding normal coordinates is carried out by comparison with the *ab initio* results [65]. The results of these calculations [65] for the ionic frequencies of the h_3 and d_3 complex are summarized in Table VI. The assignment of the experimental frequencies to the corresponding normal coordinates (see

[65] for a diagram of these motions) is straightforward only for the stretch. For the other intermolecular vibrations the comparison with the *ab initio* results makes an assignment possible (see Table VI).

1. *Intramolecular Vibrations*

The ZEKE spectra of the 1:1 complex also show a few peaks that are assigned to intramolecular (phenol-localized) vibrations. In the ZEKE spectrum of the h_3 complex via the S_1 origin (see Figs. 5 and 46), a strong feature at 450 cm^{-1} is observed. In the earlier ZEKE study [181] this band was tentatively assigned to a combination band of one quantum of the stretch with a vibrational origin at 203 cm^{-1}. The latter band, however, has now been reassigned to the third quantum of the 67-cm^{-1} fundamental (see above). Hence, the 450-cm^{-1} band is now attributed to a fundamental, which was also a possible suggestion discussed in the previous work [181]. In that work this band was possibly assigned to an intermolecular vibration since no (similarly strong) intramolecular phenol mode is seen in that frequency region and it was assumed that the frequency shifts on complexation should be small (however, see later). Searching for the corresponding d_3 band in the spectrum (shown in Fig. 5) leads to a peak at 435 cm^{-1} (in the low-energy shoulder of the second quantum of the stretch at 441 cm^{-1}). Since no high-frequency intermolecular mode with a small isotopic shift is predicted [65] from the *ab initio* calculation (see Table VI), this fundamental is now attributed to an intermolecular mode. It has also been found that for the combination bands of this particular vibration with the stretch progression the stretch frequency is noticeably reduced from 240 to 230 cm^{-1} in the case of the h_3 complex and correspondingly from 221 to 214 cm^{-1} in the d_3 complex, thus suggesting some interaction between these two vibrations. The only intramolecular vibration in that frequency region, which has the same symmetry as the intermolecular stretch, namely a' (in plane), is the 18*b* mode with a frequency of 411 cm^{-1} for the phenol cation itself [180]. Hence, an assignment of the 450 cm^{-1} band to the 18*b* vibration would imply a large frequency shift of +39 cm^{-1} on complexation. However, the 18*b* mode is the C–O bending motion, and hence a significant change in frequency on complexation as well as on deuteration is expected. Indeed, the *ab initio* calculations [65] for this vibration predict a strong increase in frequency of about 70 cm^{-1} as well as a strong deuteration shift of −25 cm^{-1}. These shifts are consistent with the experimentally observed shifts of +39 and −15 cm^{-1}. (According to *ab initio* results, the 18*b* vibration also shows a significant complexation shift of about +20 cm^{-1} in the neutral complex [175].) Furthermore, by comparison with the normal mode pictures [65] of the 18*b* vibration, it can be seen that the water molecule is noticeably involved in that motion thus explaining the strong complexation and deuteration shifts. By comparing the normal mode pictures [65] of the 18*b* vibration and the intermolecular stretch, evidence for the coupling of these two vibrations is also found: In the hydrogen-bond region the elongations are very similar, while in the phenol moiety the elongations are 180° phase shifted. The fact that the 450 cm^{-1} band is quite intense in the ZEKE spectrum of the complex (compared to the monomer ZEKE spectrum [180]), and that an intense overtone (at 895 cm^{-1}) is seen, suggests there is a large change in the geometry along this particular normal coordinate upon ionization. This finding can again be rationalized by the noticeable influence of the intermolecular hydrogen bond on that normal coordinate since the intermolecular bond length is strongly reduced on ionization. Hence, in summary, the assignment of the 450 (h_3) and 435 cm^{-1} (d_3) band to the intramolecular 18*b* vibration is supported by the *ab initio* results [65, 175] and is at present the favored assignment.

2. *Intermolecular Vibrations*

Assuming C_s symmetry for both the neutral and the ionic states, only the three totally symmetric vibrations or even-numbered overtones of the antisymmetric vibrations are expected to be seen in the ZEKE spectra since all pumped intermediate levels are totally symmetric. However, according to the assignments given in Table VI there is one antisymmetric intermolecular fundamental (the out-of-plane bending vibration) assigned to the 67 cm^{-1} band. The intensity of this band could be derived from weak Herzberg–Teller coupling. It is well known from the PE [59] and also ZEKE spectra of aromatic molecules [22, 62, 63] that Franck–Condon forbidden vibrations can be observed. However, a complete explanation for the observation of antisymmetric (a'' in C_s) intermolecular fundamentals still has to be given.

IX. CONCLUSIONS AND OUTLOOK

ZEKE spectroscopy can be expected to lead to more and more applications in chemistry and molecular physics. In this chapter we have presented a few typical examples, mainly focusing on the high-resolution spectra of molecular cations that can be obtained by ZEKE spectroscopy. The extension to radicals, anions, and hence neutrals is still an emerging field. Many laboratories across the world are contributing to the rapid emergence of this spectroscopy. This trend was very evident in the two international conferences at Kreuth in 1991 and Giens in 1993. The future will no doubt help to further clarify the nature of ZEKE Rydberg states leading to these interesting spectra and to extend applications to anion spectra and neutrals. The technique also lends itself to the study of transient species, such as radicals or complexes including the direct measurements of transition states in chemical reactions [24]. All these early examples will no doubt be followed, as in any emerging spectroscopy, by a rich body of examples, which firmly establishes this new spectroscopy.

One chemical implication of ZEKE detection arises from the particular feature of ZEKE Rydberg states: Their longevity caused by the negligible interaction between the ZEKE electron and the ion core [182, 183]. The question then arises as to what happens to the ZEKE electron if the ionic core is highly excited vibrationally or electronically, or if the ion core dissociates into a fragment ion (core) and a neutral? For complete decoupling of the ZEKE electron from the dynamics of the core, one might expect the ZEKE electron to follow the positive charge of the fragment ion core thus forming a fragment ZEKE Rydberg state; the neutral fragment carries no charge and its interaction with the ZEKE electron due to its multipole moments and polarizability should be negligible. The two-photon nonresonant ZEKE spectra of ammonia showed that the (long) lifetime of the ZEKE Rydberg states is not shortened significantly for very high vibrational excitation up to $v^+ = 9$ (corresponding to ~1 eV *internal* energy of the ion core) and it was postulated in that work that ZEKE Rydberg states can survive fragmentation of the core to form fragment ZEKE Rydberg states [182]. For van der Waals clusters some experimental evidence for the formation of a fragment ZEKE Rydberg state from the parent ZEKE Rydberg state was found in the picosecond pump–probe ZEKE spectra of anilline . . . argon [22]. Also, it was demonstrated that electronic excitation and subsequent fluorescence of the ion core leads to the formation of a ZEKE Rydberg state with an ion core in the electronic ground state that can then be pulsed-field ionized [47]. Direct experimental evidence for the survival of ZEKE Rydberg states with ion core dissociation came from experiments using a variant of the ZEKE detection scheme: Instead of detecting the electron, the ion originating from pulsed-field ionization of the ZEKE Rydberg states is detected in the experiment (this has also been termed mass analyzed threshold ionization, MATI [41, 42]). This ionization has been carried out by applying a dc field (already present at the time of photoexcitation) to spatially separate the ions from the ZEKE Rydberg state molecules. A high-voltage pulse is then used for PFI and to extract the ions so produced into a TOF mass spectrometer. For the benzene–argon (C_6H_6–Ar) van der Waals cluster, the ZEKE spectra reproduced in Fig. 47 were obtained in different mass-resolved channels. [43]. The mass

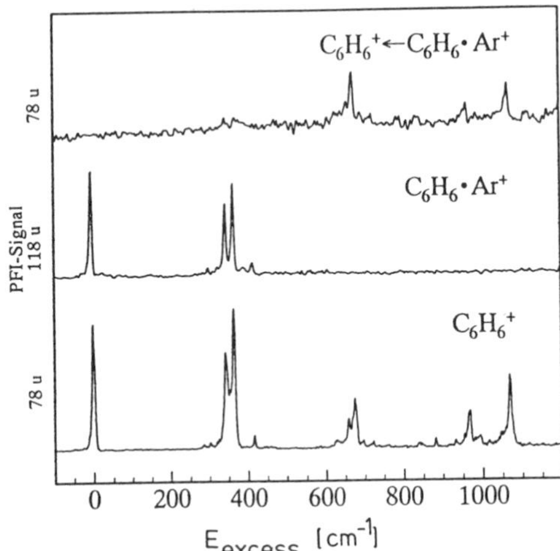

Figure 47. Mass resolved ZEKE spectra in three different mass channels. Bottom: Benzene ions produced from pulsed-field ionization of ZEKE Rydberg states of benzene. Middle: Benzene–argon ions produced from pulsed-field ionization of ZEKE Rydberg states of benzene–argon. Top: Benzene ions produced from pulsed-field ionization of fragment ZEKE Rydberg states of benzene originating from fragmentation of the benzene–argon core. (Reproduced from [43] with permission.)

resolved ZEKE signals recorded were $C_6H_6^+$ (bottom), $C_6H_6^+ \cdots Ar$ (middle), and $C_6H_6^+$ coming from the dissociation of the $C_6H_6^+ \cdots Ar$ core in a ZEKE Rydberg state. Hence, the lower frequency intramolecular vibrations (in the region of the 6^1 $(\pm\frac{3}{2})$ state, see Section II.B) seen in the C_6H_6–Ar ZEKE spectrum do not lead to dissociation of the ion complex, whereas the higher ones (around the 6^1 $(\pm\frac{1}{2})$) do; the signal loss in the $C_6H_6^+ \cdots Ar$ mass channel corresponds to signal appearing in the $C_6H_6^+$ mass channel. A similar result was found recently for mass-selected ZEKE spectra of C_6H_6: Fragmentation of the ion core led to fragment ZEKE Rydberg states that were then detected by PFI [182]. A very recent development for MATI is the use of *pulsed* separation fields to spatially separate the directly formed ions from the ZEKE Rydberg states. This detection scheme has been demonstrated to yield MATI spectra fully equivalent, in signal strength and resolution, to ZEKE spectra obtained from electron detection [184, 185].

Another important aspect for chemistry concerns the dynamics of hydrogen-bonded clusters. Attempts have been made to probe proton transfer in such systems, as shown in Fig. 48 for PhOH. . .$(NH_3)_n$, by short-pulse lasers in the femto-to-picosecond regime [186]. The combination of this pump–probe technique with ZEKE detection should be

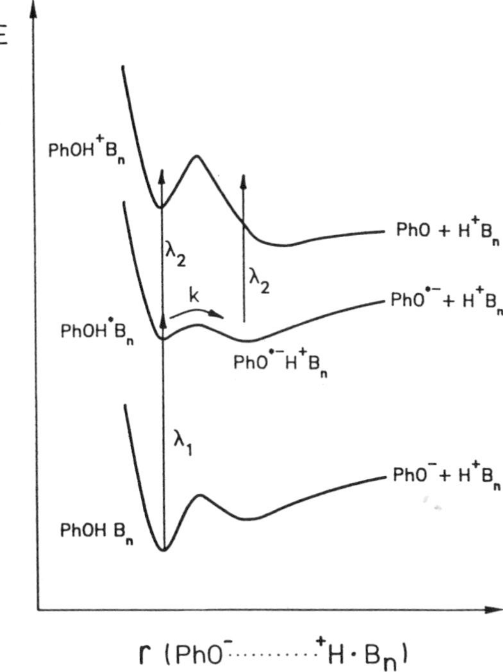

Figure 48. Schematic of pump–probe diagnostics of proton transfer in excited state of phenol–(ammonia)$_n$ clusters. (Reproduced from [184] with permission.)

particularly well suited to project out the proton-transfer coordinate from the excited S_1 state into the ion. Combined with mass-selected ZEKE detection, the proton transfer and fragmentation channel in the ion can also be monitored.

In conclusion one can say that we are now at the threshold of a rich new spectroscopy capable of providing new insights into cations, anions, and neutrals in a mass-selective fashion. Therefore, these techniques also provide information about metastable species present in reactive mixtures, clusters, van der Waals systems, and free radicals to mention only a few.

ACKNOWLEDGMENTS

Support from the Deutsche Forschungsgemeinschaft and the Commission of the European Union is gratefully acknowledged by KMD. Contributions of ERG were supported by a grant from the National Science Foundation (CHE-9307131) and a U.S. Senior Scientist Award from the

Alexander von Humboldt Foundation. The work of KW and VMK was supported by grants from the Air Force Office of Scientific Research and the Office of Health and Environmental Research of the U.S. Department of Energy. The authors KW and VMK also acknowledge use of the resources of the Jet Propulsion Laboratory/California Institute of Technology CRAY Y-MP2E/116 Supercomputer.

REFERENCES

1. K. Müller-Dethlefs, M. Sander, and E. W. Schlag, *Z. Naturforsch. A*, **39**, 1089 (1984).
2. K. Müller-Dethlefs, M. Sander, and E. W. Schlag, *Chem. Phys. Lett.*, **112**, 291 (1984).
3. G. Reiser, W. Habenicht, K. Müller-Dethlefs, and E. W. Schlag, *Chem. Phys. Lett.*, **152**, 119 (1988).
4. K. Müller-Dethlefs and E. W. Schlag, *Annu. Rev. Phys. Chem.*, **42**, 109 (1991).
5. W. Habenicht, G. Reiser, and K. Müller-Dethlefs, *J. Chem. Phys.*, **95**, 4809 (1991).
6. K. Müller-Dethlefs, *J. Chem. Phys.*, **95**, 4821 (1991).
7. G. Reiser and K. Müller-Dethlefs, *J. Phys. Chem.*, **96**, 9 (1992).
8. R. Lindner, B. Beyl, H. Sekiya, and K. Müller-Dethlefs, *Angew. Chem.*, **105**, 631 (1993); *Angew. Chem. Int. Ed. Engl.*, **32**, 603 (1993).
9. R. Lindner, H. Sekiya, and K. Müller-Dethlefs, *Angew. Chem.*, **105**, 1384 (1993); *Angew. Chem. Int. Ed. Engl.*, **32**, 1364 (1993).
10. S. Fredin, D. Gauyacq, M. Horani, Ch. Jungen, G. Lefevre, and F. Masnou-Seeuws, *Mol. Phys.*, **60**, 825 (1987).
11. M. Sander, L. A. Chewter, K. Müller-Dethlefs, and E. W. Schlag, *Phys. Rev. A*, **36**, 4543 (1987).
12. H. Rudolph, V. McKoy, and S. N. Dixit, *J. Chem. Phys.*, **90**, 2570 (1989).
13. R. G. Tonkyn, J. W. Winniczek, and M. G. White, *Chem. Phys. Lett.*, **164**, 137 (1989).
14. T. N. Kitsopoulos, I. M. Waller, J. G. Loeser, and D. M. Neumark, *Chem. Phys. Lett.*, **159**, 300 (1989).
15. I. M. Waller, T. N. Kitsopulos, and D. M. Neumark, *J. Phys. Chem.*, **94**, 2240 (1990).
16. J. M. Smith, C. Lakshiminarayan, and J. L. Knee, *J. Chem. Phys.*, **93**, 4475 (1990).
17. G. Ganteför, D. M. Cox, and A. Kaldor, *J. Chem. Phys.*, **93**, 8395 (1990).
18. G. Reiser, O. Dopfer, R. Lindner, G. Henri, K. Müller-Dethlefs, E. W. Schlag, and S. D. Colson, *Chem. Phys. Lett.*, **181**, 1 (1991).
19. F. Merkt and T. P. Softley, *Phys. Rev. A*, **46**, 302 (1992).
20. E. R. Grant and M. G. White, *Nature (London)*, **354**, 249 (1991).
21. I. Fischer, A. Strobel, J. Staecker, G. Niedner-Schatteburg, K. Müller-Dethlefs, and V. E. Bondybey, *J. Chem. Phys.*, **96**, 7171 (1992).
22. X. Zhang, J. M. Smith, and J. L. Knee, *J. Chem. Phys.*, **97**, 2843 (1992).
23. I. Fischer, A. Lochschmidt, A. Strobel, G. Niedner-Schatteburg, K. Müller-Dethlefs, and V. E. Bondybey, *J. Chem. Phys.*, **98**, 3592 (1993).
24. D. M. Neumark, *Annu. Rev. Phys. Chem.*, **43**, 153 (1992); R. B. Metz, S. E.

Bradforth, and D. M. Neumark, *Adv. Chem. Phys.*, **81** (1992); D. M. Neumark, *Acc. Chem. Res.*, **26**, 33 (1993).

25. F. Merkt and T. P. Softley, *J. Chem. Phys.*, **96**, 4149 (1992).
26. F. Merkt and T. P. Softley, *Int. Rev. Phys. Chem.*, **12**, 205 (1993).
27. J. Harrington and J. C. Weisshaar, *J. Chem. Phys.*, **97**, 2809 (1992).
28. A. Strobel, I. Fischer, J. Staecker, G. Niedner-Schatteburg, K. Müller-Dethlefs, and V. E. Bondybey, *J. Chem. Phys.*, **97**, 2332 (1992).
29. G. P. Bryant, Y. Jiang, M. Martin, and E. R. Grant, *J. Chem. Phys.*, **96**, 6875 (1992).
30. R. G. Tonkyn, R. T. Wiedmann, and M. G. White, *J. Chem. Phys.*, **96**, 3696 (1992).
31. R. G. Tonkyn, R. T. Wiedmann, E. R. Grant, and M. G. White, *J. Chem. Phys.*, **95**, 7033 (1991).
32. M.-T. Lee, K. Wang, V. McKoy, R. G. Tonkyn, R. T. Wiedmann, E. R. Grant, and M. G. White, *J. Chem. Phys.*, **96**, 7848 (1992).
33. G. F. Ganteför, D. M. Cox, A. J. Kaldor, *J. Chem. Phys.*, **96**, 4102 (1992).
34. S. T. Pratt, *J. Chem. Phys.*, **98**, 9241 (1993).
35. X. Zhang, J. M. Smith, and J. L. Knee, *J. Chem. Phys.*, **99**, 3133 (1993).
36. W. A. Chupka, *J. Chem. Phys.*, **98**, 4520 (1993); **99**, 580 (1993).
37. F. Merkt, H. H. Fielding, and T. P. Softley, *Chem. Phys. Lett.*, **202**, 153 (1993).
38. U. Even, M. Ben-Nun, and R. D. Levine, *Chem. Phys. Lett.*, **210**, 416 (1993).
39. M. C. R. Cockett and K. Kimura, *J. Chem. Phys.*, **100**, 3429 (1994).
40. S. Hillenbrand, L. Zhu, and P. Johnson, *J. Chem. Phys.*, **95**, 2237 (1991).
41. L. Zhu and P. Johnson, *J. Chem. Phys.*, **94**, 5769 (1991).
42. C. Jouvet, C. Dedonder-Lardeux, S. Martrenchard-Barra, and D. Soldagi, *Chem. Phys. Lett.*, **198**, 419 (1992).
43. H. Krause and H. J. Neusser, *J. Chem. Phys.*, **97**, 5923 (1992).
44. K. Wang, J. A. Stephens, and V. McKoy, *J. Phys. Chem.*, **98**, 460 (1994).
45. F. Merkt, *J. Chem. Phys.*, **100**, 2623 (1994).
46. F. Merkt and R. N. Zare, *J. Chem. Phys.*, **101**, 3495 (1994).
47. W. Kong, D. Rodgers, and J. W. Hepburn, *Chem. Phys. Lett.*, **203**, 497 (1993).
48. M. J. Seaton, *Rep. Prog. Phys.*, **46**, 167 (1983).
49. Ch. Jungen, *Phys. Rev. Lett.*, **53**, 2394 (1984).
50. A. Giusti-Suzor and Ch. Jungen, *J. Chem. Phys.*, **80**, 986 (1984).
51. S. N. Dixit, D. L. Lynch, V. McKoy, and W. M. Huo, *Phys. Rev. A*, **32**, 1267 (1985).
52. D. W. Turner, C. Baker, A. D. Baker, and C. R. Brundle, *Molecular Photoelectron Spectroscopy*, Wiley, London, 1970.
53. G. Herzberg, *Electronic Spectra of Polyatomic Molecules*, Van Nostrand, New York, 1966.
54. H. A. Jahn and E. Teller, *Proc. R. Soc. London A*, **161**, 220 (1937).
55. I. B. Bersuker, *The Jahn–Teller Effect and Vibronic Interactions in Modern Chemistry*, Plenum Press, New York, 1984.
56. K. Raghavachari, R. C. Haddon, T. A. Miller, and V. E. Bondybev, *J. Chem. Phys.*, **79**, 1387 (1983).
57. M. Iwasaki, K. Toriyama, and K. Nunime, *J. Chem. Soc. Chem. Commun.*, **1983**, 320.
58. T. Oka, *Rev. Mod. Phys.*, **64**, 1141 (1992).
59. E. Sekreta, K. S. Visvanathan, and J. P. Reilly, *J. Chem. Phys.*, **90**, 5349 (1989).

60. D. Rieger, G. Reiser, K. Müller-Dethlefs, and E. W. Schlag, *J. Phys. Chem.*, **96**, 12 (1992).
61. G. Reiser, D. Rieger, T. G. Wright, K. Müller-Dethlefs, and E. W. Schlag, *J. Phys. Chem.*, **97**, 4335 (1993).
62. M. C. R. Cockett, K. Okuyama, and K. Kimura, *J. Chem. Phys.*, **97**, 4679 (1992).
63. J. M. Dyke, H. Ozeki, M. Takahashi, M. C. R. Cockett, and K. Kimura, K., *J. Chem. Phys.*, **97**, 8926 (1992).
64. O. Dopfer, G. Reiser, K. Müller-Dethlefs, E. W. Schlag, and S. D. Colson, *J. Chem. Phys.*, **101**, 974 (1994).
65. P. Hobza, R. Burcl, V. Špirko, O. Dopfer, K. Müller-Dethlefs, and E. W. Schlag, *J. Chem. Phys.*, **101**, 990 (1994).
66. M. I. Al-Joboury and D. W. Turner, *J. Chem. Soc.*, **1963**, 5141; *J. Chem. Phys.*, **37**, 3007 (1962).
67. F. I. Vilesov, B. C. Kurbatov, and A. N. Terenin, *Dokl. Akad. Nauk SSSR*, **138**, 1329 (1961), Engl. transl.: *Sov. Phys. Dokl.*, **6**, 490 (1961).
68. K. Siegbahn, *Electron Spectroscopy for Chemical Analysis*, Uppsala University, Institute of Physics Report 670, Nov. 1969.
69. K. Siegbahn, C. Nordling, G. Johansson, J. Hedman, P. F. Heden, K. Gelius, T. Bergmark, L. O. Werme, T. Manne, and Y. Baer, *ESCA Applied To Free Molecules*, North Holland, Amsterdam, London 1969; K. Siegbahn, *Some Current Problems in Electron Spectroscopy*, Report Uppsala University Institute of Physics 1074, Uppsala, 1982.
70. L. Åsbrink, *Chem. Phys. Lett.*, **7**, 549 (1970).
71. A. Niehaus and M. W. Ruf, *Chem. Phys. Lett.*, **11**, 55 (1971).
72. S. Southworth, C. M. Truesdale, P. H. Kobrin, D. W. Lindle, W. D. Brewer, and D. A. Shirley, *J. Chem. Phys.*, **76**, 143 (1982).
73. S. T. Pratt, P. M. Dehmer, and J. L. Dehmer, *J. Chem. Phys.*, **87**, 3288 (1987).
74. W. G. Wilson, K. S. Viswanathan, E. Sekreta, and J. P. Reilly, *J. Phys. Chem.*, **88**, 672 (1984).
75. K. S. Viswanathan, E. Sekreta, E. R. Davidson, and J. P. Reilly, *J. Chem. Phys.*, **90**, 5078 (1986).
76. X. Song, E. Sekreta, J. P. Reilly, H. Rudolph, and V. McKoy, *J. Chem. Phys.*, **91**, 6062 (1989).
77. S. W. Allendorf, D. J. Leahy, D. C. Jacobs, and R. N. Zare, *J. Chem. Phys.*, **91**, 2216 (1989).
78. T. Baer, *Adv. Chem. Phys.*, **64**, 111 (1986).
79. R. Spohr, P. M. Guyon, W. A. Chupka, and J. Berkowitz, *Rev. Sci. Instrum.*, **42**, 1872 (1971).
80. W. B. Peatman, G. B. Kasting, and D. J. Wilson, *J. Electron. Spectrosc. Relat. Phenom.*, **7**, 233 (1975).
81. T. Baer, State Selection by Photoion-Photoelectron Coincidence. In *Gas Phase Ion Chemistry*, M. T. Bowers, Ed. Academic, New York, 1979.
82. P. M. Guyon, T. Baer, and I. Nenner, *J. Chem. Phys.*, **78**, 3665 (1983).
83. W. P. Peatman, T. B. Borne, and E. W. Schlag, *Chem. Phys. Lett.*, **3**, 492 (1969); Peatman, W. P. Ph. D. Thesis, Northwestern University, Evanston, Illinois, 1969.
84. T. Baer, W. B. Peatman, and E. W. Schlag, *Chem. Phys. Lett.*, **4**, 243 (1969).
85. H. Sekiya, R. Lindner, and K. Müller-Dethlefs, *Chem. Lett.*, **1993**, 485.

86. R. Lindner et al. *J. Chem. Phys.*, to be submitted.
87. I. Fischer, R. Lindner and K. Müller-Dethlefs, *J. Chem. Soc. Faraday Trans.*, Faraday Research Article, **90**, 2425 (1994).
88. H. Rudolph, V. McKoy, and S. N. Dixit, *Chem. Phys. Lett.*, **90**, 2570 (1988).
89. J. A. Blush, P. Chen, R. T. Wiedmann, and M. G. White, *J. Chem. Phys.*, **98**, 3557.
90. J. H. Callomon, T. M. Dunn, and I. M. Mills, Philos. Trans. R. Soc. London A, **259**, 499 (1966).
91. M. Oldani, R. Widmer, G. Grassi, and A. Bauder, *J. Mol. Struct.*, **190**, 31 (1988).
92. E. Riedle, T. Knittel, T. Weber, and H. J. Neusser, *J. Chem. Phys.*, **91**, 4555 (1989).
93. E. Riedle and J. Pliva, *Chem. Phys.*, **152**, 375 (1991).
94. A. W. Potts, W. C. Price, D. G. Streets, and T. A. Williams, *Faraday Discuss. Chem. Soc.*, **54**, 168 (1972).
95. S. R. Long, J. T. Meek, and J. P. Reilly, *J. Chem. Phys.*, **79**, 3206 (1983).
96. P. R. Bunker, *Molecular Symmetry and Spectroscopy*, Academic, New York, 1979.
97. M.-L. Junttila, J. L. Domenech, G. T. Fraser, and A. S. Pine, *J. Mol. Spectrosc.*, **147**, 513, 1991; E. Riedle, to be published, used these S_0 constants to improve the S_1 constants from [92].
98. H. C. Longuet-Higgins, U. Öpik, M. H. L. Pryce, F. R. S. Sack, and R. A. Sack, *Proc. R. Soc. London A*, **244**, 1 (1958).
99. T. A. Miller and V. E. Bondybey, *Molecular Ions: Spectroscopy, Structure and Chemistry*, North-Holland, Amsterdam, 1983 and references cited therein.
100. R. L. Whetten, K. S. Haber, and E. R. Grant, *J. Chem. Phys.*, **84**, 1270 (1986).
101. J. Eiding, R. Schneider, W. Domcke, H. Köppel, and W. von Niessen, *Chem. Phys. Lett.*, **177**, 345 (1992).
102. J. Eiding and W. Domcke, *Chem. Phys.*, **163**, 133 (1992).
103. S. L. Anderson, G. D. Kubiak, and R. N. Zare, *Chem. Phys. Lett.*, **105**, 22 (1984).
104. K. S. Viswanathan, E. Sekreta, and J. P. Reilly, *J. Phys. Chem.*, **90**, 5658 (1986).
105. S. T. Pratt, P. M. Dehmer, and J. L. Dehmer, *Chem. Phys. Lett.*, **105**, 28 (1984).
106. M. A. O'Halloran, S. T. Pratt, P. M. Dehmer, and J. L. Dehmer, *J. Chem. Phys.*, **87**, 3288 (1987).
107. A. Fujii, T. Ebata, and M. Ito, *J. Chem. Phys.*, **88**, 5307 (1988).
108. S. T. Pratt, E. F. McCormack, J. L. Dehmer, and P. M. Dehmer, *J. Chem. Phys.*, **92**, 1831 (1990).
109. S. T. Pratt, P. M. Dehmer, and J. L. Dehmer, *J. Chem. Phys.*, **92**, 262 (1990).
110. E. F. McCormack, S. T. Pratt, J. L. Dehmer, and P. M. Dehmer, *J. Chem. Phys.*, **92**, 4734 (1990).
111. E. de Beer, M. Born, C. A. de Lange, and N. P. C. Westwood, *Chem. Phys. Lett.*, **186**, 40 (1991).
112. K. Wang, J. A. Stephens, V. McKoy, E. de Beer, C. A. de Lange, and N. P. C. Westwood, *J. Chem. Phys.*, **97**, 211 (1992).
113. E. de Beer, C. A. de Lange, J. A. Stephens, K. Wang, and V. McKoy, *J. Chem. Phys.*, **95**, 714 (1991).
114. M. Braunstein, V. McKoy, S. N. Dixit, R. G. Tonkyn, and M. G. White, *J. Chem. Phys.*, **93**, 5345 (1990).
115. S. N. Dixit and V. McKoy, *Chem. Phys. Lett.*, **128**, 49 (1986).
116. K. Wang and V. McKoy, *Chem. Phys. Lett.*, **95**, 4977 (1991).

117. K. Wang and V. McKoy, to be published.

118. M.-T. Lee, K. Wang, V. McKoy, and L. E. Machado, *J. Chem. Phys.*, **97**, 3905 (1992).

119. M. Dubs, U. Bruhlmann, and J. R. Huber, *J. Chem. Phys.*, **84**, 3106 (1986).

120. J. B. Halpern, H. Zacharias, and R. Wallenstein, *J. Mol. Spectrosc.*, **79**, 1 (1980).

121. A. R. Edmonds, *Angular Momentum in Quantum Mechanics*, Princeton University, New Jersey, 1974.

122. R. R. Lucchese, G. Raseev, and V. McKoy, *Phys. Rev. A*, **25**, 2572 (1982).

123. J. M. Brown, J. T. Hougen, K. P. Huber, J. W. C. Johns, I. Kopps, H. Lefebvre-Brion, A. J. Merer, D. A. Ramsay, J. Rostas, and R. N. Zare, *J. Mol. Spectrosc.*, **55**, 500 (1975).

124. J. Xie and R. N. Zare, *J. Chem. Phys.*, **93**, 3033 (1990).

125. G. Raseev and N. Cherepkov, *Phys. Rev. A*, **42**, 3948 (1990).

126. R. S. Mulliken, *Phys. Rev.*, **59**, 873 (1941).

127. P. G. Burke, N. Chandra, and F. A. Gianturco, *J. Phys. B*, **5**, 2212 (1972).

128. W. J. Hunt and W. A. Goddard, *Chem. Phys. Lett.*, **3**, 414 (1969).

129. R. T. Wiedmann, M. G. White, K. Wang, and V. McKoy, *J. Chem. Phys.*, **98**, 7673 (1993).

130. R. T. Wiedmann, R. G. Tonkyn, M. G. White, K. Wang, and V. McKoy, *J. Chem. Phys.*, **97**, 768 (1992).

131. M. Takahashi, H. Ozeki, and K. Kimura, *Chem. Phys. Lett.*, **181**, 255 (1991).

132. W. Kong, D. Rodgers, J. W. Hepburn, K. Wang, and V. McKoy, *J. Chem. Phys.*, **99**, 3159 (1993).

133. K. Wang and V. McKoy, *J. Chem. Phys.*, **95**, 7872 (1991).

134. H. Lefebvre-Brion, *Chem. Phys. Lett.*, **171**, 377 (1990).

135. K. Wang, J. A. Stephens, and V. McKoy, *J. Chem. Phys.*, **95**, 6456 (1991).

136. M.-T. Lee, K. Wang, and V. McKoy, *J. Chem. Phys.*, **97**, 3108 (1992).

137. M. S. Child and Ch. Jungen, *J. Chem. Phys.*, **93**, 7756 (1990).

138. R. D. Gilbert and M. S. Child, *Chem. Phys. Lett.*, **187**, 153 (1991).

139. R. T. Wiedmann, M. G. White, K. Wang, and V. McKoy, *J. Chem. Phys.*, **100**, 4738 (1994).

140. R. T. Wiedmann, E. R. Grant, R. G. Tonkyn, and M. G. White, *J. Chem. Phys.*, **95**, 746 (1991).

141. a. Y.-F. Zhu, E. R. Grant, K. Wang, V. McKoy, and H. Lefebvre-Brion, *J. Chem. Phys.*, **100**, 8633 (1994);
b. Y. F. Zhu, E. R. Grant and H. Lefebvre-Brion, *J. Chem. Phys.*, **99**, 2287 (1993).

142. E. de Beer, W. J. Buma, and C. A. de Lange, *J. Chem. Phys.*, **99**, 3252 (1993).

143. C. R. Mahon, G. R. Janik, and T. F. Gallagher, *Phys. Rev. A*, **41**, 3746 (1990).

144. E. R. Grant and Ch. Jungen, unpublished results.

145. K. S. Haber, G. Bryant, Y. Jiang, E. R. Grant, and H. Lefebvre-Brion, *Phys. Rev. A*, **44**, R5331 (1991).

146. E. de Beer, B. G. Koenders, M. P. Koopmans, and C. A. de Lange, *J. Chem. Soc. Faraday Trans.*, **86**, 2038 (1990).

147. H. Lefebvre-Brion, A. Giusti-Suzor, and G. Raseev, *J. Chem. Phys.*, **83**, 1557 (1986).

148. J. Xie and R. N. Zare, *Chem. Phys. Lett.*, **159**, 399 (1989) and private communication.

149. Y. Xie, P. T. A. Reilly, S. Chilukuri, and R. J. Gordon, *J. Chem. Phys.*, **95**, 854 (1991).

150. D. K. Hsu, D. L. Monts, and R. N. Zare, *Spectral Atlas of Nitrogen Dioxide*, Academic, New York, 1978.

151. R. S. Tapper, R. L. Whetten, and E. R. Grant, *J. Phys. Chem.*, **88**, 1273 (1984).

152. M. B. Knickelbein, K. S. Haber, L. Bigio, and E. R. Grant, *Chem. Phys. Lett.*, **131**, 51 (1986).

153. F. X. Campos, Y. Jiang, and Edward R. Grant, *J. Chem. Phys.*, **93**, 2308 (1990).

154. F. X. Campos, Y. Jiang, and Edward R. Grant, *J. Chem. Phys.*, **93**, 7731 (1990).

155. G. P. Bryant, Y. Jiang, and E. R. Grant, *J. Chem. Phys.*, **96**, 4827 (1992).

156. L. Bigio and E. R. Grant, *J. Chem. Phys.*, **83**, 5361 (1985); *ibid.*, **83**, 5369 (1985).

157. J. Berkowitz, *Photoabsorption, Photoionization and Photoelectron Spectroscopy*, Academic, New York, 1979.

158. a. G. P. Bryant, Y. Jiang, and E. R. Grant, *Chem. Phys. Lett.*, **200**, 495 (1992);
b. G. P. Bryant, Y. Jiang, M. Martin, and E. R. Grant, *J. Chem. Phys.*, **101**, 7199 (1944).

159. T. J. Lee, *Chem. Phys. Lett.*, **188**, 154 (1992).

160. W. von Niessen, G. H. F. Diercksen, and L. S. Cederbaum, *Chem. Phys. Lett.*, **45**, 295 (1977).

161. M. H. Palmer, W. Moyes, M. Spiers, and J. N. A. Ridyard, *J. Mol. Struct.*, **49**, 105 (1978).

162. G. Bieri, L. Åsbrink, and W. von Niessen, *J. Electron Spectrosc. Relat. Phenom.*, **23**, 281 (1981).

163. R. S. Mulliken, *J. Chem. Phys.*, **23**, 1997 (1955).

164. (a) F. Duschinsky, *Acta Physiochem. U.R.S.S.*, **1**, 551 (1937); (b) see also: G. J. Small, *J. Chem. Phys.*, **54**, 3300 (1971).

165. X. Song, C. W. Wilkerson, J. Lucia, S. Pauls, and J. P. Reilly, *Chem. Phys. Lett.*, 174, 377 (1990).

166. R. D. Amos and J. E. Rice, *Cambridge Analytic Derivatives Package (CADPAC) 4.0*, Cambridge, 1987.

167. G. C. Eiden, F. Weinhold, and J. C. Weisshaar, *J. Chem. Phys.*, **95**, 8665 (1991).

168. D. Solgadi, C. Jouvet, and A. Tramer, *J. Phys. Chem.*, **92**, 3313 (1988).

169. C. N. R. Rao and T. Pradeep, *Chem. Soc. Rev.*, **20**, 477 (1991).

170. S. Tomoda and K. Kimura, in *Vacuum Ultraviolet Photoionization and Photodissociation of Molecules and Clusters*, C. Y. Ng, Ed., World Scientific, London, 1991.

171. K. Fuke, H. Yoshiuchi, K. Kaya, Y. Achiba, K. Sato, and K. Kimura, *Chem. Phys. Lett.*, **108**, 179 (1984).

172. K. Müller-Dethlefs, O. Dopfer, and T. G. Wright, *Chem. Rev.*, **94**, 1845 (1994).

173. R. J. Lipert and S. D. Colson, *J. Chem. Phys.*, **89**, 4579 (1988).

174. R. J. Lipert and S. D. Colson, *J. Chem. Phys.*, **93**, 135 (1989).

175. M. Schütz, T. Bürgi, S. Leutwyler, and T. Fischer, *J. Chem. Phys.*, **98**, 3763 (1993).

176. R. J. Lipert and S. D. Colson, *J. Chem. Phys.*, **92**, 3240 (1990).

177. T. G. Wright, E. Cordes, O. Dopfer, and K. Müller-Dethlefs, K., *J. Chem. Soc. Faraday Trans.*, **89**, 1609 (1993).
178. E. Cordes, O. Dopfer, T. G. Wright, and K. Müller-Dethlefs, *J. Phys. Chem.*, **97**, 7471 (1993).
179. O. Dopfer, G. Lembach, T. G. Wright, and K. Müller-Dethlefs, *J. Chem. Phys.*, **98**, 1933 (1993).
180. O. Dopfer et al., *J. Chem. Phys.*, to be submitted; O. Dopfer, Dissertation, Technische Universität München, 1944.
181. O. Dopfer, G. Reiser, R. Lindner, and K. Müller-Dethlefs, *Ber. Bunsen-Ges. Phys. Chem.*, **96**, 1259 (1992).
182. G. Reiser, W. Habenicht, and K. Müller-Dethlefs, *J. Chem. Phys.*, **98**, 8462 (1993).
183. W. G. Scherzer, H. L. Selzle, and E. W. Schlag, *Z. Naturforsch.*, **48a**, 1256 (1993).
184. H.-J. Dietrich, R. Lindner, and K. Müller-Dethlefs, *J. Chem. Phys.*, **101**, 3399 (1994).
185. R. Lindner, H.-J. Dietrich, and K. Müller-Dethlefs, *Chem. Phys. Lett.*, **228**, 417 (1994).
186. J. A. Syage and J. Steadman, *J. Chem. Phys.*, **95**, 2497 (1991); J. A. Syage, *J. Phys. Chem.*, **97**, 12523 (1993).

NEW WAYS OF UNDERSTANDING SEMICLASSICAL QUANTIZATION

P. GASPARD, D. ALONSO, AND I. BURGHARDT

Service de Chimie Physique and Centre for Nonlinear Phenomena and Complex Systems, Université Libre de Bruxelles, B-1050 Brussels, Belgium

CONTENTS

Advances in Chemical Physics, Volume XC, Edited by I. Prigogine and Stuart A. Rice.
ISBN 0-471-04234-X

I. INTRODUCTION

A. General Introduction

Thanks to laser spectroscopy, rapid progress is underway in our understanding of atoms and molecules. Modern studies have emphasized the dynamical aspects of atomic and molecular systems [1]. In this context, the relation between quantum and classical mechanics has to be reinvestigated, especially, in the light of the recent works on classical chaos [2–7].

The quantum mechanical ground state of atoms or molecules is characterized by a wave function that rapidly decays for large separations between the constituent particles, and thus lends a stability to the system that is at the basis of most material structures, such as stable molecular and supramolecular assemblies as well as solids. In the ground state, the dynamics is frozen for the internal degrees of freedom of atomic systems. This stability cannot be accounted for by classical mechanics, which has been one of the main reasons for which classical mechanics was superseded by quantum mechanics in the description of microscopic systems [8–9].

However, highly excited regimes of such systems are often characterized by physical actions that are much larger than the Planck constant and, in these semiclassical regimes, classical mechanics is a very useful guide to reach an understanding of the physicochemical processes. Indeed, the importance of quantum coherence is often decreasing upon excitation to higher energies. At low-energy excitations, the system still manifests coherent quantum beats between its discrete energy levels, which are only damped by spontaneous emission. At higher energies, however, the system becomes the stage of kinetic processes like ionization, unimolecular dissociation, bimolecular reaction, or photon absorption followed by intramolecular vibrational redistribution or isomerization and, eventually, photon emission.[1] The purpose of semiclassical methods is to find a systematic description of such phenomena at the border between classical and quantum mechanics.

[1] Most of these processes can be viewed as scattering processes of few-body systems with electrons, nuclei, and photons. However, these processes may also take place in many-body systems; for instance, in gas or liquid phases where the atom or molecule is exposed to the effects of a fluctuating environment due to collisions with other particles or with phonons. The process may then be described as a quantum stochastic process, which achieves the irreversible localization of the wave function in one of the final states with certain probabilities [10]. The description of such stochastic processes falls outside the scope of this chapter.

Since semiclassical methods carry out the quantization on the basis of our knowledge of the classical motion, they are susceptible to the difficulties caused by the nonlinearities of Hamilton's equations of classical mechanics. Over the last decades, consequences of these nonlinearities have been thoroughly studied thanks to fast modern computers, which led to the widespread recognition that trajectories (solutions of nonlinear equations) may be as irregular in time as was previously known for random processes. This phenomenon, nowadays known as chaos, has been discovered and studied in every field of natural sciences where the time evolution is governed by nonlinear equations [2–4]. Chaos turns out to be a typical behavior in classical mechanics. In this way, chaos was found in classical models of atomic and molecular systems, in particular in microscopic models of unimolecular or bimolecular chemical reactions. The inapplicability of standard WKB semiclassical methods to classically chaotic systems urged the development of new semiclassical methods that would be applicable to the typical and common chaotic dynamical regimes.

In the early 1970s, Gutzwiller [11] as well as Balian and Bloch [12, 13] derived a semiclassical expression for the trace of the resolvent of the quantum Hamiltonian operator. Through this trace formula, the quantum operator is directly related to classical periodic orbits. In 1980, Gutzwiller showed that the trace formula can be used to obtain approximate values for the quantum energy eigenvalues of classically chaotic systems that are not separable [14, 15]. In this sense, the periodic-orbit quantization method appears to be more general than the WKB method. Since then, periodic-orbit quantization has been applied to a large variety of systems, among others by the authors [16–21]. Moreover, the method provides a new way to analyze complex experimental spectra by Fourier transform (FT) from the energy domain to the time domain.

However, the trace formula derived by Gutzwiller only contains the leading term of the semiclassical expansion, and the form of the next-to-leading terms remained a challenge for a long time. Recently, Alonso and Gaspard [19, 20] derived a systematic expansion of the trace formula in powers of the Planck constant. This $\hbar$-expansion allows us to consider the semiclassical method on the same footing as the other perturbation methods of quantum mechanics. The $\hbar$-expansion and the recognition that the trace formula contains different terms, which depend on the nature and on the stability of the periodic orbits, provide the semiclassical description of new phenomena, such as diffraction effects, as well as nonlinear stability effects or bifurcation of periodic orbits besides the simple improvement of the accuracy of the quantities calculated with the semiclassical method.

The purpose of this chapter is to present some of the recent developments in the periodic-orbit quantization, particularly of classically chaotic systems, and some of its applications to atomic and molecular processes.

The plan of this chapter is the following. In section I.B, we briefly describe the history and the modern context of the development of semiclassical methods. In this way, we give several potential applications of these methods in the field of chemical physics and in related fields. Section II gives a presentation of the general formalism of time evolution in quantum mechanics and its formulation in terms of path integrals. The semiclassical limit is considered in its principle. The equations of classical mechanics and the related variational principles are derived, as well as the second variation of the action. The method by Maslov and Fedoriuk [22] for the semiclassical solution of the time-dependent initial condition Cauchy problem is summarized. We then describe the Feynman path integral and the principle of the integration method by stationary phase. Section III is devoted to a description of the phase-space structures of classical mechanics. In particular, we describe the properties associated with the stability of classical orbits and we show how chaotic behavior may appear when the classical motion becomes unstable. The relevant phase-space structures are the stable and unstable manifolds, the homoclinic tangle, the Smale horseshoe, and the related fractal repeller. Section IV describes the theory of the $\hbar$-expansion in the time domain. In the energy domain, the resummation of the $\hbar$-expansion on equilibrium points (to be compared with resummation on periodic orbits, see below) is carried out as described in Section V, where we show that the $\hbar$-expansion is essentially equivalent to standard perturbation theory by Van Vleck contact transformations. We also discuss its relationship with the semiclassical quantization of the Birkhoff normal form around equilibrium points. The limitation of this method to effectively separable systems without chaotic behavior is discussed. The resummation of the semiclassical expansion on periodic orbits is performed in Section VI, which contains the completion of the program set up by Gutzwiller for the periodic-orbit quantization of classically chaotic systems. Section VII shows how average quantities can be calculated by the periodic-orbit quantization method, particularly in view of photoabsorption cross-sections. The case of quantum billiards is described in Section VIII. An extension of the $\hbar$-expansion to matrix Hamiltonians is contained in Section IX. A comparison of the atomic and molecular Hamiltonians and their semiclassical properties is discussed in Section X. Molecular vibrograms are introduced in Section XI as a tool for the semiclassical analysis of molecular photoabsorption cross-sections that opens the way to a spectroscopy in the time domain. The theory of the molecular transition state is developed in Section XII. The periodic-orbit quantization for the hydrogen negative ion (H^-) is described in section XIII, together with a discussion of semiclassical electronic excitation in other atomic and solid-state systems. This chapter ends with Section XIV, where the main results are summarized and conclusions are drawn.

B. Brief Historical Introduction and General Background

Because of their ability to relate the quantum level of description to the classical one, semiclassical methods play a privileged role in our general understanding of atomic and molecular processes.

Semiclassical methods have been developed since the beginning of quantum mechanics because of the importance of the models based on classical particles and fields in the historical advent of Heisenberg's matrix mechanics in 1925 and of Schrödinger's wave mechanics in 1926. Let us recall that the wave theory of electrons by de Broglie and others owes much to the relation between geometric and wave optics known at that time. Earlier, the atomic model of Bohr and Rutherford led to the Bohr-Sommerfeld quantization conditions, which first appeared in an incomplete form because the Morse–Maslov indexes were unknown at that time [23]. Let us note that works in classical mechanics (in particular, by Poincaré, Bruns, and others [24]) had shown at the turn of the century that most classical systems are not integrable. This known fact shed some trouble on the methods proposed by Bohr and Sommerfeld to quantize atomic systems, as these methods implicitly assumed separability between the different internal degrees of freedom of the electronic motion around the atomic nuclei. If this is the case for hydrogen, the knowledge of divergences in the perturbation theory of the three-body problem of celestial mechanics already suggested that the helium atom with two electrons and a nucleus could not conform to the Bohr–Sommerfeld quantization without implicit or explicit approximations or further assumptions. This criticism came from Einstein in 1917 [25]. As a way to partially handle these problems, adiabatic methods were developed, particularly by Ehrenfest [26], with the concept of an effective separability between degrees of freedom evolving on very different time scales. The general context that already pointed to what is today called classical chaos led young scientists like Heisenberg to propose radical changes and to point out the necessity of founding a new quantum mechanics directly based on strictly observational quantities. The first historical jump was carried out for the harmonic oscillator but the proposed matrix mechanics turned out to be extendable to arbitrary systems, separable or not [8, 9].

The wave equation proposed by Schrödinger turns out to be extremely powerful and allows a discussion of the relation to classical mechanics along lines similar to the relation between wave optics and geometric optics, but in the multidimensional phase space rather than in the tridimensional physical phase space, as emphasized by Pauli [27]. The same year Schrödinger proposed his equation, this similarity was developed by Brillouin, Wentzel, and Kramers for systems with one degree of freedom or for larger systems that are separable and reducible to several one-degree-of-freedom systems [28].

Semiclassical methods also turned out to be crucial in statistical mechanics, in particular, in the definition of elementary phase-space cells of volume $d^f q d^f p/(2\pi\hbar)^f$. In this context, a fundamental contribution was due to Wigner in 1932, who proposed his famous representation of quantum states as real but not necessarily positive functions in the phase space of positions and momenta [29]. Systems with many fermions were also treated semiclassically by Thomas and Fermi [30].

In this way, semiclassical methods are essential tools to bridge the gap between the microscopic quantum world and the macroscopic classical world. In the following, we would like to show that their recent developments are tightly related to the progress of technologies in the mesoscopic domain. Thus, we will summarize the modern experimental context in which a rapid progress is underway in semiclassical methods.

Laser spectroscopy, atomic or molecular beams (as well as Fourier spectrometers) have allowed us to analyze spectra in detail with very high resolution, which is provided by the low temperatures achieved in supersonic jets [31]. In this way, it has been possible to study weakly bonded molecular complexes, such as Van der Waals molecules [32].

Furthermore, it is now possible to study the Rydberg states of atoms or molecules where one electron is orbiting at great distances from the ion [33]. In Rydberg states, the motion takes place at very high quantum numbers, where the classical dynamics starts to emerge from the quantum dynamics. In this context, Rydberg atoms in magnetic and electric fields are being extensively investigated, to the point where very subtle nonlinear properties typical of classical mechanics, such as bifurcations of periodic orbits, are experimentally observed in quantum spectra [34, 35]. More recently, the study of doubly excited states, where two electrons are at great distances from the nuclei, has started to develop [36]. Rydberg and doubly excited states are also studied in molecules, but the main results are probably still to be awaited [37]. Semiclassical methods are playing an essential role in this context.

In studying these highly excited states, where the constituent particles move at greater and greater distances from each other, it is important to keep the atom or molecule in isolation for longer and longer times. Along this line, several types of traps have been invented to confine neutral particles or ions that are nearly at rest. The trap may contain one or several particles. The study of the motion of these particles is becoming a field in itself, where the motion is in a first approximation classical for heavy particles, but where we must further take into account quantum effects, as well as the random forcing by the laser beams [38]. The condition of isolation in a trap is often closer to the astrophysical conditions than to experiments in the gas phase or in a supersonic jet so that they make possible the laboratory study of special reactions occurring in interstellar space.

Generally, Rydberg states are stable or metastable assemblies of the constituent particles with very long lifetimes. In contrast, the vibrational motion of molecules takes place on a much shorter time scale of the order of 10–1000 fs. The advent of femtosecond laser pulses has only recently opened this new time domain [39]. Thus, the study of vibrational motions and, in particular, those taking place in the course of chemical reactions are nowadays possible. These new techniques give us the hope of understanding the details of how chemical reactions occur. Semiclassical methods are particularly useful in this context, since they allow the calculation of several quantities of interest like the lifetimes of the metastable states of the molecular transition complex, as we will show in Section XII.

In larger systems, such as electronic semiconductor devices, semiclassical methods are currently experiencing a special interest in the study of electronic conductance in nanometric ultrapure semiconductor devices [40]. With lithography and epitaxy techniques, it is possible today to build electronic circuits of predesigned shapes where the electrons are in ballistic motion between the walls of the device at a low temperature of a few Kelvin [41]. In a first approximation, it may be assumed that the potential rises so steeply at the walls that the electron moves like in a billiard where the wave function vanishes at the walls. Irregular scattering phenomena of the electron in such billiards have been shown to be in relation to universal electric conductance fluctuations as well as to Ericson fluctuations in nuclear physics [42, 43]. Similar fluctuations may exist in atomic or molecular scattering processes [44].

At higher temperatures, there is the transition between the ballistic motion of the electron, which is essentially a one-body phenomenon, and the more standard Ohm's regime, which is thermal and of many-body character [45].

Other applications concern the electronic motion on surfaces. Electron tunneling microscopy has reached an unprecedented capacity in manipulating atoms on surfaces. On the other hand, surfaces are known to sustain electronic motions in the form of charge density waves. By manipulating atoms it is possible to construct, out of only a few atoms, a scatterer for the surface electronic waves [46]. The electrons undergo S-wave scattering on the atoms that can be considered as impurities if they are randomly distributed on the surface. This new work is opening up the study of electronic motion on an even smaller scale than in the layers of nanometric semiconductor circuits.

In a similar direction, it is possible today to build nanomachines, such as rotors, bearings, or gears, from semiconductor or diamond materials [47]. The motion of the nanometric mechanical parts of these devices is affected by the various possible atomic and molecular forces that cause friction. Semiclassical methods may be relevant to consider when quantum effects become nonnegligible.

Let us also mention the application of the semiclassical methods to electromagnetic microwave cavities that are experiencing a current rise in interest in the field of classically chaotic systems [48], as well as to the acoustics of sound cavities or scattering [49].

A further domain where semiclassical methods are important is the study of quantum systems coupled to a fluctuating environment. In these cases, the system contains a large number of degrees of freedom. We may cite proton- or electron-transfer reactions in liquids or at interfaces, as well as superconducting or superfluid devices. In this context, current research is using several methods, particularly the elimination of the irrelevant degrees of freedom of the reservoir in the Feynman path integral, and the use of the semiclassical approximation in certain regimes, as proposed by several authors, such as Leggett, Chandler, Miller, and Voth [50].

The preceding large but partial survey of the mesoscopic domain of chemical physics and related fields indicates the broadness of the range of potential applications for semiclassical methods. Semiclassical methods should not substitute wave function methods, but they turn out to be complementary. If wave function methods can achieve high-precision results, semiclassical methods, on the other hand, often provide a general description of the laws underlying series and distributions of energy levels, of lifetimes, or of other observables. Semiclassical methods are of special interest in the study of chemical reactions, where they offer a unifying approach for understanding several aspects of reactive and dynamical processes, as recently emphasized by Marcus [51].

II. QUANTUM TIME EVOLUTION, PATH INTEGRALS,AND SEMICLASSICAL LIMIT

A. Propagator and Green Operator

Microscopic nonrelativistic systems are described by Schrödinger's equation

$$i\hbar\, \partial_t \psi = \hat{\mathscr{H}} \psi \tag{2.1}$$

where the Hamiltonian operator is assumed to be time independent and of the form

$$\hat{\mathscr{H}} = \sum_{j=1}^{f} -\frac{\hbar^2}{2m_j} \frac{\partial^2}{\partial q_j^2} + V(q_1, \ldots, q_f) \tag{2.2}$$

for a system with f degrees of freedom. After rescaling the position variables $\sqrt{m_j} q_j \rightarrow q_j$, the masses m_j can be set equal to unity.

Since Schrödinger's equation is linear, the time evolution can be written in terms of the evolution operator as

$$\psi(T) = \hat{K}(T)\psi(0) = \exp\left(-\frac{i}{\hbar} \hat{\mathscr{H}} T\right) \psi(0) \tag{2.3}$$

On the other hand, we may introduce the resolvent of the Hamiltonian (also called the Green operator) as the Laplace transform or semisided Fourier transform of the propagator

$$\hat{G}(E) \equiv \frac{1}{E - \hat{\mathscr{H}}} = \frac{1}{i\hbar} \int_0^\infty dT \exp\left(\frac{i}{\hbar} ET\right) \hat{K}(T) \tag{2.4}$$

where the last equality holds if $\operatorname{Im} E > 0$ [52].

Reciprocally, the time evolution can be expressed as a complex integral of the resolvent. The choice of the integration contour determines whether the unitary group, the forward semigroup (corresponding to forward-time propagation), or the backward semi-group will be selected according to [53]

$$\hat{K}(T) = \frac{1}{2\pi i} \int_{C_+ + C_-} dE \exp\left(-\frac{i}{\hbar} ET\right) \hat{G}(E) \tag{2.5}$$

$$\theta(\pm T)\hat{K}(T) = \frac{1}{2\pi i} \int_{C_\pm} dE \exp\left(-\frac{i}{\hbar} ET\right) \hat{G}(E) \tag{2.6}$$

where $\theta(\cdot)$ denotes the Heaviside function and the different contours are given in Fig. 1(*a*).

Analytic continuation in the complex energy surface may simplify the expression of the time evolution. The analytic continuation decomposes the integrals [Eqs. (2.5 and 2.6)] into the contribution of the different singularities of the resolvent [54]. In many systems of interest, the singularities are of three types:

1. Branch points at real energies $\{E_c\}$ are the energy thresholds where the channels $\{c\}$ are opened. In most cases, the branch points are of order two because the energy is related to the momentum by $E = E_c + p^2/(2M)$, whereas the wave function is analytic in the momentum variable. Accordingly, two Riemann sheets are associated with each branch point and the complex energy surface is

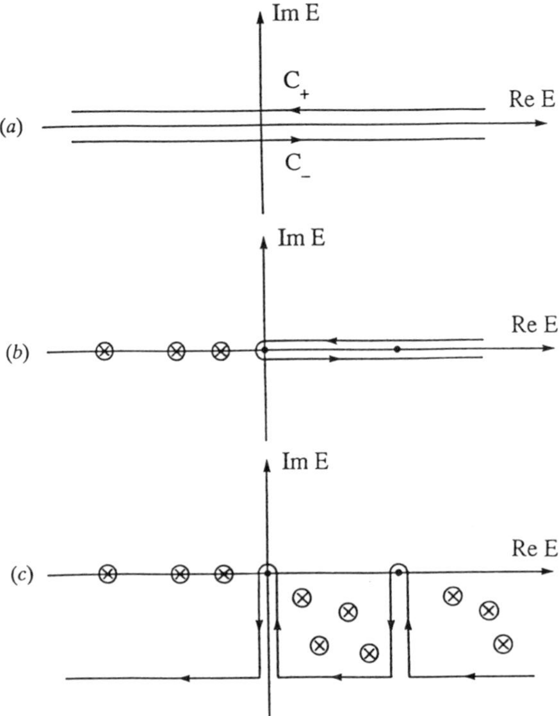

Figure 1. Integration contours in the Riemann surface of complex energies, which are entering into the definitions of the propagators of the unitary group [Eq. (2.5)] and of the forward and backward semigroups [Eq. (2.6)]: (*a*) Contours $C_\pm$ of the forward and backward semigroups [Eq. (2.6)]; (*b*) Distortion of $C_+ + C_-$ to obtain the group propagator in the form of Eq. (2.7); (*c*) Distortion of C_+ to obtain the forward semigroup propagator in the form of Eq. (2.8).

therefore composed of several Riemann sheets, which are connected at the branch points E_c.

2. Poles $\{E_b\}$ are located on the real energy axis below the lowest energy threshold E_0. These poles are associated with the bound states of the system.
3. Poles $\{E_{cr} = \varepsilon_{cr} - i\Gamma_{cr}/2\}$ are located in the lower half of the second Riemann sheet, for a given Riemann surface composed of two sheets. These poles determine the scattering resonances that correspond to metastable states, as encountered, in particular, in atomic ionization or molecular dissociation processes. Another pole, located in the upper half of the same sheet, is associated with each of the former poles, and represents an antiresonance.

In the case of the unitary group, the cuts that are associated with the continuous energies of the channels are chosen on the real energy axis above each branch point E_c. The contour can be distorted, as shown in Fig. 1(b), so that Eq. (2.5) is decomposed into a sum of terms for each bound state and a sum of integrals for each channel. The energy spectrum has both discrete and continuous components so that the evolution operator can be written as [54]

$$\hat{K}(T) = \sum_b |\varphi_b\rangle\, e^{-iE_bT/\hbar}\langle\varphi_b| + \sum_c \int_{E_c}^{\infty} dE|\varphi_{cE}\rangle\, e^{-iET/\hbar}\langle\varphi_{cE}| \tag{2.7}$$

where the quantities φ_b are the bound-state eigenfunctions corresponding to the discrete energies E_b. The quantities φ_{cE} are the scattering eigenfunctions in the different channels c of energy threshold E_c.

On the other hand, the integral in Eq. (2.6) can be distorted into a contour extending into the second of each pair of sheets. Two different semigroups are selected by the condition that the scattering part of the wave function [Eq. (2.6)] vanishes for either $t \to +\infty$ or $t \to -\infty$. These two semigroups correspond, respectively, to an analytic continuation toward either negative or positive values of $\operatorname{Im} E$.

For the analytic continuation to $\operatorname{Im} E < 0$, the cuts may be chosen parallel to the axis $\operatorname{Im} E$ and extend to negative values, which uncovers the poles located at the resonances, as shown in Fig. 1(c). For the forward semigroup ($T > 0$), Eq. (2.6) becomes

$$\begin{aligned}\theta(+T)\hat{K}(T) = \sum_b |\varphi_b\rangle\, e^{-iE_bT/\hbar}\langle\varphi_b| + \sum_{cr} |\varphi_{cr}\rangle\, e^{-i\varepsilon_{cr}T/\hbar}\, e^{-\Gamma_{cr}T/2\hbar}\langle\tilde{\varphi}_{cr}| \\ + \frac{\hbar}{2\pi}\sum_c e^{-iE_cT/\hbar}\int_0^{\infty} ds\, e^{-sT}[\hat{G}(E_c + 0 - i\hbar s) \\ - \hat{G}(E_c - 0 - i\hbar s)]\end{aligned} \tag{2.8}$$

The third term is at the origin of the long-time tails where the wave function decays algebraically as $|\psi(T)|^2 \sim T^{-3\nu_c}$, ν_c being the number of fragments in the channel c. Let us remark that this power-law decay is simply the result of the quasifree propagation of the fragments flying away from the interaction region [55].

By contrast, the poles contribute to exponential decays. If a resonance is isolated either because of its large spacing with respect to neighboring resonances or because of the initial condition of the wavepacket, the wave function decays as $|\psi(T)|^2 \sim \exp(-\Gamma_{cr}T/\hbar)$, so that the poles or resonances correspond to metastable states of the system with lifetimes $\tau_{cr} = \hbar/\Gamma_{cr}$. Recently, a theory has been developed by Böhm and others [56, 57], where the complex energies of the resonances are considered as eigenvalues of the generator of the forward semigroup. The corresponding eigenstates are known as Gamow vectors. The generator of the semigroup is no longer a Hermitian operator although it is formally equal to the Hamiltonian. It is acting on a space of distributions that differs from the usual Hilbert space. According to this theory of so-called rigged Hilbert spaces, the metastable states can be interpreted as eigenvectors of the time evolution operator. As a consequence of the non-Hermiticity of the extended Hamiltonian, the left eigenstate $\langle \tilde{\varphi}_{cr}|$ differs from the complex conjugate of the right eigenstate $|\varphi_{cr}\rangle$.

B. Trace and Level Density

Our purpose in this section is to calculate the real or complex energy eigenvalues $\{E_b, E_{cr}\}$ of some Hamiltonian operator $\hat{\mathcal{H}}$. If the energy eigenvalues are the poles of the Green operator they are also the poles of its trace [58]. This relationship can be established by considering the level density of a bounded system, which is given by the distribution

$$n(E) = \sum_{n=0}^{\infty} \delta(E - E_n) = \operatorname{tr} \delta(E - \hat{\mathcal{H}}) = -\frac{1}{\pi} \operatorname{Im} \operatorname{tr} \frac{1}{E - \hat{\mathcal{H}} + i0} \tag{2.9}$$

where E is real. The level density is the derivative of the staircase function $N(E) = \text{Number}\{E_n < E\}$.

Similarly, the resonances of a scattering system can be obtained as the poles of a trace function such as

$$g(E) = \operatorname{tr}\left(\frac{1}{E - \hat{\mathcal{H}}} - \frac{1}{E - \hat{\mathcal{H}}_{\text{as}}}\right) \tag{2.10}$$

where $\hat{\mathcal{H}}_{\text{as}}$ is some reference Hamiltonian whose asymptotic behavior coincides with $\hat{\mathcal{H}}$ but whose resonances are known. This subtraction is necessary for the trace function to be defined; otherwise the trace is infinite because the scattering system has an infinite extension in space [13]. In Eq. (2.10), the energy is complex, as in Fig. 1. Because the Green operator is related to the propagator by Eq. (2.4), it turns out that the above traces of the Green operator can be evaluated by a Laplace or semisided Fourier transform of the trace of the propagator. Sections II.D and E will be devoted to the semiclassical estimation of these traces, but we will first describe several relevant observable quantities.

C. Observable Quantities

We may distinguish two basic kinds of experiments: (1) interaction of one or several laser beams with a molecular beam, gas, liquid, or solid; (2) interaction of two molecular beams [1]. In the first type of experiment, the molecules are interacting with the electromagnetic field and exchange photons with it, in addition to their internal dynamics. In the second kind of experiments, the molecules interact between themselves and the interaction with the electromagnetic field is secondary.

1. *Interaction with Light*

In this case, we must add to the free-system Hamiltonian the Hamiltonians of the electromagnetic field and of its interaction with the system. The treatment of such Hamiltonians can be found in several references [59]. It is not our purpose in this chapter to enter further into such developments.

The absorption of monochromatic light by the system is the simplest type of experiment. The cross-section for the absorption of one photon $\hbar\omega$ is known to be given by [59, 60]

$$\kappa_\omega = \frac{\omega}{6\hbar c \varepsilon_0}(1 - e^{-\beta\hbar\omega}) \int_{-\infty}^{+\infty} dt \, e^{-i\omega t} \langle \hat{\mathbf{d}}(0) \cdot \hat{\mathbf{d}}(t) \rangle \tag{2.11}$$

where $\langle \cdot \rangle = \mathrm{tr}(\hat{\rho} \, \cdot)$ is the statistical average over a density matrix for a canonical equilibrium ensemble of atoms or molecules at an inverse temperature β. The operator c is the light velocity and ε_0 is the vacuum permittivity. The operator $\hat{\mathbf{d}}(t)$ is the electric dipole operator at time t given by

$$\hat{\mathbf{d}}(t) = \exp\left(+\frac{i}{\hbar}\hat{\mathcal{H}}t\right) \hat{\mathbf{d}} \exp\left(-\frac{i}{\hbar}\hat{\mathcal{H}}t\right) \tag{2.12}$$

where $\hat{\mathcal{H}}$ is the free-system Hamiltonian. This result is obtained by perturbation theory using the weakness of the coupling to the electromagnetic field.

Works by Heller et al. [61] have shown that it is particularly enlightening to consider not only the photoabsorption cross-section in the energy domain but also the correlation function in the time domain that appears in the Fourier transform of the absorption spectrum. This method is standard by now and can be extended to other observable quantities related to the exchange of more than one photon. In this context, we have the processes with spontaneous emission during reaction; the scattering of light whose cross-section is given in the energy domain by the Kramers–Heisenberg–Dirac relation and in the time-domain by integrals of four-time correlation functions of the electric dipole; as well as other multiphotonic processes which are typical of nonlinear optics [59, 61].

The newly developed femtosecond laser techniques emphasize the time-dependent description of the molecule-light interaction [39]. It is in principle required to know the precise form of the electromagnetic pulse that may influence the driving of the molecular motion. However, if we assume that the time length of the pulse is much shorter than the resulting molecular dynamics, we may assume that the pulse prepares an initial wavepacket that will thereafter follow a time evolution under the interaction-free molecular Hamiltonian. With such a scheme, we are coming closer to the second type of experiments. Indeed, the dissociation following the photoabsorption can be viewed as a half-collision that is controlled by the interaction-free molecular Hamiltonian [62]. Thus, we are close to a description in terms of the scattering matrix.

2. *Scattering between Atomic Particles*

The second basic class of experiments deals with collisions between atoms or molecules, such as

$$\mathrm{A} + \mathrm{B} \rightarrow \mathrm{C} + \mathrm{D}$$

The atoms and molecules possess internal degrees of freedom that may be activated either in the reactant or the product species. For each particular ingoing or outgoing channel, we have the Hamiltonian describing the free flight of the species, that is obtained as some limits of the total Hamiltonian. The collision is described by the scattering matrix that connects the ingoing wave function to the outgoing one at a certain energy E. For large separations between the particles, the total outgoing eigenfunction at energy E therefore has the form [63]

$$\Psi_E^{(+)} \underset{r\to\infty}{\cong} \varphi_{\mathbf{n}_i}(\xi_i)\frac{e^{-ik_i r}}{r} - \sum_{\mathbf{n}_f} S_{\mathbf{n}_i\mathbf{n}_f}(k_i, k_f)\varphi_{\mathbf{n}_f}(\xi_f)\frac{e^{+ik_f r}}{r} \tag{2.13}$$

where $\varphi_{\mathbf{n}}(\xi)$ denotes the asymptotic wave function expressed with respect to the internal degrees of freedom ξ of quantum numbers $\mathbf{n}$. Among them, we find the quantum numbers of rotation and vibration of the separated atoms or molecules A, B, C, D. The parameter r is the relative distance between the particles in free flight, while k represents the corresponding momentum. The quantity S is the scattering matrix at the energy $E = E(\mathbf{n}_i, k_i) = E(\mathbf{n}_f, k_f)$.

Scattering Resonances. It is known that the S-matrix is related to the Green operator, which is the resolvent of the total Hamiltonian $\hat{G}(z) = (z - \hat{\mathcal{H}})^{-1}$. By this relation, several of the properties of the S-matrix can also be obtained from the properties of the Green operator [54, 56]. This is the case for the scattering resonances corresponding to the metastable states of the collisional transition complex. The scattering resonances are determined by the poles of the S-matrix or of the Green operator in the surface of complex energy $z = E$, as discussed above [54]. We may therefore obtain the scattering resonances as the poles of the trace of the Green operator tr $\hat{G}(z)$. Because the metastable states are transient and, thus, localized in a finite region of position space, these states are intrinsic to the system, which explains that they are determined by the total Hamiltonian (or its resolvent) and, consequently, appear at the level of the S-matrix.

Cross-Sections and Product Distributions. On the other hand, there are other properties of the collision that cannot be determined from tr $\hat{G}(z)$, such as the total and differential cross-sections, as well as the rotational or vibrational distributions of the products of the collision. Those distributions are sensitively dependent not only on the dynamics in the region of the collisional transition complex but also on the dynamics in the ingoing and exit channels that may produce distortions in the wave function over a distance that is larger than the extension of the metastable states. For those quantities, we need to calculate the elements of the scattering matrix $S_{\mathbf{n}_i\mathbf{n}_f}(k_i, k_f)$ explicitly. These elements can also be related to elements of the Green operator $\hat{G}(z)$ between the initial and final states, but not simply to its trace.

Average Time Delay. Wigner [64] was among the first to observe that the existence of metastable states would cause a time delay in the scattering of a wavepacket with respect to a situation without metastable states. Wigner showed that the time delay is related to the derivative of the phase shift due to the collision with respect to the energy. The concept of time delay has been studied in more general cases since then, and it has been shown [58] that the average time delay at energy E is given in terms of the scattering matrix or of the Green operator by

$$\mathcal{T}(E) = \frac{\hbar}{i} \operatorname{tr} \frac{d}{dE} \ln \hat{S}(E)$$

$$= -2\hbar \operatorname{Im} \operatorname{tr}\left(\frac{1}{E - \hat{\mathcal{H}} + i0} - \frac{1}{E - \hat{\mathcal{H}}_0 + i0}\right) \tag{2.14}$$

$\hat{\mathcal{H}}$ is the total Hamiltonian and the average time delay is calculated with respect to the asymptotically free Hamiltonian $\hat{\mathcal{H}}_0$, which plays the role of a reference Hamiltonian to calculate both the S-matrix and the time delay. The time delay has been the object of recent interest, especially in the context of time delay in tunneling, for instance, in electronic circuits [64].

Reaction Rates. Rates are of special importance when we want to relate the microscopic quantities to the time evolution of macroscopic systems, especially during chemical reactions near or far from equilibrium [65]. It is not our purpose to describe the kinetic processes in the condensed phases, but the same problem of definition of the rate constants already appears for isolated unimolecular or bimolecular reactions. At this level, the reaction rate can be defined in terms of the scattering resonances of the S-matrix. The scattering resonances are the poles of the S-matrix at complex energies $E_r = \varepsilon_r - i\Gamma_r/2$, where the half-width Γ_r determines the lifetime $\tau_r = \hbar/\Gamma_r$ of the corresponding metastable state. Accordingly, a wavepacket around the energy ε_r of the resonance will decay over the lifetime τ_r. This decay occurs in either the unimolecular or the bimolecular reaction if the former is considered as the half-collision related to the latter.

For nonisolated molecules or atoms, the reaction rates may be defined in very different manners. We may start from the point of view that we know *a priori* the location of the reactant and product species in configuration space, which leads to the consideration of fluxes across predefined hypersurfaces. On the other hand, there is the alternative point of view that the metastable state (this time of the many-body system) should be defined, as for an isolated system, from the intrinsic dynamics of the many-body Hamiltonian. Reaction rate theory is apparently progressing between these two points of view [66].

This overview of different experimental situations shows how the observational quantities are related to the Green operator, and thus to the time evolution of the quantum system. In the following sections, we will therefore focus on the problem of time evolution and its semiclassical limit. We first give a short introduction to the quasiclassical method, and then proceed to develop the semiclassical propagators.

D. Weyl–Wigner Representation and the Quasiclassical Method

The semiclassical method, which is the subject of this chapter, is a method used to determine the quantum mechanical eigenvalues and wave functions from information on the classical orbits. The quasiclassical method is another method used to study the regimes intermediate between the quantum and the classical ones. This method is complementary to the semiclassical method [29, 67]. It is based on the idea of expansion of the physical observables in a series in the Planck constant. For this reason, the quasiclassical method is closer to classical dynamics than to quantum dynamics and presents difficulties in problems where the quantum mechanical phases and wave functions are important. In Section VI.A we will use the quasiclassical method to derive the Thomas–Fermi–Weyl–Wigner average level density.

The quasiclassical method uses the Wigner–Weyl representation [67–69]. Operators $\hat{A}$ are transformed into functions over the position-momentum space $(\mathbf{q}, \mathbf{p})$ according to

$$A_{\mathrm{W}}(\mathbf{q}, \mathbf{p}) \equiv \int d^f x \exp\left(\frac{i}{\hbar} \mathbf{p} \cdot \mathbf{x}\right) \left\langle \mathbf{q} - \frac{\mathbf{x}}{2} \middle| \hat{A} \middle| \mathbf{q} + \frac{\mathbf{x}}{2} \right\rangle \tag{2.15}$$

Each operation on observables can be expressed in the Weyl–Wigner representation. In particular, the trace is

$$\operatorname{tr} \hat{A} = \int \frac{d^f q d^f p}{(2\pi\hbar)^f} A_{\mathrm{W}}(\mathbf{q}, \mathbf{p}) \tag{2.16}$$

The equation of motion for the function A_{W} is

$$\partial_t A_{\mathrm{W}} = \frac{2}{\hbar} \mathcal{H}_{\mathrm{W}} \sin\left(\frac{\hbar \hat{\Lambda}}{2}\right) A_W \qquad \text{with} \qquad \hat{\Lambda} = \overleftarrow{\partial}_{\mathbf{p}} \cdot \overrightarrow{\partial}_{\mathbf{q}} - \overleftarrow{\partial}_{\mathbf{q}} \cdot \overrightarrow{\partial}_{\mathbf{p}} \tag{2.17}$$

In many cases, the Weyl–Wigner representation of the Hamiltonian operator coincides with the classical Hamiltonian function, $\mathcal{H}_{\mathrm{W}} = \mathcal{H}_{\mathrm{cl}}$. The generator of the evolution can be expanded in a series in the Planck constant together with the function A_{W} to obtain the solution of the equation of motion in the form

$$A_{\mathrm{W}}(t) = A_{\mathrm{W}}^{(0)}(t) + \hbar^2 A_{\mathrm{W}}^{(2)}(t) + \mathcal{O}(\hbar^4) \tag{2.18}$$

The first term $A_{\mathrm{W}}^{(0)}(t)$ obeys the classical Liouville equation $\partial_t A_{\mathrm{W}}^{(0)} = \{A_{\mathrm{W}}^{(0)}, \mathcal{H}_{\mathrm{W}}\}$, while the next terms obey similar but inhomogeneous equations [69]. The symbol $\{\cdot, \cdot\}$ is the classical Poisson bracket. This series is an asymptotic series in $\hbar$. Here the time dependency is determined by the classical evolution operator $\exp(-t\hat{L})$ with the Liouville generator $\hat{L} = \{\mathcal{H}_{\mathrm{W}}, \cdot\}$. Correlation functions can then be expressed as the classical correlation functions plus corrections in powers of the Planck constant. This quasiclassical scheme differs from the semiclassical scheme in that it is unable to reproduce quantization effects, such as the formation of eigenenergies, but it is very useful in thermal systems of statistical mechanics.

E. Short-Wavelength Asymptotics of Schrödinger's Equation

1. *From Schrödinger's to Hamilton's Equations*

In this section we would like to describe semiclassical methods in general, as they developed since the beginning of quantum mechanics. We will focus on the semiclassical methods applied to the general problem of solving an initial value problem of the Schrödinger equation of the form

$$i\hbar\, \partial_t \psi = \left(-\frac{\hbar^2}{2}\nabla^2 + V\right)\psi \tag{2.19}$$

Writing the wave function ψ in terms of its modulus and its phase, which are real quantities,

$$\psi = \sqrt{\rho}\, \exp \frac{i}{\hbar} \Upsilon \tag{2.20}$$

where $\rho = |\psi|^2$ is the probability density, we obtain two coupled equations for the two fields

$$\begin{aligned} &\partial_t \Upsilon + \frac{1}{2}(\nabla\Upsilon)^2 + V = \frac{\hbar^2}{2\sqrt{\rho}}\nabla^2\sqrt{\rho} \\ &\partial_t \rho + \nabla \cdot (\rho \nabla \Upsilon) = 0 \end{aligned} \tag{2.21}$$

The second equation is the continuity equation of probability conservation, which in this case holds in position space where the velocity or momentum is given by $\mathbf{v} = \mathbf{p} = \nabla\Upsilon$.

In the classical limit $\hbar \to 0$, we suppose that the quantum density and phase tend to classical correspondent quantities

$$\lim_{\hbar\to 0} \Upsilon(\mathbf{q}, t; \hbar) = W(\mathbf{q}, t)$$
$$\lim_{\hbar\to 0} \rho(\mathbf{q}, t; \hbar) = D(\mathbf{q}, t) \tag{2.22}$$

which obey the equations

$$\partial_t W + \tfrac{1}{2}(\nabla W)^2 + V = 0$$
$$\partial_t D + \nabla \cdot (D\nabla W) = 0 \tag{2.23}$$

We see that the first equation is the Hamilton–Jacobi equation of classical mechanics for the classical action $W(\mathbf{q}, t)$. The second one is the continuity equation for the classical probability density $D(\mathbf{q}, t)$ in position space [70, 71].

The purpose of the semiclassical method is to express the quantum wave function in terms of the classical quantities D and W. To this end, the quantum corrections due to the right-hand side of Eq. (2.21) must be properly taken into account in a systematic semiclassical method. Let us make a remark about the form of those corrections. At the level of the real quantities ρ and Υ, the $\hbar$-corrections appear as series in even powers $\hbar^{2n}$ of the Planck constant due to Eq. (2.21). If the wave function is written as the exponential of another function, such as $\psi = \exp(\eta)$, we see that the function η can be expanded in powers of $\hbar$. Except for a constant phase, each term that is odd in $\hbar$ is imaginary and contributes to the phase of ψ while each term that is even in $\hbar$ is then real and contributes to the modulus ρ.

In terms of the classical quantities, the wave function becomes

$$\psi = \phi\sqrt{D}\exp\frac{i}{\hbar}W \tag{2.24}$$

where ϕ is a factor that is the ratio between the quantum wave function ψ and its semiclassical approximant $\sqrt{D}\exp(iW/\hbar)$. The equation of motion for this factor is

$$\frac{d\phi}{dt} = \partial_t\phi + \nabla W \cdot \nabla\phi = \frac{i\hbar}{2\sqrt{D}}\nabla^2(\phi\sqrt{D}) \tag{2.25}$$

which has to be solved along the classical trajectories. The factor ϕ can be expanded in powers of $i\hbar$ as

$$\phi = \sum_{n=0}^{\infty} (i\hbar)^n \phi_n \tag{2.26}$$

which reduces Eq. (2.25) to a hierarchy of coupled equations

$$\begin{aligned} \frac{d\phi_0}{dt} &= 0 \qquad n = 0 \\ \frac{d\phi_n}{dt} &= \frac{1}{2\sqrt{D}} \nabla^2(\phi_{n-1}\sqrt{D}) \qquad n = 1, 2, 3, \ldots \end{aligned} \tag{2.27}$$

where ϕ_0 is a constant along each classical orbit as expected [22].

The solution to these equations is based on the knowledge of the classical problem and on the solution of the Hamilton–Jacobi equation. Thus, we will now shortly comment on the classical problem, and come back to the determination of the factor ϕ in Section II.E.4.

The characteristics of the Hamilton–Jacobi equation are given by the classical trajectories in position space [71]. From the solution of the classical Hamiltonian equations in phase space

$$\begin{aligned} \dot{\mathbf{q}} &= \frac{\partial \mathcal{H}}{\partial \mathbf{p}} \\ \dot{\mathbf{p}} &= -\frac{\partial \mathcal{H}}{\partial \mathbf{q}} \end{aligned} \tag{2.28}$$

the action is given by

$$W = \int (\mathbf{p} \cdot d\mathbf{q} - \mathcal{H}\, dt) \tag{2.29}$$

The solution of Hamilton's equation is unique if the initial positions and momenta are given. However, this single-valuedness is lost in position space due to the following mechanism. The structures formed in position space by the trajectories are projections of the phase-space structures. If we have families of trajectories forming manifolds in phase space, their projections may have all sorts of singularities in position space that are known as catastrophes [73]. The most common is the fold of a manifold that produces a caustic line in position space. Two sheets of the manifold exist on one side of the caustic and zero on the other side. Accordingly, there may exist several trajectories passing through the same position. This multivaluedness is a major cause of difficulties in the semiclassical methods, but these difficulties have been overcome today, particularly by the works of Morse [72], Maslov [74], and others who provided the methods to match the semiclassical wave functions on the different sides of the caustic lines.

2. Classical Probability Conservation and the Van Vleck–Morette Matrix

After the above comments on the Hamilton–Jacobi equation, let us come back to the classical continuity equation. One of the best ways to find the solution to this equation, from the viewpoint of quantum mechanics, has been given by Berry and Mount [75]. We consider the Cauchy problem where the initial state is the state $|\mathbf{q}_0\rangle$ where the particle is located at the position $\mathbf{q}_0$. The solution to this problem is known as the propagator $K(\mathbf{q}, \mathbf{q}_0, t) = \langle \mathbf{q}|\hat{K}(t)|\mathbf{q}_0\rangle$, which satisfies Eq. (2.19) as well as the initial condition

$$\lim_{t\to 0} K(\mathbf{q}, \mathbf{q}_0, t) = \delta(\mathbf{q} - \mathbf{q}_0) \tag{2.30}$$

The propagator represents the probability amplitude to propagate from $\mathbf{q}_0$ to $\mathbf{q}$ in the time t. Its absolute square gives the probability density for such a propagation

$$\rho(\mathbf{q}, \mathbf{q}_0, t) = |K(\mathbf{q}, \mathbf{q}_0, t)|^2 = |\phi|^2 D(\mathbf{q}, \mathbf{q}_0, t) \tag{2.31}$$

where D is the classical density and ϕ becomes a constant in the classical limit. The semiclassical solution of this problem goes through the search for all the classical trajectories from $\mathbf{q}_0$ to $\mathbf{q}$ in time t. Several solutions to this problem may exist, as illustrated by the case of a reflection in a mirror that provides a second solution besides the direct solution. If the initial state is localized at $\mathbf{q}_0$, its wave function is given in the momentum representation by

$$\langle \mathbf{p}_0|\mathbf{q}_0\rangle = \frac{1}{(2\pi\hbar)^{f/2}} \exp\left(-\frac{i}{\hbar}\mathbf{p}_0 \cdot \mathbf{q}_0\right) \tag{2.32}$$

Accordingly, the probability for the initial momentum to take a value in the cell $(\mathbf{p}_0, \mathbf{p}_0 + d\mathbf{p}_0)$ is uniform and given by

$$|\langle \mathbf{p}_0|\mathbf{q}_0\rangle|^2 \, d\mathbf{p}_0 = \frac{d\mathbf{p}_0}{(2\pi\hbar)^f} \tag{2.33}$$

Let us suppose that $\mathbf{p}_0$ is the initial momentum of one of the trajectories joining $\mathbf{q}_0$ to $\mathbf{q}$ in the time t. Then, the probability mass [Eq. (2.33)] will be the contribution to the probability mass centered around the position $\mathbf{q}$

at time t, so that we have

$$|K(\mathbf{q}, \mathbf{q}_0, t)|^2 \, d\mathbf{q} = |\phi|^2 D(\mathbf{q}, \mathbf{q}_0, t) \, d\mathbf{q} \underset{\hbar \to 0}{\simeq} \frac{d\mathbf{p}_0}{(2\pi\hbar)^f} \tag{2.34}$$

The classical density D is therefore proportional to the Jacobian determinant of the transformation from the initial momentum $\mathbf{p}_0$ to the final position $\mathbf{q}$, while $|\phi_0|$ is given by $(2\pi\hbar)^{-f/2}$. Since the equation for D is linear, the proportionality may be turned into an equality. Moreover, since the classical action $W = W(\mathbf{q}, \mathbf{q}_0, t)$ defines a canonical transformation from the initial variables $(\mathbf{q}_0, \mathbf{p}_0)$ to the final variables $(\mathbf{q}, \mathbf{p})$ given by

$$\mathbf{p} = \frac{\partial W}{\partial \mathbf{q}} \qquad \mathbf{p}_0 = -\frac{\partial W}{\partial \mathbf{q}_0} \tag{2.35}$$

we get

$$D(\mathbf{q}, \mathbf{q}_0, t) = \det \frac{\partial \mathbf{p}_0}{\partial \mathbf{q}} = \det\left(-\frac{\partial^2 W}{\partial \mathbf{q} \, \partial \mathbf{q}_0}\right) \tag{2.36}$$

This determinant is called the Van Vleck–Morette determinant [76, 77]. It is possible to show that D is a solution of the classical continuity equation (2.25), which can be rewritten in the useful form

$$\frac{d}{dt} \ln D = -\nabla^2 W = -\mathrm{tr} \frac{\partial^2 W}{\partial \mathbf{q} \, \partial \mathbf{q}} \tag{2.37}$$

with the total derivative defined by $d/dt = \partial_t + \mathbf{p} \cdot \nabla$. Indeed, if we introduce the Jacobian matrix

$$\mathbf{D} = -\frac{\partial^2 W}{\partial \mathbf{q} \, \partial \mathbf{q}_0} \tag{2.38}$$

which is known as the Van Vleck–Morette matrix, we can relate the derivative of the determinant to a trace involving the derivative of the matrix:

$$\frac{d}{dt} \ln D = \mathrm{tr} \, \mathbf{D}^{-1} \cdot \frac{d\mathbf{D}}{dt} \tag{2.39}$$

by using the well-known formula that $\ln \det \mathbf{A} = \mathrm{tr} \ln \mathbf{A}$ for a matrix $\mathbf{A}$.

The Jacobian matrix obeys the equation

$$\frac{d\mathbf{D}}{dt} = -\frac{\partial^2 W}{\partial \mathbf{q}\, \partial \mathbf{q}} \cdot \mathbf{D} \tag{2.40}$$

which can be obtained by differentiation of the Hamilton–Jacobi equation with respect to $\mathbf{q}_0$ and $\mathbf{q}$. The continuity equation (2.37) results from Eqs. (2.39) and (2.40).

3. *Jacobi–Hill Equation and Linear Stability*

The continuity equation as well as Eq. (2.40) for the Jacobian matrix $\mathbf{D}$ are partial differential equations based on the knowledge of the scalar field given by the action W. The partial differential equation (2.40) can be transformed into an equation to be solved along a given classical trajectory for the inverse matrix

$$\mathbf{J} = \mathbf{D}^{-1} = \frac{\partial \mathbf{q}}{\partial \mathbf{p}_0} \tag{2.41}$$

which relates the final position to the initial momentum. An equation for $\mathbf{J}$ is easy to obtain if we refer back to Newton's equations, which govern the trajectories $\mathbf{q} = \mathbf{q}_{\rm cl}(t; \mathbf{q}_0, \mathbf{p}_0)$

$$\ddot{\mathbf{q}} = -\frac{\partial V}{\partial \mathbf{q}} \tag{2.42}$$

By differentiating Newton's equations with respect to the initial momentum, we obtain

$$\frac{d^2}{dt^2}\mathbf{J} = -\frac{\partial^2 V}{\partial \mathbf{q}\, \partial \mathbf{q}} [\mathbf{q}_{\rm cl}(t)] \cdot \mathbf{J} \tag{2.43}$$

If we refer to the motion in position space as the motion of a fluid whose velocity field is given by $\mathbf{v} = \mathbf{p} = \nabla W$, Eq. (2.40) is the equation of motion of $\mathbf{D} = \mathbf{J}^{-1}$ under its Eulerian form, while Eq. (2.43) is the Lagrangian form of this equation. Furthermore, let us remark that, contrary to the matrix $\mathbf{J}$ itself, its determinant J does not obey a closed equation except for one-degree-of-freedom systems. Indeed, the matrix coincides with its determinant if $f = 1$. However, as soon as $f = 2$, the second time derivative of J involves the elements of $\mathbf{J}$ and $\dot{\mathbf{J}}$ in a complicated form that cannot be simplified to an expression involving only J and $\dot{J}$.

A few words are now in order about the behavior of those Jacobians near caustics. As we mentioned above, the single-valuedness of the solution is lost on a caustic line. Indeed, there appears a degeneracy in the initial momentum leading to a given position that is located on a caustic line such that there exist several nearby values of $\delta\mathbf{p}_0$ corresponding to the same $\delta\mathbf{q}$. Since both are related by $\mathbf{J}$, the rank of the Jacobian matrix decreases on the caustic line and its determinant vanishes, $J = \det \mathbf{J} = 0$. As a consequence, the Van Vleck–Morette determinant $D = J^{-1}$ diverges on the caustic line. These points of an orbit, where the determinant D is diverging, are called conjugate points, that is, points that are conjugated to the initial point. The multiplicity of the conjugate point is the number of zero eigenvalues of the matrix $\mathbf{J} = \mathbf{D}^{-1}$ at the conjugate point. The number of conjugate points along the trajectory from $\mathbf{q}_0$ to $\mathbf{q}$ counted with their multiplicity is an integer called the Morse index [72, 74], which determines the sign of the Van Vleck–Morette determinant

$$\nu_\ell = \frac{1}{\pi}[\arg \det \mathbf{J}_\ell(t) - \arg \det \mathbf{J}_\ell(0)] \tag{2.44}$$

Using the first determination of the square root, we have $\sqrt{D_\ell} = |D_\ell|^{1/2} \exp(-i\pi\nu_\ell/2)$ for that particular orbit ℓ. The integer ν_ℓ plays an essential role in the semiclassical method as soon as different trajectories contribute to a given amplitude. We shall come back to this integer below. Before proceeding, a further relationship is important to establish.

Equation (2.43) is also the equation for any perturbation $\delta\mathbf{q}(t)$ with respect to the classical trajectory $\mathbf{q}_{\text{cl}}(t)$

$$\hat{\mathcal{D}} \cdot \delta\mathbf{q}(t) = 0 \tag{2.45}$$

with the real symmetric Sturm–Liouville operator

$$\hat{\mathcal{D}} = -\mathbf{1}\frac{d^2}{dt^2} - \frac{\partial^2 V}{\partial\mathbf{q}\,\partial\mathbf{q}}[\mathbf{q}_{\text{cl}}(t)] \tag{2.46}$$

which controls the linear stability around the classical orbit. This operator plays a very important role in the following considerations. Equation (2.45) is known as the Jacobi–Hill equation and its solutions are known as Jacobi fields [78]. The Jacobi–Hill equation is solved here from the initial conditions $\mathbf{J}(0) = 0$ and $\dot{\mathbf{J}}(0) = \mathbf{1}$. The operator $\hat{\mathcal{D}}$ is a tensor with two indexes in position space.

Thus we reach the general conclusion that the amplitude of the semiclassical wave function is related to the linear stability of the classical trajectories. In this sense, the semiclassical method is very sensitive to any form of dynamical instability, such as bifurcations of classical trajectories or dynamical chaos.

4. *Semiclassical Propagator*

By gathering the preceding results, the propagator is given as the following sum over all orbits ℓ joining $\mathbf{q}_0$ to $\mathbf{q}$ in the time t, assuming that the initial and final points are not located on caustics

$$K(\mathbf{q}, \mathbf{q}_0, t) = \sum_{\ell} [\phi_{0\ell} + i\hbar\phi_{1\ell} + (i\hbar)^2\phi_{2\ell} + \cdots] \times \left|\det\left(-\frac{\partial^2 W_\ell}{\partial \mathbf{q}\, \partial \mathbf{q}_0}\right)\right|^{1/2} \exp\left[\frac{i}{\hbar} W_\ell(\mathbf{q}, \mathbf{q}_0, t) - i\frac{\pi}{2}\nu_\ell\right] \tag{2.47}$$

The Morse indexes ν_ℓ are determined by the changes of sign of the Jacobian determinant D along the classical orbits. Maslov [22, 74] developed an alternative method to calculate the integers ν_ℓ in terms of topological indexes of the Lagrangian manifolds supporting the wave functions. The Lagrangian manifolds are manifolds formed by continuous ensembles of trajectories in phase space [71]. Since these Lagrangian manifolds are phase-space objects it turns out that the indexes can be calculated if consistency is guaranteed between the wave functions in the position and momentum representations. These representations are related by a Fourier transform. The advantage of going from one representation to the other is that the caustic lines generated by a projection of the manifold onto position space are not related to the caustic lines generated by projection onto momentum space. Therefore, the singularities appearing at the caustic lines are specific to a given representation, and we can locally avoid them by going to another representation. In this way, the phases between the different local wave functions associated with the different sheets of a manifold of trajectories can be determined, and the global wave function can thus be constructed by a matching procedure [22, 74]. The topological indexes of Maslov turn out to be identical to the Morse indexes in many cases so that these indexes are also called Maslov indexes. We will not give further details for the moment because Section II.F will present another more direct method for the determination of the phases, which is close to the original method used by Morse.

In the limit where $t \to 0$, there is no caustic line between the initial and final points so that the Morse index vanishes. By using the initial condition [Eq. (2.30)], we can fix the factor

$$\phi_{0\ell} = \frac{1}{(2\pi i\hbar)^{f/2}} \tag{2.48}$$

which is constant along each orbit. Beyond the leading term, the $\hbar$-corrections are obtained as a solution of Eqs. (2.27) along the classical trajectories.

A simple illustration is given by the free particle for which the classical action and density are given by

$$W(\mathbf{q}, \mathbf{q}_0, t) = \frac{1}{2t}(\mathbf{q}-\mathbf{q}_0)^2 \qquad D(\mathbf{q}, \mathbf{q}_0, t) = \frac{1}{t^f} \tag{2.49}$$

Since the density is independent of position, Eqs. (2.27) give vanishing $\hbar$-corrections so that the semiclassical solution is the exact quantum solution. The factor ϕ_0 can be determined from the initial condition [Eq. (2.30)]. Accordingly, the free-particle propagator becomes

$$K(\mathbf{q}, \mathbf{q}_0, t) = \frac{1}{(2\pi i\hbar t)^{f/2}} \exp\left[\frac{i}{2\hbar t}(\mathbf{q}-\mathbf{q}_0)^2\right] \tag{2.50}$$

as expected.

The preceding method is simple and allows us to derive several important expressions. However, caustics are causing such difficulties that the formulation has a restricted applicability. In particular, the formulation gives a too important role to the determinant D, which becomes singular each time the trajectory meets a caustic. At caustics, the method of matching, which we mentioned above, uses quite different equations than the preceding ones, like the Fourier transform to the momentum representation. This method becomes difficult to apply when we want to calculate the $\hbar$-corrections. For those reasons, it is advantageous to reformulate the problem in terms of Feynman path integrals. The expressions are perhaps heavier to manipulate but the relevant quantities appear more explicitly.

F. Feynman Path Integrals and the Semiclassical Method

1. *Propagator as a Path Integral*

The path integrals that were introduced by Feynman have become a standard method in quantum mechanics [79]. This formulation is very useful in understanding the relationship between quantum and classical mechanics. The present section will be devoted to the description of this connection.

Because of the linearity of the time evolution, the propagation over a time interval T can be decomposed into N successive propagations over small intervals $\Delta t = T/N$. In the position representation, we have

$$K(\mathbf{q}, \mathbf{q}_0, T) = \langle \mathbf{q}|\hat{K}(T)|\mathbf{q}_0\rangle = \langle \mathbf{q}|\hat{K}(\Delta t)^N|\mathbf{q}_0\rangle$$

$$= \int d\mathbf{q}_1\, d\mathbf{q}_2 \cdots d\mathbf{q}_{N-1} \prod_{n=0}^{N-1} \langle \mathbf{q}_{n+1}|\hat{K}(\Delta t)|\mathbf{q}_n\rangle \qquad (2.51)$$

where we have inserted $N-1$ closure relations $\int d\mathbf{q}|\mathbf{q}\rangle\langle\mathbf{q}| = \hat{I}$, and where we identified $\mathbf{q} = \mathbf{q}_N$. The propagator is thus given by a multiple integral. We now need the expression of the propagators over Δt. Assuming that the time interval is small enough we can expand using

$$e^{\hat{A}+\hat{B}} = e^{\hat{A}} e^{\hat{B}} e^{-\frac{1}{2}[\hat{A}, \hat{B}]} \cdots \qquad (2.52)$$

where the dots denote exponentials of commutators of higher degrees than two between the operators $\hat{A}$ and $\hat{B}$. Applying this formula to a Hamiltonian that separates into kinetic and potential energies, $\hat{\mathcal{H}} = \hat{\mathbf{p}}^2/2 + V(\hat{\mathbf{q}})$, we obtain

$$\langle q_{n+1}|\hat{K}(\Delta t)|\mathbf{q}_n\rangle = \langle \mathbf{q}_{n+1}| \exp\Big(-\frac{i}{\hbar}\hat{\mathcal{H}}\Delta t\Big)|\mathbf{q}_n\rangle$$

$$= \frac{1}{(2\pi i\hbar\Delta t)^{f/2}} \exp\Big[\frac{i}{\hbar} L\Big(\frac{\mathbf{q}_{n+1} - \mathbf{q}_n}{\Delta t}, \mathbf{q}_n\Big)\Delta t + \mathcal{O}(\Delta t^2)\Big] \qquad (2.53)$$

in terms of the classical Lagrangian

$$L(\dot{\mathbf{q}}, \mathbf{q}) = \frac{\dot{\mathbf{q}}^2}{2} - V(\mathbf{q}) \qquad (2.54)$$

In Eq. (2.53), the equality is obtained up to terms of order Δt^2 in the exponential, which are due to the commutator in Eq. (2.52). These terms are of interest if the above procedure is applied to the numerical evaluation of the propagation, in which case the precision of the numerical integration can be improved by including the next terms. However, those higher terms turn out to disappear as higher order infinitesimals in the limit $\Delta t \to 0$. Inserting Eq. (2.53) into the multiple integral Eq. (2.51), we get

$$K(\mathbf{q}, \mathbf{q}_0, T) = \int d\mathbf{q}_1\, d\mathbf{q}_2 \cdots d\mathbf{q}_{N-1} (2\pi i \hbar \Delta t)^{-Nf/2}$$

$$\times \exp\left[\frac{i}{\hbar} \sum_{n=0}^{N-1} L\left(\frac{\mathbf{q}_{n+1} - \mathbf{q}_n}{\Delta t}, \mathbf{q}_n\right) \Delta t + \mathcal{O}(\Delta t)\right] \quad (2.55)$$

In the continuum limit $N \to \infty$, the phase becomes the classical action along the path

$$W[\mathbf{q}(t)] = \int_0^T L(\dot{\mathbf{q}}, \mathbf{q})\, dt \quad (2.56)$$

The path integral formulation is very close to an explicit recipe of numerical integration [80]. In this sense, the method is particularly useful.

Besides the numerical evaluation of the multiple integral, there are few systems that are integrable analytically such as the harmonic oscillator. Numerical integration of multiple integrals is extremely time consuming. For that reason, it is often advantageous to use a semiclassical approach where the integrals are performed by the method of stationary phases.

2. *Stationary-Phase Integration*

The principle of this method is the following [81]. The integrand is an oscillatory kernel. The oscillations become faster and faster as the Planck constant is formally considered to become smaller and smaller ($\hbar \to 0$). The regions of the integration domain where the integrand is rapidly oscillating give nearly vanishing contributions to the integral. However, there may be some points where the oscillations slow down and whose vicinity thus contributes to the integral. Examples of such critical points are encountered where the phase becomes stationary when its first derivative vanishes so that the phase is quadratic near the critical point. The integral is then transformed into an imaginary Gaussian integral in a first approximation. Higher order approximations can be obtained by expansion of the phase and integrand in Taylor series around the critical point, which allows moments of the imaginary Gaussian integral to appear. This expansion generates a series in powers of the Planck constant. For a one-variable integral with an arbitrary function $\phi(x)$

having one stationary point, where $\partial_x \phi(x_0) = 0$, we have

$$\int_{-\infty}^{+\infty} e^{(i/\hbar)\phi(x)}\, dx = e^{(i/\hbar)\phi_0} \left(\frac{2\pi i \hbar}{\phi_0^{(2)}} \right)^{1/2} \left[1 - \frac{i\hbar \phi_0^{(4)}}{8\phi_0^{(2)\,2}} + \frac{5i\hbar \phi_0^{(3)\,2}}{24\phi_0^{(2)\,3}} + \mathcal{O}(\hbar^2) \right] \tag{2.57}$$

with the notation $\phi_0^{(n)} = \partial_x^n \phi(x_0)$.

A stationary point, where $\phi^{(1)}(x_0) = 0$ but $\phi^{(2)}(x_0) \neq 0$, is not the only example of a critical point, but it is certainly the most common one, as a quadratic extremum is generic. We may also find cubic or higher order stationary points as well as critical points where the phase is either discontinuous or coincides with the border of the integration domain. In this way, there is a whole zoology of critical points that require specific integration techniques [81]. Besides the critical points in the real integration domain, the integrals may be continued into complex domains where other critical points may be encountered. The expansion of the integral around these complex critical points is done by the steepest descent method. Complex critical points contribute to terms that are exponentially small in the Planck constant, i.e., of the form $\exp(-\mathrm{Cst}/\hbar)$. Nevertheless, it may happen that the corresponding terms have a large numerical value for a finite value of $\hbar$ so that the complex critical points may give a larger contribution than the real critical points. This is the case for tunneling phenomena where no real critical point exists, but where the tunneling amplitude is nonvanishing.

If we apply the stationary-phase method to the multiple integral [Eq. (2.55)], we obtain the critical points as solutions of the stationary-phase condition $\partial W / \partial \mathbf{q}_n = 0$, which gives a discrete version of Newton's equation (2.42). In the time-continuous limit where the time step $\Delta t = T/N$ vanishes, the stationary-phase condition is nothing else than Hamilton's variational principle, which requires the vanishing of the first variation of the action along the classical orbit with boundary conditions corresponding to a propagation from $\mathbf{q}_0$ to $\mathbf{q}$ in time T:

$$\frac{\delta W}{\delta \mathbf{q}} = 0 \qquad \text{with} \qquad \mathbf{q}(0) = \mathbf{q}_0 \qquad \text{and} \qquad \mathbf{q}(T) = \mathbf{q} \tag{2.58}$$

In order to perform the path integral, we also need information on the second variation of the action functional. This second variation gives us the quadratic form at the basis of the imaginary Gaussian integral. Equation (2.57) then suggests that $\hbar$-corrections are related to the higher order variations [82, 83]

$$W[\mathbf{q}_{\text{cl}} + \delta\mathbf{q}] = W_{\text{cl}} + \delta W + \delta^2 W + \delta^3 W + \delta^4 W + \cdots \tag{2.59}$$

Section II.F.3 is concerned with the second variation of the action in the classical context while the higher variations are treated later when we go back to the quantum path integral.

3. *Second Variation of the Action and Jacobi–Hill Equation*

The knowledge of the second variation is essential because it determines the type of extremum for the classical orbit. We therefore give some details about it and show how it is related to the linear stability of the orbit and to the quantities obtained in Section II.E. The second variation is given by a quadratic Lagrangian $\mathscr{L}$, which is now dependent on time, since the quadratic form is evaluated along the periodic orbit [78]

$$\delta^2 W = \int_0^T \mathscr{L}(\delta\dot{\mathbf{q}}, \delta\mathbf{q}, t)\, dt \tag{2.60}$$

with

$$\mathscr{L} = \frac{1}{2}\delta\dot{\mathbf{q}}^2 - \frac{1}{2}\delta\mathbf{q}^{\mathrm{T}} \cdot \frac{\partial^2 V}{\partial\mathbf{q}\,\partial\mathbf{q}}[\mathbf{q}_{\text{cl}}(t)] \cdot \delta\mathbf{q} \tag{2.61}$$

where T denotes the transpose. The equation of motion for the variation $\delta\mathbf{q}$ under this Lagrangian is nothing but the Jacobi–Hill equation (2.45)–(2.46). With the boundary conditions $\delta\mathbf{q}(0) = \delta\mathbf{q}(T) = 0$, we have a Sturm–Liouville problem for the operator $\hat{\mathscr{D}}$, which is given by Eq. (2.46) over the time interval $(0, T)$. The operator $\hat{\mathscr{D}}$ is real symmetric so that it has real eigenvalues μ_n corresponding to real eigenfunctions $\mathbf{u}_n$ forming a complete basis on which it is possible to expand any variation satisfying the boundary conditions,

$$\delta\mathbf{q} = \sum_{n=0}^{\infty} a_n \mathbf{u}_n \tag{2.62}$$

with

$$\hat{\mathscr{D}} \cdot \mathbf{u}_n = \mu_n \mathbf{u}_n \qquad \mathbf{u}_n(0) = \mathbf{u}_n(T) = 0 \quad \text{and} \quad \int_0^T \mathbf{u}_m^{\mathrm{T}}(t) \cdot \mathbf{u}_n(t)\, dt = \delta_{mn} \tag{2.63}$$

The second variation becomes diagonal in this new basis [78]

$$\delta^2 W = \frac{1}{2} \sum_{n=1}^{\infty} \mu_n a_n^2 \tag{2.64}$$

Therefore, the sign of the quadratic form of the second variation is determined by the number of negative eigenvalues of the operator $\hat{\mathcal{D}}$.

Let us consider the simple one-dimensional example of the harmonic oscillator for which the Jacobi–Hill operator is

$$\hat{\mathcal{D}} = -\frac{d^2}{dt^2} - \omega^2 \tag{2.65}$$

Here the Jacobi–Hill operator is a scalar quantity. The eigenvalues and eigenfunctions are given by

$$\mu_n = \left(\frac{\pi n}{T}\right)^2 - \omega^2 \qquad \text{and} \qquad u_n(t) = \sqrt{\frac{2}{T}} \sin\frac{\pi n t}{T} \tag{2.66}$$

Figure 2 shows how the eigenvalues vary with the length of the time interval T. For small values of T, all the eigenvalues are positive so that the second variation has a minimum. There appears an unstable direction at the time $T = \pi/\omega$ that corresponds to the first conjugate point. Successive conjugate points are met at each half-period of the oscillator $T_n = n\pi/\omega$, where one extra eigenvalue becomes negative. Since the

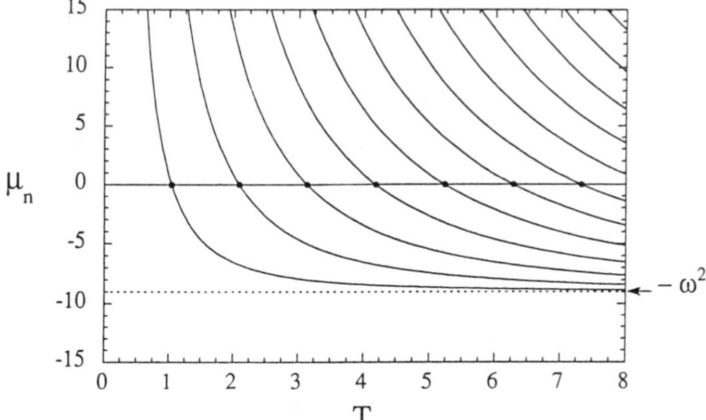

Figure 2. Eigenvalues μ_n of the Jacobi–Hill operator of linear stability versus the time interval T of the trajectory for a harmonic oscillator as given by Eq. (2.66).

equation $\hat{\mathscr{D}} \cdot \delta\mathbf{q} = 0$ admits a nontrivial solution at those particular times ($T = T_n$) satisfying the boundary condition $\delta\mathbf{q}(T_n) = 0$, the rank of the Jacobian matrix $\mathbf{J}$, solution of Eq. (2.43), decreases at these times (T_n). At the other times, when $T \neq T_n$, there exists no such solution and $\hat{\mathscr{D}} \cdot \delta\mathbf{q} \neq 0$. This reasoning shows that the conjugate points where $J = 0$ correspond to the times T where one eigenvalue of the Sturm–Liouville problem crosses the line $\mu = 0$. We will see the consequence of this change of sign on the path integral below. We also remark that the change of sign is due to the focusing character of the harmonic potential. The preceding reasoning shows that there is no conjugate point at short times after the initial condition. For an inverted harmonic potential, where $\omega^2 = -\lambda^2$, the eigenvalues $\mu_n = (\pi n/T)^2 + \lambda^2$ always remain positive and there is no conjugate point.

The decrease of the eigenvalues μ_n for the increasing time interval seen in Fig. 2 is very general since we can show that the variations of the eigenvalues are always negative or zero

$$\frac{\partial \mu_n}{\partial T} = -\dot{\mathbf{u}}_n^2(T) \tag{2.67}$$

for the solutions of the Sturm–Liouville problem [Eq. (2.63)]. This last result [Eq. (2.67)] is obtained by considering Eq. (2.64) with $a_n = 1$ and $a_m = 0$ for $m \neq n$. Both sides of Eq. (2.64) are then differentiated with respect to T and the integral Eq. (2.60), which defines $\delta^2 W$, is integrated by parts, using the equation and boundary conditions [Eq. (2.63)] for the function $\delta\mathbf{q} = \mathbf{u}_n(t; T)$ which, as a function of t, has furthermore a parametric dependency on T. The expression is then simplified by using the orthonormality condition [Eq. (2.63)], which is also differentiated with respect to T. Finally, we use the fact that $\partial_T \mathbf{u}_n(T; T) = -\partial_t \mathbf{u}_n(T; T)$, which results from the boundary condition $\mathbf{u}_n(T; T) = 0$, which holds for all values of T.

We can compare the preceding eigenvalue problem for the Jacobi–Hill operator to the quantum mechanical determination of the bound states in a potential given by the local frequencies of the physical potential along the classical orbit. For large values of the number $n \to \infty$, the eigenvalues increase as $\mu_n \simeq (\pi n/T)^2$. The number of negative eigenvalues will depend not only on T but also on the number of oscillatory frequencies around the classical orbit. We can obtain an evaluation of the number of eigenvalues using the Thomas–Fermi–Weyl–Wigner semiclassical approximation to the classical potential given by the local eigenvalues of the second derivative of the physical potential

$$\det\left\{\frac{\partial^2 V}{\partial \mathbf{q}^2}[\mathbf{q}_{\text{cl}}(t)] - \sigma^{(i)}(t)\mathbf{1}\right\} = 0 \tag{2.68}$$

where the index i labels the f different eigenvalues. We obtain the average approximation of the staircase function $N(\mu) = \text{Number}\{\mu_n < \mu\}$ for the eigenvalues of the Jacobi–Hill operator,

$$N_{\text{av}}(\mu) = \sum_{i=1}^{f} \int_0^T \frac{dt}{2\pi} \theta[\mu + \sigma^{(i)}(t)] \sqrt{\mu + \sigma^{(i)}(t)} \tag{2.69}$$

where $\theta(\cdot)$ is the Heaviside function. This result only provides a crude estimation of the number of eigenvalues below μ, but it can help in cases where the behavior is well pronounced.

4. *Classical Green Function as Resolvent of the Jacobi–Hill Operator*

Associated with each classical orbit, we have a classical Green function corresponding to the Jacobi–Hill operator

$$\hat{\mathcal{D}} \cdot \mathcal{G}(t, t') = \mathbf{1}\delta(t - t') \tag{2.70}$$

which is an $f \times f$ matrix function of the times t and t'. Here $\mathbf{1}$ is the $f \times f$ identity matrix. Equation (2.70) can be rewritten as

$$-\frac{d^2}{dt^2} \mathcal{G}_{ij}(t, t') - \frac{\partial^2 V}{\partial q_i \, \partial q_k} [\mathbf{q}_{\text{cl}}(t)] \mathcal{G}_{kj}(t, t') = \delta_{ij}\delta(t - t') \tag{2.71}$$

with a summation over k. Knowing the eigenvectors and eigenvalues, the classical Green operator is obtained as

$$\mathcal{G}(t, t') = \sum_{n=0}^{\infty} \mathbf{u}_n(t) \frac{1}{\mu_n} \mathbf{u}_n^{\mathrm{T}}(t') \tag{2.72}$$

For a trajectory of the harmonic oscillator between the times $t = 0$ and $t = T$, the corresponding Green function solution of Eq. (2.71) with the Dirichlet boundary conditions $\mathcal{G}(0, t') = \mathcal{G}(T, t') = 0$ is

$$\begin{aligned} \mathcal{G}(t, t') &= \frac{\sin \omega t \sin \omega(T - t')}{\omega \sin \omega T} \qquad (t < t') \\ \mathcal{G}(t, t') &= \frac{\sin \omega(T - t) \sin \omega t'}{\omega \sin \omega T} \qquad (t > t') \end{aligned} \tag{2.73}$$

which is a scalar quantity here. For the inverted harmonic potential with $\omega^2 = -\lambda^2$, the Green function is obtained by replacing ω by $\pm i\lambda$ in Eq. (2.73). These Green functions are shown in Fig. 3 for comparison. We will see below that this classical Green function plays an essential role in the calculation of the $\hbar$-corrections.

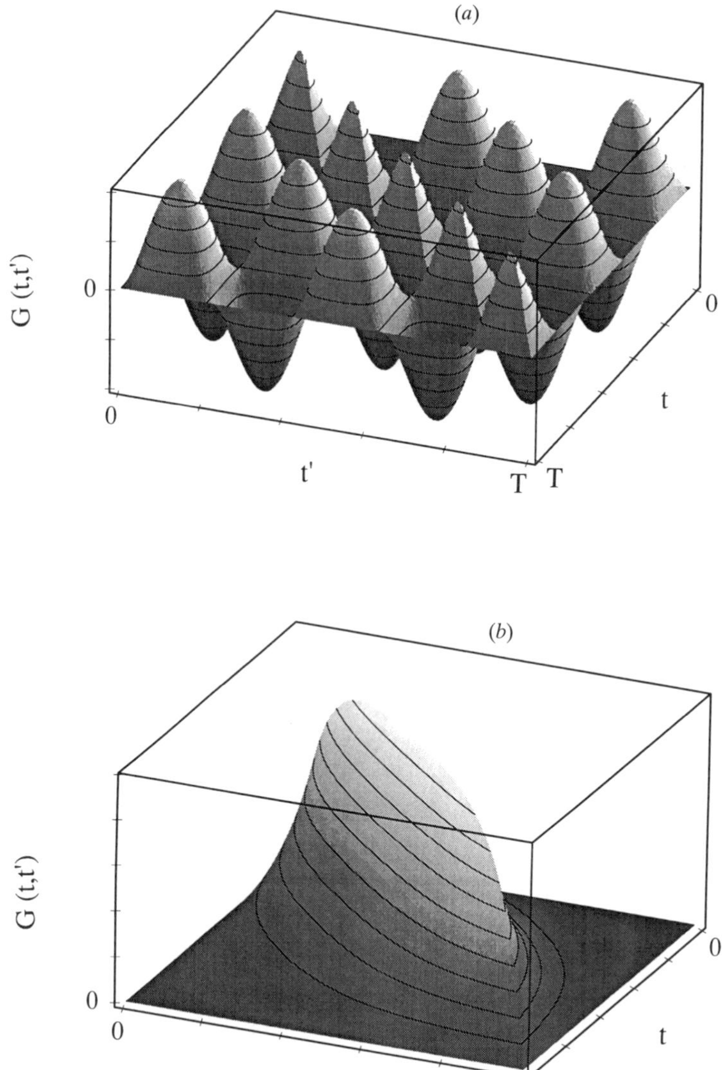

Figure 3. (*a*) Classical Green function (2.73) of a trajectory of a harmonic oscillator of frequency $\omega = 5$ over a time $T = 5$ versus the times t and t'. (*b*) Same as (*a*) but for an unstable inverted harmonic potential for which the classical Green function is given by replacing ω by $i\lambda$ in Eq. (2.73) with $T = 5$ and $\lambda = 1$. In this case, we observe the exponential damping of the Green function away from the diagonal.

5. *ħ-Expansion of the Propagator*

After these considerations on the second variation of the action, we would like to return to the multiple integral [Eq. (2.55)]. As we already mentioned in Section II.E, there may exist several classical orbits $\mathbf{q}_\ell(t)$, which are going from $\mathbf{q}_0$ to $\mathbf{q}$ during the time t, and each of these stationary solutions contributes to the propagator as already obtained in Section II.E. As in the simple example of Eq. (2.57), the integrand is expanded in Taylor series around each classical orbit. We denote the deviation of the path with respect to the classical orbit ℓ by $\xi^a = \{q_{in} - q_i^{(\mathrm{cl})}(n\Delta t)\}_{n=1}^{N-1}$. An index such as a stands for a double index $a=(i,n)$ where the first index i refers to the ith component of the position $\mathbf{q}$ $(i=1,\ldots,f)$ while the second index n refers to the time $t=n\Delta t$ $(n=1,\ldots,N-1)$. The propagator becomes [83, 84]

$$K(\mathbf{q},\mathbf{q}_0,T) = \sum_\ell \left(\frac{N}{2\pi i\hbar T}\right)^{Nf/2} \exp\left(\frac{i}{\hbar}W_{\mathrm{cl}}\right) \int d^{(N-1)f}\xi \, \exp\left(\frac{i}{2\hbar} W_{,ab}\xi^a\xi^b\right)$$
$$\times \left[1 + \frac{i}{6\hbar}W_{,abc}\xi^a\xi^b\xi^c + \frac{i}{24\hbar}W_{,abcd}\xi^a\xi^b\xi^c\xi^d + \mathcal{O}(\xi^5/\hbar)\right.$$
$$\left. - \frac{1}{72\hbar^2}W_{,abc}W_{,def}\xi^a\xi^b\xi^c\xi^d\xi^e\xi^f + \mathcal{O}(\xi^7/\hbar^2) + \mathcal{O}(\xi^9/\hbar^3)\right] \tag{2.74}$$

where $W_{,a_1\cdots a_j}$ denotes the partial derivatives $\partial^j W/\partial\xi^{a_1}\cdots\partial\xi^{a_j}$ evaluated at the classical solution $\xi^a = 0$, and W_{cl} is the classical action of the orbit ℓ. The matrix of second derivatives $D_{ab} = (\mathbb{D})_{ab} = W_{,ab}$ of the action is a $(N-1)f \times (N-1)f$ matrix of the form

$$\mathbb{D} = \frac{1}{\Delta t}\begin{pmatrix} \mathbf{A}_1 & -\mathbf{1} & 0 & 0 & \cdots & 0 & 0 & 0 \\ -\mathbf{1} & \mathbf{A}_2 & -\mathbf{1} & 0 & \cdots & 0 & 0 & 0 \\ 0 & -\mathbf{1} & \mathbf{A}_3 & -\mathbf{1} & \cdots & 0 & 0 & 0 \\ \vdots & \vdots & \vdots & \vdots & \ddots & \vdots & \vdots & \vdots \\ 0 & 0 & 0 & 0 & \cdots & -\mathbf{1} & \mathbf{A}_{N-2} & -\mathbf{1} \\ 0 & 0 & 0 & 0 & \cdots & 0 & -\mathbf{1} & \mathbf{A}_{N-1} \end{pmatrix} + \mathcal{O}(\Delta t^2) \tag{2.75}$$

where the $f \times f$ matrices $\mathbf{A}_n$ have the elements

$$A_{ij}(n\Delta t) = 2\delta_{ij} - \Delta t^2 \frac{\partial^2 V}{\partial q_i\,\partial q_j}[\mathbf{q}_{\mathrm{cl}}(n\Delta t)] \tag{2.76}$$

with $i, j = 1, \ldots, f$. The symbol **1** denotes the $f \times f$ identity matrix. In Eq. (2.75), the corrections in Δt^2 are negligible since we will take the limit $\Delta t \to 0$ at the end of the calculation and keep only the expressions forming standard integrals over $T = N\Delta t$. We observe that the matrix $\mathbb{D}$ is the discrete version of the Jacobi–Hill operator [Eq. (2.46)] according to the correspondence $\hat{\mathscr{D}} \leftrightarrow \mathbb{D}/\Delta t$. Since we also have the correspondence $\delta(t-t') \leftrightarrow \delta_{nn'}/\Delta t$, we conclude that the inverse of the second derivative matrix is given according to $(\mathbb{D}^{-1})^{ab} = \mathscr{G}_{ij}(m\Delta t, n\Delta t)$ [in the limit $\Delta t \to 0$ with $a = (i, m)$, $b = (j, n)$, $i, j = 1, \ldots, f$, and $m, n = 1, \ldots, N-1$] in terms of the classical Green function $\mathscr{G}(t, t')$ solution of Eq. (2.71) with the boundary conditions [83, 84]

$$\mathscr{G}_{ij}(0, t') = \mathscr{G}_{ij}(T, t') = 0 \tag{2.77}$$

The eigenvalues of the second derivative matrix $\mathbb{D}$ are also in correspondence with those of the Jacobi–Hill operator $\hat{\mathscr{D}}$. For the example of the harmonic oscillator, we obtain the eigenvalues of the matrix $\mathbb{D}$ as $\tilde{\mu}_n = -\omega^2\Delta t + 2[1 - \cos(n\pi\Delta t/T)]/\Delta t$ with $n = 1, 2, \ldots, N-1$. We observe that we have the approximate equality $\tilde{\mu}_n \simeq \mu_n \Delta t$ for the low-lying eigenvalues, which is a general result. In particular, the number of negative eigenvalues of $\hat{\mathscr{D}}$ and $\mathbb{D}$ will be equal in the limit where Δt is small enough. As a consequence, the number of negative eigenvalues of $\mathbb{D}$ is precisely given by the Morse index, which was determined in the preceding section.

The higher derivatives of the action are given by

$$\frac{\partial^m W}{\partial q_{i_1}(t) \cdots \partial q_{i_m}(t)} = -\Delta t \frac{\partial^m V}{\partial q_{i_1} \cdots \partial q_{i_m}} [\mathbf{q}_{\text{cl}}(t)] + \mathscr{O}(\Delta t^2) \tag{2.78}$$

and zero otherwise. With these ingredients, we can perform the integrals.

The first term in Eq. (2.74) contains the imaginary Gaussian integral

$$\int d^M\xi \exp\left(\frac{i}{2\hbar} D_{ab}\xi^a\xi^b\right) = \left[\frac{(2\pi i\hbar)^M}{\det \mathbb{D}}\right]^{1/2} \tag{2.79}$$

with $M = (N-1)f$. The higher-order terms are given in terms of the inverse of the matrix $\mathbb{D}$. The odd powers of ξ give a vanishing contribution when the domain of integration extends over the real numbers. On the other hand, the integrals containing even powers of ξ are

$$\int d^M\xi \, \exp\left(\frac{i}{2\hbar} D_{ab}\xi^a\xi^b\right) \xi^{c_1}\xi^{c_2}\cdots\xi^{c_{2L-1}}\xi^{c_{2L}}$$

$$= \left[\frac{(2\pi i\hbar)^M}{\det \mathbb{D}}\right]^{1/2} (i\hbar)^L[(\mathbb{D}^{-1})^{c_1c_2}\cdots(\mathbb{D}^{-1})^{c_{2L-1}c_{2L}} + \cdots] \tag{2.80}$$

where all the $(2L-1)!!$ terms that are obtained by grouping the indexes two by two appear.

The determinant of the matrix $\mathbb{D}$ can be calculated by comparison with the results of Section II.E as

$$\det \mathbb{D} = \frac{1}{\Delta t^{Nf}} \det\left(-\frac{\partial^2 W}{\partial \mathbf{q}\, \partial \mathbf{q}_0}\right)^{-1} \tag{2.81}$$

in the limit $\Delta t \to 0$. Therefore, the integrals (2.79) and (2.80) exist as long as the time T does not coincide with a conjugate point along the orbit, which guarantees $\det \mathbb{D} \neq 0$.

Moreover, the phase is now fixed without ambiguity in the calculation of the imaginary Gaussian integral. Indeed, the integral is given by the first determination of the square root. Equation (2.79) is obtained by diagonalization of the matrix $\mathbb{D}$ so that we are left with $(N-1)f$ integrals of the form

$$\int_{-\infty}^{+\infty} d\tilde{\xi}_n \exp\left(\frac{i}{2\hbar}\tilde{\mu}_n \tilde{\xi}_n^2\right) = \left(\frac{2\pi\hbar}{|\tilde{\mu}_n|}\right)^{1/2} e^{i\frac{\pi}{4}} e^{-i\frac{\pi}{2}\tilde{\nu}_n} \tag{2.82}$$

where the phase is fixed because the integral is convergent and its value is known. The index $\tilde{\nu}_n$ is equal to 1 if $\tilde{\mu}_n$ is negative and 0 otherwise. The sum $\nu_\ell = \Sigma_{n=1}^{(N-1)f} \tilde{\nu}_n$ is equal to the Morse index of the orbit ℓ as discussed above.

We have thus established the relation to the previous section, but as a byproduct of the path integral approach we can calculate the $\hbar$-corrections. These corrections are given by the higher integrals [Eq. (2.80)], which involve the inverse of the matrices $\mathbb{D}$, that is, the classical Green functions $\mathcal{G}(t, t')$ along each classical orbit. When the results are gathered the propagator is given by

$$K(\mathbf{q}, \mathbf{q}_0, T) = \sum_\ell \frac{1}{(2\pi i\hbar)^{f/2}} \left| \det\left(-\frac{\partial^2 W_\ell}{\partial \mathbf{q}\, \partial \mathbf{q}_0} \right) \right|^{1/2}$$

$$\times \exp\left[\frac{i}{\hbar} W_\ell(\mathbf{q}, \mathbf{q}_0, T) - i\frac{\pi}{2}\nu_\ell \right]$$

$$\times \left\{ 1 + \frac{i\hbar}{8} \int_0^T dt \frac{\partial^4 V(t)}{\partial q_i\, \partial q_j\, \partial q_k\, \partial q_l} \mathcal{G}_{ij}(t,t)\mathcal{G}_{kl}(t,t) \right.$$

$$+ \frac{i\hbar}{24} \int_0^T dt\, dt' \frac{\partial^3 V(t)}{\partial q_i\, \partial q_j\, \partial q_k} \frac{\partial^3 V(t')}{\partial q_l\, \partial q_m\, \partial q_n} [3\mathcal{G}_{ij}(t,t)\mathcal{G}_{kl}(t,t')$$

$$\left. \times \mathcal{G}_{mn}(t',t') + 2\mathcal{G}_{il}(t,t')\mathcal{G}_{jm}(t,t')\mathcal{G}_{kn}(t,t')] + \mathcal{O}(\hbar^2) \right\} \qquad (2.83)$$

Diagrams are associated with these integrals as follows [83–87]. A vertex with m legs is associated here with each $-\partial^m V(t)/\partial q_{i_1} \cdots \partial q_{i_m}$. A line is associated with each Green function $\mathcal{G}_{ij}(t, t')$, which joins two free legs either of the same vertex or between two different vertices. Integrals are then performed over the times of the different vertices. These diagrams have the following properties. In the Taylor expansion [Eq. (2.83)], we find terms with V vertices and ℓ legs. The number ℓ of legs is always even otherwise the term is vanishing. If there are $\ell = 2L$ legs there are $(2L-1)!!$ different manners to join them two by two with L lines. Hence, many diagrams are generated but many of them are identical and, as a consequence, have the same numerical value. The sign of the numerical weight of a diagram with V vertices is $(-)^V$. The difference $L - V$ gives the power of the Planck constant to which the diagram contributes.

In the series, we find connected and disconnected diagrams. A very important property is that the series in the Planck constant involving all the diagrams can be transformed into the exponential of a series involving only the connected diagrams. Accordingly, the propagator can be written as

$$K(\mathbf{q}, \mathbf{q}_0, T) = \sum_\ell \frac{1}{(2\pi i\hbar)^{f/2}} \left| \det\left(-\frac{\partial^2 W_\ell}{\partial \mathbf{q}\, \partial \mathbf{q}_0} \right) \right|^{1/2}$$

$$\times \exp\left[\frac{i}{\hbar} W_\ell(\mathbf{q}, \mathbf{q}_0, T) - i\frac{\pi}{2}\nu_\ell + \sum_{n=1}^{\infty} (i\hbar)^n C_{\ell n}(\mathbf{q}, \mathbf{q}_0, T) \right] \qquad (2.84)$$

The first coefficient is given explicitly by

$$C_1 = \frac{1}{8}\int_0^T dt \frac{\partial^4 V(t)}{\partial q_i\, \partial q_j\, \partial q_k\, \partial q_l}\, \mathcal{G}_{ij}(t,t)\mathcal{G}_{kl}(t,t)$$

$$+ \frac{1}{24}\int_0^T dt\, dt' \frac{\partial^3 V(t)}{\partial q_i\, \partial q_j\, \partial q_k} \frac{\partial^3 V(t')}{\partial q_l\, \partial q_m\, \partial q_n} [3\mathcal{G}_{ij}(t,t)\mathcal{G}_{kl}(t,t')\mathcal{G}_{mn}(t',t')$$

$$+ 2\mathcal{G}_{il}(t,t')\mathcal{G}_{jm}(t,t')\mathcal{G}_{kn}(t,t')] + \mathcal{O}(\hbar^2)\Big\} \tag{2.85}$$

The diagram associated with C_1 is shown in Fig. 4 together with that of the $\hbar^2$-correction C_2.

The previous method and the appearance of the Feynman diagrams of Fig. 4 are typical of $\hbar$-expansions. The integrals corresponding to these diagrams are well behaved in problems with a smooth potential $V(\mathbf{q})$ (Coulomb potentials require an extra regularization as discussed in a following section). By contrast, the calculation of the $\hbar$-corrections may be problematic if the Eqs. (2.27) have to be integrated, in view of the divergence of the classical density D on caustic lines. In this regard, the method of classical Green functions avoids a severe difficulty.

In the following sections, we will apply the above expansion to the calculation of the trace of the propagator and to the determination of the eigenvalues of the Schrödinger equation. However, first we will describe the classical structures we should expect in phase space.

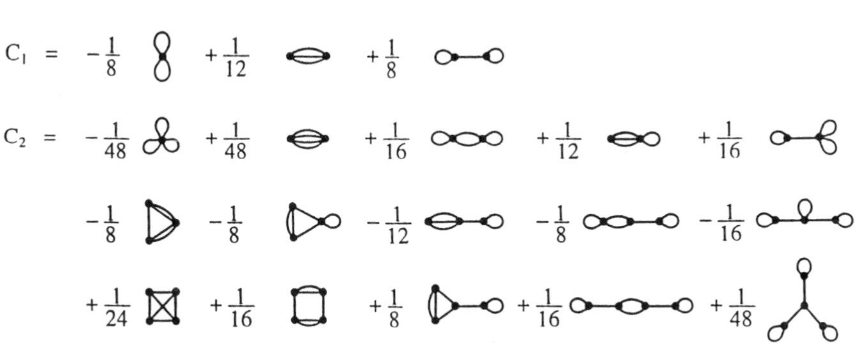

Figure 4. Feynman diagrams contributing to the coefficients C_1 and C_2 of the $\hbar$- and $\hbar^2$-corrections, respectively, in the case of smooth Hamiltonians.

III. CLASSICAL DYNAMICS

The phase-space structures are the feature of the pointlike classical dynamics of the trajectories derived from Hamilton's equations, which can be written in the form

$$\dot{\mathbf{X}} = \mathbf{F}(\mathbf{X}) = \Sigma \cdot \frac{\partial \mathcal{H}}{\partial \mathbf{X}} \tag{3.1}$$

where $\mathbf{X} = (\mathbf{q}, \mathbf{p})$ are the positions and momenta, $\mathcal{H}(\mathbf{X})$ is the Hamiltonian function, and

$$\Sigma = \begin{pmatrix} 0 & \mathbf{1} \\ -\mathbf{1} & 0 \end{pmatrix} \tag{3.2}$$

is an antisymmetric $2f \times 2f$ matrix at the basis of the symplectic structure of phase space. Cauchy's theorem guarantees the uniqueness of the solutions of Eq. (3.1) from given initial conditions: $\mathbf{X}_t = \Phi^t(\mathbf{X}_0)$. As a consequence of the Hamiltonian form, we have $\operatorname{div} \mathbf{F} = 0$ so that the flow is conservative and phase-space volumes are preserved. Moreover, energy is conserved if the Hamiltonian is time independent: $\mathcal{H} = E$. Therefore, the flow takes place in $2f - 1$ dimensions. Extra constants of motion may further reduce the dimension of the effective phase space.

A. Linear Stability in Phase Space

The preceding sections have shown the importance of linear stability of the trajectories. Infinitesimal perturbations $\delta\mathbf{X}$ around a trajectory $\mathbf{X}_t$ of Eq. (3.1) are solutions of the equation [88]

$$\delta\dot{\mathbf{X}} = \mathbf{L}(t) \cdot \delta\mathbf{X} \tag{3.3}$$

with

$$\mathbf{L}(t) = \frac{\partial \mathbf{F}}{\partial \mathbf{X}}(\mathbf{X}_t) = \Sigma \cdot \frac{\partial^2 \mathcal{H}}{\partial \mathbf{X}^2}(\mathbf{X}_t) \tag{3.4}$$

In the case of a Hamiltonian of the form $\mathcal{H} = \mathbf{p}^2/2 + V(\mathbf{q})$, we have that $\delta\mathbf{X} = (\delta\mathbf{q}, \delta\mathbf{p})$ and

$$\mathbf{L} = \frac{\partial \mathbf{F}}{\partial \mathbf{X}} = \begin{pmatrix} 0 & \mathbf{1} \\ -\dfrac{\partial^2 V}{\partial \mathbf{q}^2} & 0 \end{pmatrix} \tag{3.5}$$

so that the equations for the infinitesimal perturbations are

$$\begin{aligned} \delta\dot{\mathbf{q}} &= \delta\mathbf{p} \\ \delta\dot{\mathbf{p}} &= -\frac{\partial^2 V}{\partial\mathbf{q}^2}\cdot\delta\mathbf{q} \end{aligned} \tag{3.6}$$

By eliminating $\delta\mathbf{p}$, we recover the Jacobi–Hill equation (2.45)–(2.46), which shows the relationship between the problems of linear stability in position space (Section II) and in phase space (Section III).

Because Eq. (3.3) is linear, its solutions are given as a linear combination of the initial perturbations

$$\delta\mathbf{X}(t) = \mathbf{M}(t)\cdot\delta\mathbf{X}(0) \tag{3.7}$$

where $\mathbf{M}(t)$ is a $2f\times 2f$ matrix solution of

$$\dot{\mathbf{M}} = \mathbf{L}\cdot\mathbf{M} \tag{3.8}$$

from the initial conditions $\mathbf{M}(0)=\mathbf{1}$. The Lyapunov exponents of the trajectory are then defined by [89, 90]

$$\lambda(\mathbf{X}_0;\mathbf{e}) = \lim_{t\to\infty}\frac{1}{t}\ln|\mathbf{M}(t;\mathbf{X}_0)\cdot\mathbf{e}| \tag{3.9}$$

where the vector $\mathbf{e}=\delta\mathbf{X}/|\delta\mathbf{X}|$ is the direction of the initial perturbation $\delta\mathbf{X}$. The unit vector $\mathbf{e}$ belongs to the space that is tangent to the phase space, that is, to the space of the infinitesimal perturbations $\delta\mathbf{X}$. *A priori*, the Lyapunov exponent depends both on the trajectory given by the initial condition $\mathbf{X}_0$ and on the direction of the initial perturbation, $\mathbf{e}$. After the limit $t\to\infty$ is taken it turns out that the Lyapunov exponent only takes its values in a discrete set called the Lyapunov spectrum: $\lambda_1\geq\lambda_2\geq\cdots\geq\lambda_\nu$, with multiplicities $m_1, m_2, \ldots, m_\nu$ such that $\Sigma_{i=1}^{\nu} m_i = 2f$. The main Lyapunov exponent is obtained for unit vectors $\mathbf{e}$ which are arbitrary, except that they should not belong to a linear submanifold of the tangent space that is of dimension $2f-m_1$ and of zero measure. Next, one chooses a unit vector that is confined to precisely this submanifold, and thus the second largest Lyapunov exponent λ_2 is found. The next Lyapunov exponents are obtained successively in the same way. In ergodic systems, the dependency on $\mathbf{X}_0$ disappears for almost all trajectories according to the multiplicative ergodic theorem established by Oseledec, which generalizes the Birkhoff ergodic theorem [89, 90].

By differentiating the equation of motion [Eq. (3.1)] with respect to time, we obtain

$$\ddot{\mathbf{X}} = \frac{\partial \mathbf{F}}{\partial \mathbf{X}}(\mathbf{X}_t) \cdot \dot{\mathbf{X}} \tag{3.10}$$

so that the direction of the flow $\dot{\mathbf{X}}$ is a solution of the linear system [Eq. (3.3)] and we have $\dot{\mathbf{X}}(t) = \mathbf{M}(t) \cdot \dot{\mathbf{X}}(0)$. If we take the initial unit vector parallel to the direction of the flow as $\mathbf{e} = \dot{\mathbf{X}}(0)/|\dot{\mathbf{X}}(0)|$, we have

$$\mathbf{M}(t) \cdot \mathbf{e} = \mathbf{M}(t) \cdot \frac{\dot{\mathbf{X}}(0)}{|\dot{\mathbf{X}}(0)|} = \frac{1}{|\dot{\mathbf{X}}(0)|} \dot{\mathbf{X}}(t) = \frac{1}{|\dot{\mathbf{X}}(0)|} \mathbf{F}[\mathbf{X}(t)] \tag{3.11}$$

If the trajectory stays away from the equilibrium points the vector field remains nonvanishing so that the logarithm in the definition [Eq. (3.9)] keeps a finite value and the limit vanishes. We conclude that any trajectory different from an equilibrium point always has one vanishing Lyapunov exponent. Because of energy conservation, another Lyapunov exponent of such a trajectory is vanishing, which corresponds to a unit vector pointing outside the energy shell as $\partial \mathcal{H}/\partial \mathbf{X}$.

Because of the special Hamiltonian form [Eq. (3.1)] of the vector field, the matrix $\mathbf{M}(t)$ turns out to be a symplectic matrix satisfying [71, 91]

$$\mathbf{M}^{\mathrm{T}}(t) \cdot \Sigma \cdot \mathbf{M}(t) = \Sigma \tag{3.12}$$

which is proved be differentiation of both sides with respect to the time and by the use of $\Sigma \cdot \mathbf{L} \cdot \Sigma = \mathbf{L}^{\mathrm{T}}$, which follows from the definition [Eq. (3.4)] of $\mathbf{L}$ for the case of Hamiltonian systems. As a consequence, the Lyapunov exponents fall in pairs $\{\lambda_i, -\lambda_i\}_{i=1}^{f}$ [71, 91].

For Hamiltonian systems $\mathcal{H} = \mathbf{p}^2/2 + V$ with two degrees of freedom ($f = 2$), Eckhardt and Wintgen [92] have shown how to simplify the linear stability problem by going to a special coordinate system moving with the trajectory. The perturbation is decomposed onto a set of four mutually orthogonal vectors according to

$$\delta \mathbf{X}(t) = \mathbf{A}(t) \cdot \mathbf{x}(t) \tag{3.13}$$

where $\mathbf{x} = (x_{\parallel}, x_q, x_{\perp}, x_p)$ are the new coordinates and where the matrix

A has the form [92]

$$\mathbf{A} = \begin{pmatrix} \dot{q}_1 & -\dot{q}_2/r & -\dot{p}_1/r^2 & -\dot{p}_2/r \\ \dot{q}_2 & \dot{q}_1/r & -\dot{p}_2/r^2 & \dot{p}_1/r \\ \dot{p}_1 & \dot{p}_2/r & \dot{q}_1/r^2 & -\dot{q}_2/r \\ \dot{p}_2 & -\dot{p}_1/r & \dot{q}_2/r^2 & \dot{q}_1/r \end{pmatrix} \tag{3.14}$$

with $r = |\dot{\mathbf{X}}| = (\dot{q}_1^2 + \dot{q}_2^2 + \dot{p}_1^2 + \dot{p}_2^2)^{1/2}$. The first column is the vector that is parallel to the flow while the second column is chosen to be perpendicular to the first column without exchanging positions with momenta. The third column is a vector that points outside the energy shell and is proportional to the matrix Σ applied to the first column. Similarly, the fourth column is proportional to Σ applied to the second column. At time t, the perturbation in the new coordinates is given by

$$\mathbf{x}(t) = \mathbf{A}^{-1}(t) \cdot \mathbf{M}(t) \cdot \mathbf{A}(0) \cdot \mathbf{x}(0) \tag{3.15}$$

Accordingly, the new variables are governed by the equation $\dot{\mathbf{x}} = \mathbf{L}' \cdot \mathbf{x}$ with a matrix of the form

$$\mathbf{L}' = \mathbf{A}^{-1} \cdot (\mathbf{L} \cdot \mathbf{A} - \dot{\mathbf{A}}) = \begin{pmatrix} 0 & -d & e & f \\ 0 & a & f & b \\ 0 & 0 & 0 & 0 \\ 0 & -c & d & -a \end{pmatrix} \tag{3.16}$$

where the elements are obtained by a symbolic manipulation program according to

$$a = [\dot{q}_1\dot{p}_1(V_{11} - 1) + \dot{q}_2\dot{p}_2(V_{22} - 1) - (\dot{q}_1\dot{p}_2 + \dot{q}_2\dot{p}_1)V_{12}]/r^2 \tag{3.17}$$

$$b = [(\dot{q}_1^2 + \dot{p}_2^2)(V_{11} + 1) + (\dot{q}_2^2 + \dot{p}_1^2)(V_{22} + 1) + 2(\dot{q}_1\dot{q}_2 - \dot{p}_1\dot{p}_2)V_{12}]/r^2 \tag{3.18}$$

$$c = [(\dot{q}_1^2 + \dot{q}_2^2)(V_{11} + V_{22}) + 2(\dot{p}_1^2 + \dot{p}_2^2)]/r^2 \tag{3.19}$$

$$d = (\dot{q}_1\dot{p}_2 - \dot{q}_2\dot{p}_1)(V_{11} + V_{22} - 2)/r^{3/2} \tag{3.20}$$

$$e = [(\dot{q}_1^2 - \dot{p}_1^2)(1 - V_{11}) + (\dot{q}_2^2 - \dot{p}_2^2)(1 - V_{22}) + 2(\dot{p}_1\dot{p}_2 - \dot{q}_1\dot{q}_2)V_{12}]/r^2 \tag{3.21}$$

$$f = [(\dot{q}_1\dot{q}_2 + \dot{p}_1\dot{p}_2)(V_{11} - V_{22}) - (\dot{q}_1^2 + \dot{p}_1^2 - \dot{q}_2^2 - \dot{p}_2^2)V_{12}]/r^{3/2} \tag{3.22}$$

using the definition of $\mathbf{A}$ and $\ddot{q}_i = -V_{ij}q_j$ with $V_{ij} = \partial^2 V/\partial q_i\, \partial q_j$.

The form of the matrix [Eq. (3.16)] implies that, if the initial perturbation is within the energy shell, that is, $x_\perp(0) = 0$, then it remains there for all times, $x_\perp(t) = 0$. Moreover, the components x_q and x_p formed a closed system. The component $x_\parallel$ is driven by the three others so that it remains constant if the three other components are initially set to zero, as expected for the direction parallel to the flow. The nontrivial Lyapunov exponents are therefore determined by the behavior of the two components x_q and x_p so that the linear stability problem is reduced to the integration of the 2×2 matrix system

$$\dot{\mathbf{m}} = \mathbf{l} \cdot \mathbf{m} \tag{3.23}$$

with

$$\mathbf{l} = \begin{pmatrix} a & b \\ -c & -a \end{pmatrix} \tag{3.24}$$

along the orbit.

We now describe several types of trajectories and their linear stability:

1. *The Equilibrium Points*. For these special trajectories, the variables stay constant: $\mathbf{X}(t) = \mathbf{X}_s$. Hence, they correspond to the zeros of the vector field where $\mathbf{F}(\mathbf{X}_s) = 0$. For a Hamiltonian of the form $\mathcal{H} = \mathbf{p}^2/2 + V(\mathbf{q})$, the equilibrium points satisfy $\mathbf{p}_s = 0$ and $\partial V/\partial \mathbf{q}(\mathbf{q}_s) = 0$. They occur at special energy values $E_s = V(\mathbf{q}_s)$. The linear stability is controlled by the eigenvalues and eigenvectors of the gradient of the vector field at the equilibrium point

$$\det\left[\frac{\partial \mathbf{F}}{\partial \mathbf{X}}(\mathbf{X}_s) - \xi \mathbf{1}\right] = 0 \tag{3.25}$$

which reduces to

$$\det\left[\frac{\partial^2 V}{\partial \mathbf{q}^2}(\mathbf{q}_s) - \sigma \mathbf{1}\right] = 0 \tag{3.26}$$

for our particular Hamiltonians with $\sigma = -\xi^2$. As a consequence of the symmetry of the second derivative matrix of the potential, the eigenvalues ξ are either pure real or imaginary: $\xi = \pm\lambda$ or $\pm i\omega$. The imaginary eigenvalues correspond to the directions where the motion is oscillatory, and the real eigenvalues to the unstable directions. In this latter case, which occurs at saddle equilibrium points, the eigenvalues $\xi = \pm\lambda$ are the Lyapunov exponents associated with the unstable direction of the orbit, while the two other Lyapunov exponents are vanishing. In general, the Lyapunov exponents are related to the stability eigenvalues by $\lambda_i = \text{Re}\,\xi_i$.

2. *The Periodic Orbits*. For periodic orbits, there exist real numbers called periods, for which the trajectory comes back to its initial condition $\mathbf{X}(t+T) = \mathbf{X}(t)$. The smallest among those periods is the prime period T_p of the periodic orbit p. The other periods are then given as multiples of the prime period $T = rT_p$, with $r = 1, 2, \ldots$. The linear stability is determined by the matrix [Eq. (3.7)] over one prime period

$$\det[\mathbf{M}(T_p) - \Lambda\mathbf{1}] = 0 \tag{3.27}$$

According to Eq. (3.12), the matrix $\mathbf{M}$ is symplectic so that its eigenvalues fall in quadruplets as Λ, Λ^*, Λ^{-1}, and Λ^{-1*}. By the same reasoning we gave in the context of Eq. (3.11), and by energy conservation, two among the eigenvalues are necessarily equal to 1, $\Lambda = 1$, which correspond to the directions parallel to the flow and perpendicular to the energy shell. For two-degree-of-freedom systems, we have three kinds of periodic orbits: (1) hyperbolic without reflection if the two nontrivial eigenvalues are real and positive, $\Lambda > 1 > \Lambda^{-1} > 0$; (2) hyperbolic with reflection if they are real but negative, $\Lambda < -1 < \Lambda^{-1} < 0$; (3) elliptic if they are complex on the unit circle, $\Lambda = \exp(\pm 2\pi i \rho)$, where ρ defines the rotation or winding number. In higher-degree-of-freedom systems, there appears the fourth possibility that the eigenvalues are forming a genuine quadruplet of complex numbers outside the unit circle [71, 91].

The nontrivial eigenvalues are also given by the eigenvalues of the monodromy matrix, which is the linearized return map in a Poincaré surface of section that is transverse to the periodic orbit [93]. If the surface of section is defined by the equation $\mathbf{S}(\mathbf{X}) = 0$, the full return map is governing the successive intersections $\mathbf{X}_n = \mathbf{X}(t_n)$ of a trajectory with the surface of section. For a general flow, the successive times t_n when the crossings occur are not necessarily multiples of a constant as in periodically forced systems so that we have

$$\mathbf{X}_{n+1} = \phi(\mathbf{X}_n) = \mathbf{\Phi}^{\mathcal{T}(\mathbf{X}_n)}(\mathbf{X}_n) \tag{3.28}$$

$$t_{n+1} = t_n + \mathcal{T}(\mathbf{X}_n) \tag{3.29}$$

where $\mathcal{T}(\mathbf{X})$ is called the ceiling function. At the fixed point corresponding to the intersection of the periodic orbit with the surface of section, the matrix $\partial\phi/\partial\mathbf{X}$ admits the same eigenvalues as the matrix $\mathbf{M}(T_p)$. In two-degree-of-freedom systems, we may construct the Poincaré return map in some two-dimensional surface of section. In some intrinsic coordinates of this surface, the return map can be written as $\mathbf{y}_{n+1} = \varphi(\mathbf{y}_n)$, and the linearized map near the fixed point of the periodic orbit is then given by the monodromy matrix $\partial\varphi(\mathbf{y}_p)/\partial\mathbf{y} = \mathbf{m}_p = \mathbf{m}(T_p)$, which is solution of Eq. (3.23). A similar monodromy matrix $\mathbf{m}_p$ exists in higher-degree-of-freedom systems.

For periodic orbits, the Lyapunov exponents are given by

$$\lambda_i = \frac{1}{T_p} \ln|\Lambda_i| \tag{3.30}$$

in terms of the eigenvalues of $\mathbf{M}(T_p)$ or, for the non-zero Lyapunov exponents, of $\mathbf{m}(T_p)$.

The preceding considerations form the phase-space version of the problem of linear stability, which we already described in position space in Section II. As we showed at the beginning of this section both formulations are equivalent and give the same results.

When the stability eigenvalues cross the unit circle, the periodic orbit changes its stability and, in general, undergoes a bifurcation where other orbits may emerge or disappear. The different types of bifurcations that an equilibrium point or a periodic orbit may undergo have been classified elsewhere [94]. At these bifurcations, the topology of the phase-space structures is changing. In particular, it is through bifurcations that the flow can become chaotic with an infinite number of periodic orbits if the original flow was not chaotic. We will now turn to these topological aspects.

Energy is the principal bifurcation parameter in a time-independent Hamiltonian system. Topology changes with energy. Among the quantities of particular interest are the periods of the periodic orbits. Therefore, we will compare different systems by their period-energy diagram, which is important for semiclassical quantization.

One of the main tools used to study dynamical systems is the Poincaré surface of section where the flow induces a return mapping (see above). If the surface of section is transverse to the flow, a typical trajectory will have successive crossings that will form a sequence of points in the Poincaré surface. The dimensional reduction provided by the Poincaré section simplifies the analysis of the geometry of the flow.

Among the manifolds that can be formed in the phase space, a special role is played by the stable and unstable manifolds associated with the stable or unstable directions of either equilibrium points or periodic orbits [95]. They are invariant manifolds that are separating the trajectories with different behaviors forward or backward in time so that these manifolds can be used to classify the trajectories. The stable manifold is formed by all the trajectories that are converging to the periodic orbit for $t \to +\infty$. On the other hand, the unstable manifold is the set of trajectories converging to the periodic orbit for $t \to -\infty$. Therefore, they are necessarily tangent to the corresponding eigenvectors of the linear stability matrix.

The trajectories that are at the intersections between the stable and unstable manifolds of a periodic orbit of saddle type are the so-called homoclinic orbits [2, 95]. The importance of homoclinic orbits for chaotic behavior was pointed out in the works by Poincaré and by Birkhoff [2, 24]. According to a theorem established by Birkhoff, the presence of a homoclinic orbit associated with an unstable periodic orbit implies the existence of an uncountable set of nonperiodic trajectories which contains a dense and countable set of periodic orbits of arbitrarily long periods. This topological property is the signature of what is today called chaos. Homoclinic orbits of the Birkhoff type have been discovered and studied in a number of conservative and dissipative systems, particularly in celestial mechanics. We refer the reader to the vast literature on the subject [2, 3, 95].

B. Examples of Mechanical Systems

Classical mechanics concerns not only the smooth Hamiltonian systems whose time evolution is ruled by the first-order differential equations (3.1) but also a variety of more general systems in which the particles encounter hard walls where they undergo elastic collisions. It is natural to refer to these general systems as mechanical systems. Between the elastic collisions, the particles may be exposed to electric or magnetic forces. Billiards are special cases of these mechanical systems where the motion is free between the collisions [96]. During elastic collisions, energy is conserved and the reflected ray is related to the incident ray according to the laws of geometrical optics.

Let us remark that the concept of hard walls is an idealization of a very steep potential, such as occurs in elastic collisions of ultracold neutrons on solid surfaces. In microscopic and mesoscopic systems, where the dimensions of the system become comparable to the atomic dimensions, this idealization may turn out to be too crude an approximation.

Nevertheless, we will see that a number of important properties are already captured by the billiard approximation [97].

We may also distinguish between bounded and open systems. In bounded systems, the phase space is compact and the trajectories remain forever at finite distances. On the other hand, the trajectories may escape to large distances in open systems of scattering type, which are of special importance for the modeling of ionization, dissociation, or reaction processes [97]. In spite of the fact that a large portion of the trajectories are escaping to large distances in scattering systems, the pointlike character of classical dynamics allows for the existence of trajectories that remain forever at finite distances. These trapped trajectories may be stationary, periodic, quasiperiodic, or nonperiodic with both types of stability. The trapped trajectories correspond to the formation of a complex. Due to quantum mechanical tunneling, this complex is always metastable although classically there may exist trapped orbits that are absolutely stable in positive measure phase-space regions of one- and two-degree-of-freedom systems. The phenomenon called Arnold diffusion prevents the existence of such absolutely stable trapped orbits in higher-degree-of-freedom systems [2]. Nevertheless, in this latter case, there may still exist marginally stable or unstable trapped orbits.

Let us give a few examples of mechanical systems.

1. Hamiltonian Systems

1. *Harmonic Oscillator*. For the Hamiltonian $\mathcal{H} = (1/2)p^2 + (\omega^2/2)q^2$, there exists a unique fixed point at $q = 0$ and energy $E_0 = 0$. The periods, $T = rT_p = 2\pi r/\omega$, are independent of the energy in this special case. The period-energy diagram is shown in Fig. 5. All the orbits of phase space are actually periodic orbits that are multiples of a single prime periodic orbit. Therefore, the number of periods below T grows linearly with T.
2. *Quartic Oscillator*. Fig. 6 depicts the period-energy diagram for the Hamiltonian $\mathcal{H} = (1/2)p^2 - (a/2)q^2 + (b/4)q^4$. When $a, b > 0$, this potential has three equilibrium points, two of which are stable, at $q = \pm\sqrt{a/b}$ and $E_0 = -a^2/(4b)$, while the other is unstable at $q = 0$ and $E_1 = 0$. The periods of the orbits are doubly degenerate for $E_0 < E < E_1$. Due to the anharmonicity of the potential, the periods increase with energy below E_1 but decrease with energy above E_1. A bifurcation occurs at E_1. With increasing energy, there is an infinite number of periodic orbits that correspond to a fixed value T of the period. However, the periods are transverse to the period axis near E_0 in agreement with our expectation that the harmonic approximation is valid at low energy.

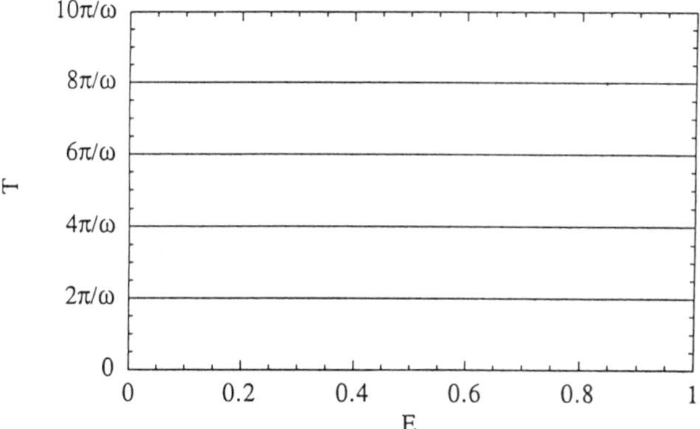

Figure 5. Period-energy diagram for the harmonic oscillator $\mathscr{H} = (1/2)p^2 + (\omega^2/2)q^2$.

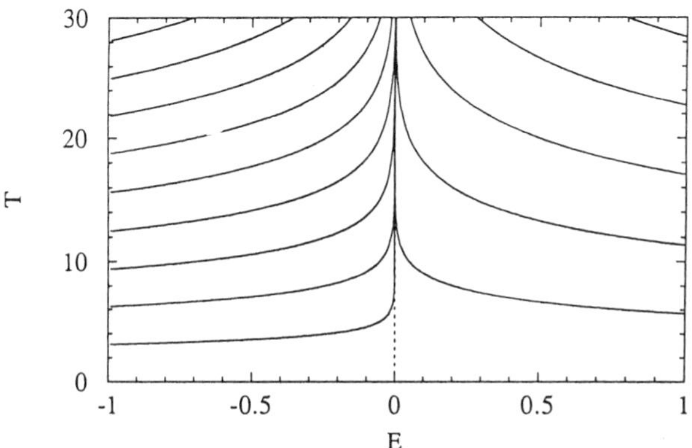

Figure 6. Period-energy diagram for the quartic oscillator $\mathscr{H} = (1/2)p^2 - (a/2)q^2 + (b/4)q^4$ with $a = 2$ and $b = 1$.

3. Morse Oscillator. For the Morse Hamiltonian [98]

$$\mathscr{H} = \frac{p^2}{2} + D[1 - \exp(-\alpha q)]^2 \tag{3.31}$$

the system is bounded below the dissociation energy $E = D$ but is unbounded for positive energies. The system is exactly solvable in classical and quantum mechanics. The bounded and periodic trajectories are given by

$$q(t) = \frac{1}{\alpha} \ln\left\{ \frac{1}{\sin^2\theta} [1 - \cos^2\theta \cos(\omega_0 t \sin\theta + \phi)] \right\} \tag{3.32}$$

for $\cos^2\theta = E/D < 1$, where $\omega_0 = \alpha\sqrt{2D}$ is the frequency of the harmonic vibrations near the equilibrium point $q = 0$, while ϕ is an arbitrary phase. The energy-dependent periods are

$$T = \frac{2\pi r}{\omega_0 \sqrt{1 - (E/D)}} \tag{3.33}$$

where r is the number of repetitions of the prime period. The corresponding period-energy diagram is plotted in Fig. 7, where we see the divergence of the period near the dissociation energy $E = D$. Moreover, above the dissociation energy $(E > D)$, there exist unbounded trajectories of scattering type, which are obtained by replacing θ by $i\theta$ in Eq. (3.32).

In the preceding one-degree-of-freedom systems, all the orbits are stable and there is no chaos. However, this situation changes as soon as we go to systems with two degrees of freedom.

4. *The Two-Electron Atomic Systems*. These systems are composed of two electrons orbiting around a nucleus of charge Z and are therefore isoelectronic with Helium [99–101]. In a classical model of the dynamics, we may imagine different configurations for the

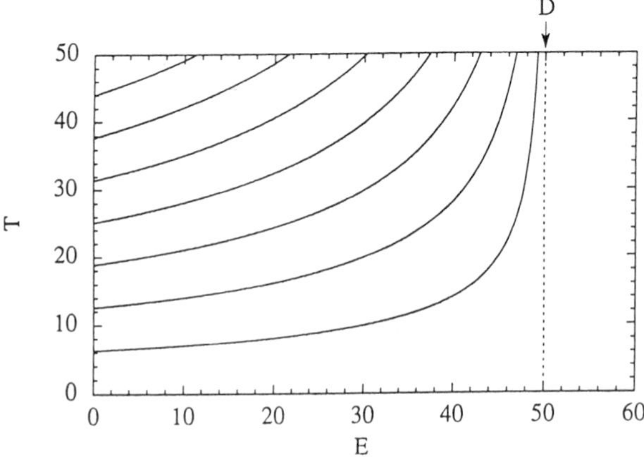

Figure 7. Period-energy diagram [Eq. (3.33)] for the Morse potential [Eq. (3.31)] with $\omega_0 = 1$ and $D = 50$.

motion of the electrons. The configuration where both electrons move collinearly on opposite sides of the proton seems to play a particular role in the sector of zero total angular momentum, especially in the hydrogen negative ion for which $Z = 1$. The Hamiltonian of this two-degree-of-freedom system is

$$\mathscr{H} = \frac{p_1^2 + p_2^2}{2} - \frac{1}{r_1} - \frac{1}{r_2} + \frac{1}{Z(r_1 + r_2)} \tag{3.34}$$

As in pure Coulomb systems, the periods scale as $T_p(E) = 0.5(-E)^{-3/2}\tilde{S}_p$ with energy, where the constants $\tilde{S}_p$ are the reduced actions of the periodic orbits at energy $E = -1$ in atomic units. The period-energy diagram of the hydrogen negative ion (H^-) is given in Fig. 8. This system is known to be classically chaotic and there exists a symbolic coding of its periodic orbits, as described in Section XIII. Contrary to the previous examples, there is no energy value for which the phase space of H^- is compact so that most trajectories escape to infinity, which corresponds to autoionization. As a consequence, there exists a fractal repeller containing all the trajectories that remain bounded in phase space.

5. *The Hénon–Heiles Hamiltonian*. This abstract model has been studied as a prototype of nonlinear coupling between two oscillators with chaotic behavior resulting from the coupling [102]. The model was first proposed in the context of celestial mechanics but its

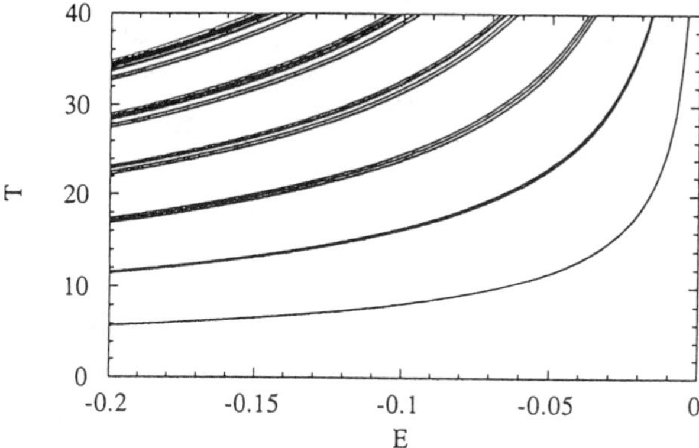

Figure 8. Period-energy diagram for the H^- ion in the collinear eZe configuration (in atomic units).

quantization has been carried out later. The Hamiltonian is

$$\mathcal{H} = \frac{p_x^2 + p_y^2}{2} + \frac{x^2 + y^2}{2} + x^2 y - \frac{1}{3} y^3 \tag{3.35}$$

The potential is represented in Fig. 9. The potential has a threefold symmetry under the group C_{3v}. At the energy $E = 0$, there is one stable equilibrium point at the origin, $x = y = 0$, with eigenvalues $\omega_1 = \omega_2 = 1$. At the energy $E = \frac{1}{6}$, there are three saddle equilibrium points at $(x, y) = (0, 1)$, $(+\sqrt{3}/2, -1/2)$, and $(-\sqrt{3}/2, -1/2)$, with eigenvalues $\omega = \sqrt{3}$ and $\lambda = 1$. The nonlinearity is cubic in the potential that is therefore unbounded. As a consequence, all quantum states below $E = \frac{1}{6}$ are metastable with a lifetime determined by tunneling. However, the motion in the bounded part of the potential is classically absolutely stable, without the possibility

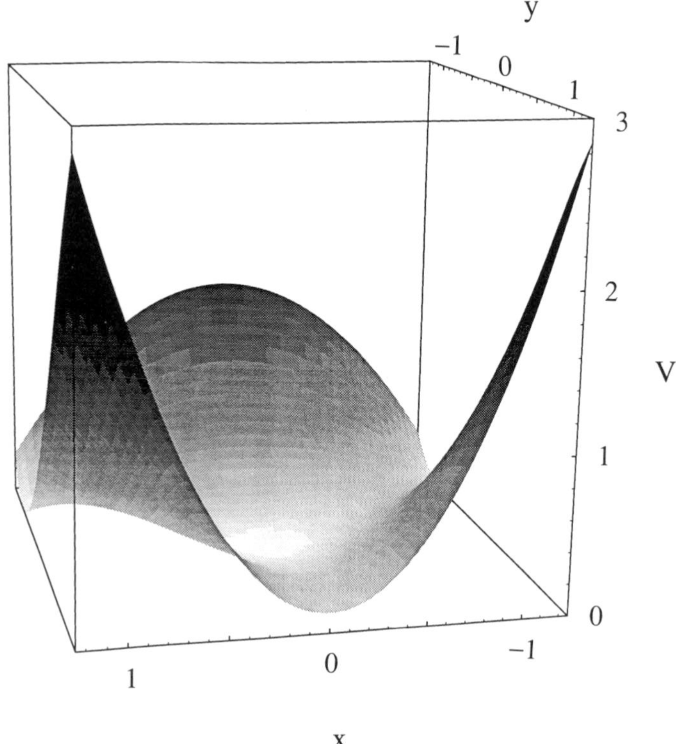

Figure 9. The Hénon–Heiles potential [Eq. (3.35)].

of escaping to large distances. This model has been intensely studied numerically and phase portraits can be found elsewhere [2]. Let us mention that the bounded motion appears to be regular (i.e., quasiperiodic) up to an energy $E \simeq 0.11$, where a global chaotic zone emerges after very complex bifurcations of the periodic and quasiperiodic motions. The area of the chaotic zone rapidly increases beyond $E \simeq 0.11$. Finally, at $E = 0.1666, \ldots$, the motion becomes unbounded and most trajectories escape to infinity through the saddle points. Above this dissociation energy, the trapped motion remains chaotic with quasiperiodic islands progressively disappearing as the energy increases [102]. This scenario is typical of a transition to chaos followed by a transition to chaotic dissociation in classical systems with two degrees of freedom or more. We will encounter this scenario under different forms in the following sections.

Many other classically chaotic Hamiltonians have been studied in the literature [103].

2. *Hamiltonian Mappings*

Hamiltonian or symplectic mappings are generated by successive crossings of the trajectories in the Poincaré surface of section. Such mappings are in general given by a nonlinear symplectic transformation of the intrinsic variables of the surface of section onto themselves, that is, a transformation whose gradient satisfies Eq. (3.12). Such a nonlinear transformation should be obtained by integration of the flow. However, we may also proceed by modelization and assume a certain form for the mapping. In the same context, there exist simple periodically driven Hamiltonian systems that admit an analytic Poincaré mapping. This applies in particular to the periodically kicked systems with Hamiltonians of the form [104]

$$\mathcal{H}(\mathbf{q}, \mathbf{p}) = \mathcal{H}_0(\mathbf{q}, \mathbf{p}) + TV(\mathbf{q}) \sum_{n=-\infty}^{+\infty} \delta(t - nT) \tag{3.36}$$

where $\mathcal{H}_0$ is the Hamiltonian of a known or integrable system and V is a potential controlling the amplitudes of the kicks given to the particle at each period T. We consider the Poincaré mapping from the variables $(\mathbf{q}_n, \mathbf{p}_n)$ before the nth kick to the variables before the $(n+1)$th kick. At the nth kick, the momentum is modified by $\Delta\mathbf{p}_n = -T\, \partial V/\partial \mathbf{q}_n$. If we denote the generating function of the motion between two kicks under the free Hamiltonian $\mathcal{H}_0$ by $F_0(\mathbf{q}_n, \mathbf{q}_{n+1}, T)$, then the generating function of the Poincaré mapping is equal to

$$F(\mathbf{q}_n, \mathbf{q}_{n+1}) = F_0(\mathbf{q}_n, \mathbf{q}_{n+1}, T) - TV(\mathbf{q}_n) \tag{3.37}$$

so that the action of a trajectory over N kicks is $W = \Sigma_{n=0}^{N-1} F(\mathbf{q}_n, \mathbf{q}_{n+1})$ and the classical motion results from its variation δW. In the special case of a particle in free flight between the kicks ($\mathcal{H}_0 = \mathbf{p}^2/2$), the Poincaré mapping becomes

$$\begin{aligned} \mathbf{p}_{n+1} &= \mathbf{p}_n - T\frac{\partial V}{\partial \mathbf{q}_n} \\ \mathbf{q}_{n+1} &= \mathbf{q}_n + T\mathbf{p}_{n+1} \end{aligned} \tag{3.38}$$

In the limit $T \to 0$, the mapping converges to the flow corresponding to the Hamiltonian $\mathcal{H} = \mathbf{p}^2/2 + V(\mathbf{q})$, which is often referred to as the integrable limit if $\mathcal{H}$ is integrable. The linearized mapping is given by

$$\mathbf{M} = \begin{pmatrix} \mathbf{1} & \mathbf{1} - \dfrac{\partial^2 V}{\partial \mathbf{q}^2} \\ \mathbf{1} & -\dfrac{\partial^2 V}{\partial \mathbf{q}^2} \end{pmatrix} \tag{3.39}$$

which satisfies Eq. (3.12) so that it is symplectic.

Let us give several examples of such mappings:

1. *Kicked Morse Oscillator*. For this system, which was considered by Tersigni et al., [105], $\mathcal{H}_0$ is the Morse Hamiltonian [Eq. (3.31)] and $V = f[1 - \exp(-\alpha q)]^2$. The mapping shows a large variety of features from elliptic islands to chaotic behavior. Since the Morse oscillator may lead to unbounded motion, the mapping is of scattering type with the possibility of escape at large distances.
2. *Standard Map*. This archetype of chaos is generated by $\mathcal{H}_0 = p^2/2$ and $V = K \cos q$ with $q \in [0, 2\pi)$ [106]. The motion of this system remains bounded along the angle variable q but the momentum may become unbounded above the threshold at $K \simeq 0.97$ for global chaos. Its properties have been reviewed elsewhere [2].
3. *Hénon Map*. This other archetype of chaos is generated by $\mathcal{H}_0 = p^2/2$ and $V = d(q^3/3 - q)$ with q real [107]. The motion may become unbounded in position. The particle is continuously accelerated at large distances so that this map is not convenient to study chaotic scattering processes. To avoid this inconvenience, Gaspard and Rice [108] introduced the following mappings.

4. *One-Channel Scattering Map with a Minimum*. This mapping is generated by $\mathcal{H}_0 = p^2/2$ and $V = d[1 - \exp(-q)]^2$ with q real. The motion may become unbounded in position but the particle reaches a free flight when $q \to +\infty$. The behavior of the trapped orbits is very similar to the one in the Hénon map with qualitatively the same sequence of bifurcations as the parameter d increases.
5. *One-Channel Scattering Map with a Minimum and a Maximum*. This mapping is generated by $\mathcal{H}_0 = p^2/2$ and $V = d \exp(-q)(q^2 + q + 1)$ with q real. The behavior of this map is similar to the one of Example 4 except that here there is an extra maximum point at finite distance. Phase portraits, as well as homoclinic tangles of the stable and unstable manifolds, have been numerically constructed elsewhere [93, 97].
6. *Two-Channel Scattering Map*. This mapping is generated by $\mathcal{H}_0 = p^2/2$ and $V = \exp(-q^2/2)(\alpha + \beta q + \gamma q^2/2)$ with q real. Contrary to the two preceding examples, the particle can escape in free motion either for $q \to +\infty$ or for $q \to -\infty$. This model has been introduced by Burghardt and Gaspard [21] in the study of direct dissociation of triatomic molecules, where two channels of dissociation exist in a collinear configuration; the model will be discussed in detail in Section XII.

3. *Billiards*

Besides Hamiltonian flows and maps, billiards have been the subject of many recent works [96, 109, 110]. There exist billiards in the plane, such as the Bunimovich stadium, the Sinai billiard, or the Lorentz gas, as well as in higher dimensions, such as three-dimensional Lorentz billiards or the gases composed of hard spheres or ellipsoids. A billiard may be integrable like the circular and elliptic ones. We speak of a hyperbolic billiard if all its periodic orbits are unstable. Billiards may be bounded or open. In the latter case, the billiard is a scattering system and its area is infinite. As we mentioned earlier, two-dimensional billiards are also models of ballistic electronic circuits at low temperatures and of nanometric sizes [40–43].

Because the motion is free between the bounces, the trajectories are generated by the following action function, which is similar to the case of mappings

$$W = \sum_{n=0}^{N-1} \ell(s_n, s_{n+1}) \tag{3.40}$$

where s_n denotes the length of perimeter along the border of the billiard where the nth bounce happened. The quantity $\ell(s_n, s_{n+1})$ is the length of the path between the bounces at s_n and s_{n+1}. The variable that is canonically conjugate to s_n is the velocity parallel to the wall of the billiard, $p_n = \sin\phi_n$, where $\pi/2 \leq \phi_n \leq 3\pi/2$ is the angle between the incident ray and the normal to the wall that points toward the exterior of the walls. The coordinates $(s_n, \sin\phi_n)$ are called the Birkhoff coordinates of the billiard [110]. All the billiards share the same dependency of the periods on the energy. Indeed, the periods $\{T_{rp}\}$ are determined by the total lengths $\{L_p\}$ of the prime periodic orbits and by the velocity according to

$$T_{rp} = rT_p = r\frac{L_p}{\sqrt{2ME}} \tag{3.41}$$

where M is the mass of the particle and $r = 1, 2, \ldots$ is the integer counting the repetitions of the prime orbit p.

In the following list, we give several examples of billiards.

1. *Circle Billiard.* Even in the absence of chaos, the spectrum of the periodic orbits may present peculiarities as in the simple case of the circle billiard [111]. A periodic orbit of this billiard will close onto itself after n bounces off the wall and after m turns around the center. The number m may differ from $m = 1$ in the case of orbits forming stars. Accordingly, the periodic orbits are labeled by two integers m and n, such that $n \geq 2m$. Their lengths are $L_{mn} = 2n\sin(\pi m/n)$ for a circle of unit radius. Because of the rotational symmetry, the periodic orbits form continuous families so that each of the preceding lengths has the degeneracy of the one-dimensional continuum. Another peculiarity comes from the accumulation of series of periodic orbits at the multiples m of the perimeter when $n \to \infty$, a phenomenon referred to as the whispering gallery. Moreover, as in the disk scatterers, every point inside the billiard is an equilibrium point of marginal stability that forms a (two-dimensional) continuum. The period-energy diagram is plotted in Fig. 10.
2. *Stadium Billiard.* This famous billiard is known to be chaotic as proved by Bunimovich [112]. The origin of chaos in the stadium is interesting because its walls are focusing rather than defocusing. Dynamical instability and positive Lyapunov exponents finally arise because of an excess of focusing due to the fact that the two focusing semicircles have been separated with respect to each other. On the other hand, the system is mixed if the two semicircles are put closer to each other, as in the lemon billiard [61]. As mentioned

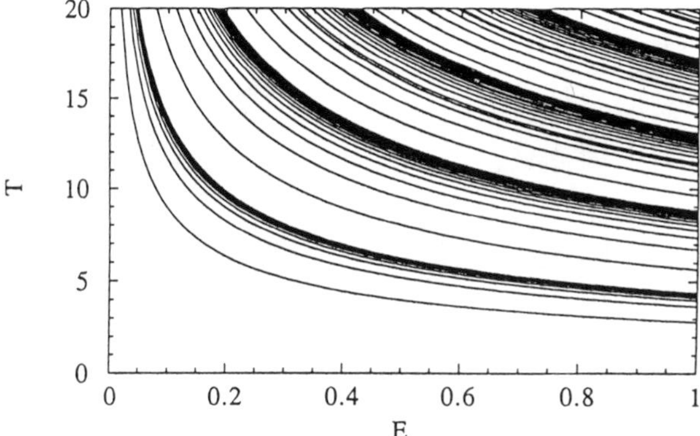

Figure 10. Period-energy diagram for the circle billiard with a unit mass.

above, the stadium is chaotic but not hyperbolic since bouncing between the parallel rectilinear walls generates a continuous family of periodic orbits they are not unstable. As in the circle billiard, the periodic orbits of the stadium billiard accumulate on multiples of the whispering gallery orbit. However, this particular orbit is infinitely unstable due to the discontinuity of the curvature of the wall at the matching points between the semicircles and the rectilinear walls. The bounded stadium may be transformed into a scattering system if the two rectilinear walls are replaced by two semiinfinite channels.

3. *Disk Scatterers*. These billiards are used as models of unimolecular dissociations as well as of diffusion [113–115]. For these billiards, a point particle undergoes elastic collisions with hard disks that are fixed in the plane. The defocusing character of the collisions on disks induces a dynamical instability so that all the trapped orbits are unstable and these systems are hyperbolic. Since these billiards are of the scattering type, their trapped orbits form a repeller of zero measure in phase space. We may consider a scatterer formed by a varying number of disks from the one-disk scatterer to the many-disk scatterer, which is equivalent to a Lorentz gas if the disks form a regular lattice. Increasing the number of disks is interesting because we observe the transition to chaos between the two- and the three-disk scatterers (we suppose that the disks form an equilateral triangle in the three-disk scatterer). Indeed, the two-disk scatterer is nonchaotic since its repeller contains a single prime periodic orbit, whereas the three-disk scatterer is chaotic and its repeller is fractal. Figure 11 shows several examples of disk

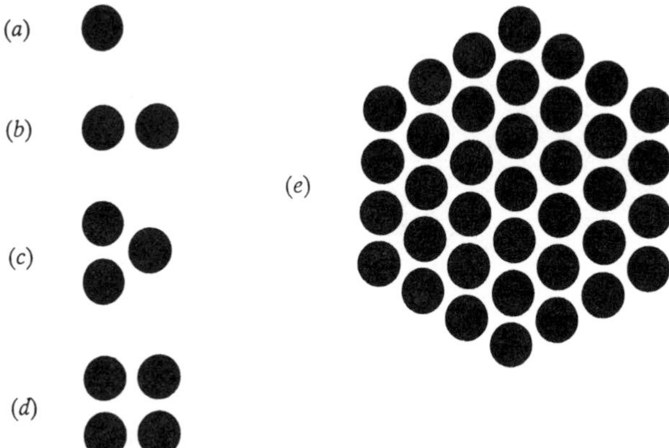

Figure 11. Different configurations of disk scatterers: (*a*) one-disk scatterer; (*b*) two-disk scatterer; (*c*) three-disk scatterer; (*d*) four-disk scatterer; (*e*) Lorentz-type disk cluster.

scatterers. The period-energy diagrams of the two- and three-disk scatterers are depicted in Fig. 12. For many geometric configurations of the N-disk scatterers, where each disk is "visible" from the others, the trapped orbits are in 1:1 correspondence with a symbolic dynamics whose alphabet is formed with the N labels of the disks. The only constraint on the succession of the symbols $\{\omega_n\}$ is then given by the rule $\omega_{n+1} \neq \omega_n$. In this case, the number of periodic orbits in the repeller increases as $(N-1)^n$ with the number n of bounces. However, when some disks are hiding others, which is the case when they are close to each other, extra constraints may appear and we talk about "pruning" of periodic orbits [115]. In this other case, the number of periodic orbits grows more slowly with the number of bounces. In both cases, we have an exponential proliferation of periodic orbits that is at the origin of the difficulties to quantize chaos.

In the following sections, we present the symbolic dynamics and the Smale horseshoe as well as the thermodynamic formalism whose purpose is to understand and characterize this exponential proliferation and its geometric and probabilistic implications.

C. Smale Horseshoe and Symbolic Dynamics

We would like to introduce the notion of symbolic dynamics in the context of the horseshoe map, as proposed by Smale in 1967 [116]. This map represents a model for a hyperbolic chaotic system. The map is constructed geometrically (see below) by a stretching and folding process,

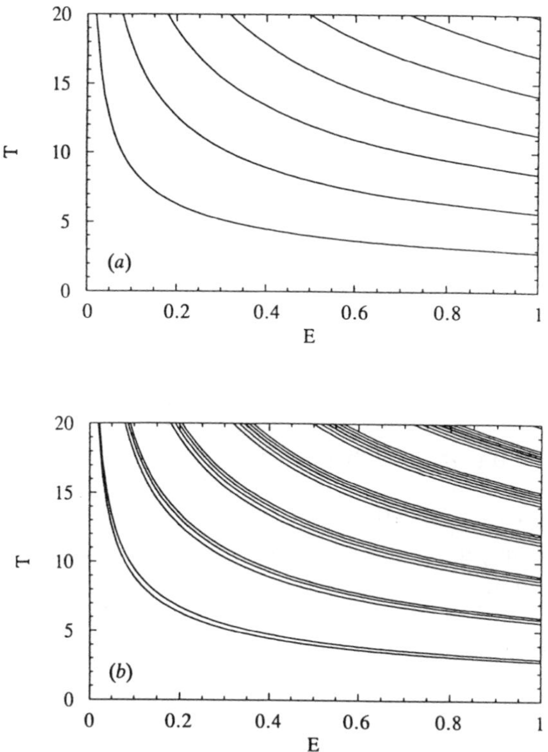

Figure 12. Period-energy diagram for (*a*) the two-disk and (*b*) the three-disk scatterers with a unit mass.

which schematically represents the way in which the stable and unstable manifolds of a hyperbolic fixed point form homoclinic intersections, thus providing a pathway to chaos. First, we will give a short description of the simplest horseshoe map, and then associate with it a symbolic dynamics. The symbolic dynamics allows a classification of the trajectories belonging to the invariant set, which will turn out to be the classification underlying the cycle expansion method (see Section VI). We will then show how the Smale horseshoe and some of its generalizations appear in some of the aforementioned mechanical systems.

1. *Horseshoe Map*

The horseshoe map is illustrated in Fig. 13. By stretching and folding process [117], regions that initially represent vertical slices within the area D are mapped to horizontal slices, while other regions that were initially

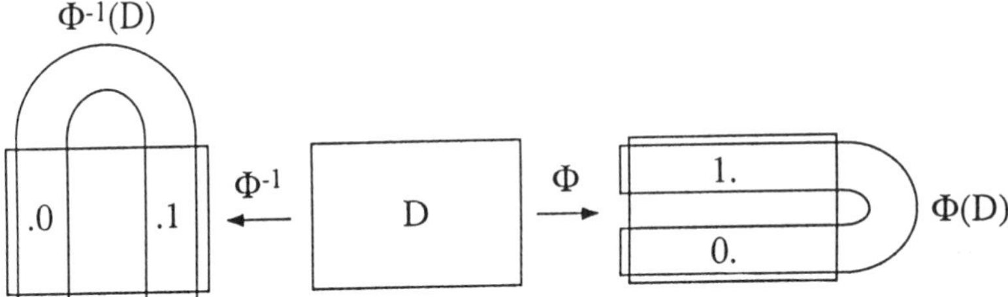

Figure 13. Formation of a Smale horseshoe under action of the mapping Φ on the domain D and definition of the generating partition {0·, 1·} at the intersections of D and of its image Φ(D). The other generating partition {·0, ·1} at the intersections of D and of its preimage Φ^{-1}(D) is also shown.

confined to D are mapped to the outside of the area. The cells generated by the intersection of D with the image $\Phi(D)$ may be labeled by the symbols 0· and 1·; analogously the cells originating from the intersection of D with its preimage $\Phi^{-1}(D)$ are labeled ·0 and ·1. The intersection of these cells then represents the area of phase space that remains confined to the original area D over a backward and forward iteration. The invariant set of trajectories that remain trapped within D over an infinite number of iterations is represented by the bi-infinite intersections of the cells with all their forward and backward iterates. The invariant set of the horseshoe map is hyperbolic and nonattracting. The property of hyperbolicity refers to the fact that all trajectories are of the saddle type, that is, the tangent space of each point of the invariant set can be uniquely decomposed into stable and unstable subspaces. The nonattracting property is reflected in the fact that the invariant set is a Cantor set, and is of zero Lebesgue measure (i.e., it occupies a vanishing phase-space area). In fact, the invariant set of the horseshoe map is referred to as a repeller, which is characterized by exponential escape dynamics [117]. Thus the horseshoe map represents a model for the chaotic dynamics in unstable physical systems, for example, the molecular transition state that will be discussed in detail in Section XII.

2. *Symbolic Dynamics*

The pair of cells ·0 and ·1 is called the generating partition. The intersections with the nth forward or backward iterates of this partition correspond to 2^n cells labeled by strings of n symbols 0 or 1. The size of the cells decreases rapidly as $2^{-n/D_{\mathrm{H}}}$, where D_{H} is the Hausdorff dimension $(0 < D_{\mathrm{H}} < 2)$ of the Cantor set [117]. In the limit $n \to \infty$ the cells correspond to single points $\mathbf{X}$, so that each point of the invariant set is in 1 : 1 correspondence with a bi-infinite sequence, with a dot denoting the time zero:

$$\mathbf{X} \leftrightarrow \cdots \omega_{-4}\omega_{-3}\omega_{-2}\omega_{-1} \cdot \omega_0\omega_1\omega_2\omega_3 \cdots \tag{3.42}$$

Now this symbol sequence not only labels a point of the invariant set, but also represents an encoding of the history of this point under the repeated application of the horseshoe map. This history may be followed from iteration to iteration by applying a shift operation to the symbol sequence, that is, by shifting the decimal point to the right (corresponding to Φ) or to the left (corresponding to Φ^{-1}). The application of the horseshoe map to the invariant set can thus be represented in terms of the shift operation on the bi-infinite symbol sequence.

If the reference to time zero, that is, the dot in Eq. (3.42), is removed from the symbol sequence, we have a bi-infinite sequence that is in 1:1 correspondence with the trajectories of the invariant set

$$\boldsymbol{\omega} = \cdots \omega_{-4}\omega_{-3}\omega_{-2}\omega_{-1}\omega_0\omega_1\omega_2\omega_3 \cdots \tag{3.43}$$

The periodic orbits are given by the repetition of the same string

$$\boldsymbol{\omega} = [\omega_0\omega_1 \cdots \omega_{n-1}]^{\infty} \tag{3.44}$$

We remark that a dotted string like $[\omega_0 \cdots \omega_{n-1}]^{m-}\omega_0\omega_1 \cdot \omega_2 \cdots \omega_{n-1}[\omega_0 \cdots \omega_{n-1}]^{m+}$ for instance, which is obtained by truncating Eq. (3.44) into a finite sequence of symbols, corresponds to a cell of phase space that contains the corresponding periodic orbit Eq. (3.44).

In regard to the classification of the periodic orbits, the periodic orbits labeled with one-symbol sequences, which correspond to the fixed points of the map, are referred to as the *fundamental* periodic orbits, since they generate all the other periodic orbits of the repeller by topological combination [118].

Let us also mention that a symbolic dynamics is characterized by the so-called topological entropy that is the rate of exponential proliferation of the periodic orbits for increasing periods [119]

$$\text{Number}\{p : T_p < T\} \approx \frac{\exp(h_{\text{top}}T)}{h_{\text{top}}T} \tag{3.45}$$

For the Smale horseshoe, the topological entropy per iteration is equal to $\ln 2$. In flows, a distinction must be established between the topological entropy per iteration and the one per unit time, since the time between two crossings in the Poincaré surface of section may vary from one crossing to another [see Eq. (3.29)].

3. *Examples*

Figure 14 depicts a Smale horseshoe with a symbolic dynamics $\{0, 1\}$ appearing in the one-channel scattering map with a minimum and a maximum, corresponding to Example 5 in Section III.B.2 [93].

A similar Smale horseshoe exists in a slightly different form in the three-disk scatterer [115], as well as in the three-hill scatterer [120]. Figure 15 shows the set of initial conditions on the borders of the disks that are trapped for one backward and one forward collision. The initial conditions appear in the Birkhoff coordinates. In this set, we find 12 cells labeled as $\omega_{-1} \cdot \omega_0 \omega_1$ in terms of three symbols $\{a, b, c\}$, which are the labels of the three disks. We find differences with respect to the Smale horseshoe: the first one is that the folding is absent but is replaced by a double reflection on the two opposite disks combined with an escape of the other trajectories that are not reflected on disks. This difference is minor since the folding or the reflection have the same effect of doubling the number of cells at each new backward or forward collision. The second difference concerns the symbolic dynamics that has three rather than two symbols. In spite of the three symbols, the topological entropy per collision is still equal to $\ln 2$. Indeed, the particle has only two possibilities after each collision on a disk so that the symbols follow each

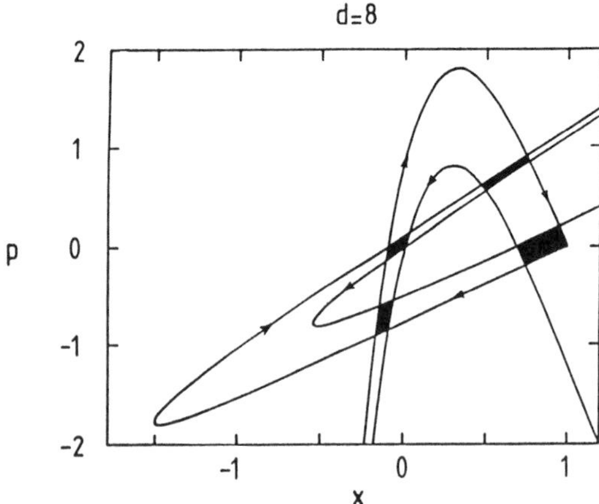

Figure 14. Smale horseshoe appearing in the (q, p) plane of the Hamiltonian mapping [Eq. (3.36)] with $\mathcal{H}_0 = p^2/2$ and $V = d\exp(-q)(q^2 + q + 1)$, when $d = 8$ (Example 5 of Section III.B.2). The repeller exists at the intersections of both horseshoes, of their images, and of their preimages. (Note that $q = x$.)

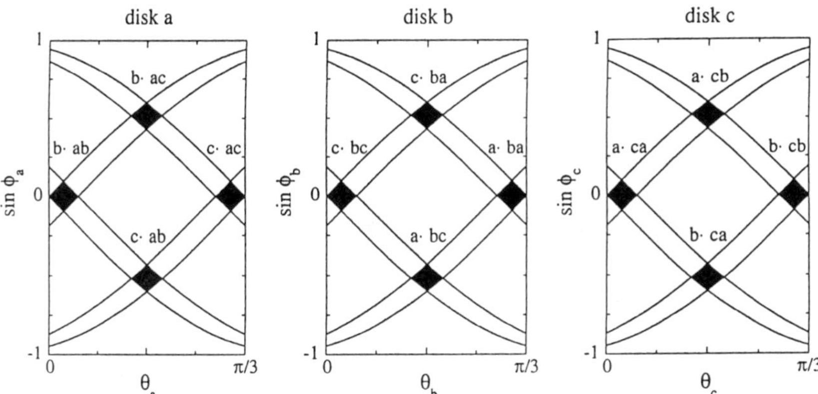

Figure 15. Horseshoe-type structures in the three-disk scatterer. The parameters θ_i are the positions of the collisions along the border of the disks, while the parameters ϕ_i are the angles between the incident ray and the normal at the point of collision. Note that the number of cells multiplies by a factor 2 at each new collision.

other in Eq. (3.43) with the constraint $\omega_n \neq \omega_{n+1}$. Moreover, if we use the threefold symmetry of the equilateral three-disk scatterer we can reduce the symbolic dynamics to the one of the Smale horseshoe.

A binary symbolic dynamics $\{+, -\}$ also exists in the two-electron atomic systems (like the hydrogen negative ion and the helium atom) in the collinear configuration where both electrons are on opposite sides of the nucleus [99, 100]. In these cases, a generating partition can be constructed from the structure of the trajectories issued from triple collisions of both electrons with the nucleus ($r_1 = r_2 = 0$). Triple collisions are highly unstable, and therefore very rare, but they nevertheless control the global behavior of these Coulomb systems.

4. *Generalizations*

The horseshoe is the simplest example of a symbolic dynamics. More complicated symbolic dynamics arise in a variety of situations. One generalization applies when the stretching of the domain D is followed by a more complicated folding, for instance, into three rather than two branches. In this case, we need three symbols $\{0, 1, 2\}$ instead of two so that the topological entropy is now equal to $\ln 3$. This situation arises in the Hamiltonian mapping of the above Example 6, which is relevant to the theory of the molecular transition complex in triatomic molecules, as we will describe in Section XII [21].

In the four-disk scatterer, an extra symbol is also necessary [121]. In this case, we have the four symbols labeling the disks but a similar constraint on two consecutive symbols exists that limits the topological entropy per collision to ln 3. Figure 16 shows the generating partition of such a symbolic dynamics. Further comments on this point will be given in Section XIII.

There are other generalizations of the symbolic dynamics if more than one constraint exists on the way the symbols follow each other. These constraints may be understood as grammatical rules in the construction of the symbolic sequences from the alphabet, that is, from the set of possible symbols $\omega \in \{0, 1, \ldots, M-1\}$ [118].

The preceding examples illustrate the ubiquity of the Smale horseshoe and of its variants in classical scattering systems where it appears naturally. The nonattractivity does not constitute a contradiction in scattering systems. However, Smale horseshoes also exist in bounded systems but as subsets of trajectories embedded in the whole phase space. In particular, the Birkhoff homoclinic theorem asserts the existence of Smale horseshoes in the vicinity of the homoclinic orbit as a subset of trajectories [2, 95]. Such a result may be very useful in our knowledge of the regularity of the motion but does not provide a complete description of all the trajectories. In contrast, in the extreme scattering situations described above, the Smale horseshoe contains the whole set of trapped orbits and provides a complete description of the trapped dynamics. These situations are referred to as full or complete chaos and they are the simplest ones to handle.

In the following section, we will see how the symbolic dynamics can be combined with information on the stabilities of the periodic orbits in order to characterize the classical chaos in a quantitative way.

D. Thermodynamic Formalism

We suppose that the trajectories of our system are in 1:1 correspondence with bi-infinite sequences [Eq. (3.42)] of a symbolic dynamics. Moreover, we may define an invariant measure μ, which we assume to be normalized. Let us denote by $\mu(\omega_0 \cdots \omega_{n-1})$ the probability of the particle belonging to one of the cells $\omega_0 \cdots \omega_{n-1}$. These probabilities are supposed to vanish when $n \to \infty$, otherwise there is a set of orbits that are not distinguished by the symbolic dynamics.

Now, we consider a physical observable $A(\mathbf{X})$ defined over phase space and we want to calculate its average. If we assume that the invariant probability measure μ is the natural ergodic measure, the ergodic theorem suggests

$$\langle A \rangle_\mu = \lim_{T\to\infty} \frac{1}{T} \int_0^T A(\mathbf{\Phi}^t \mathbf{X}_0)\, dt = \lim_{n\to\infty} \sum_{\omega_0 \cdots \omega_{n-1}} A_{\omega_0 \cdots \omega_{n-1}} \mu(\omega_0 \cdots \omega_{n-1}) \tag{3.46}$$

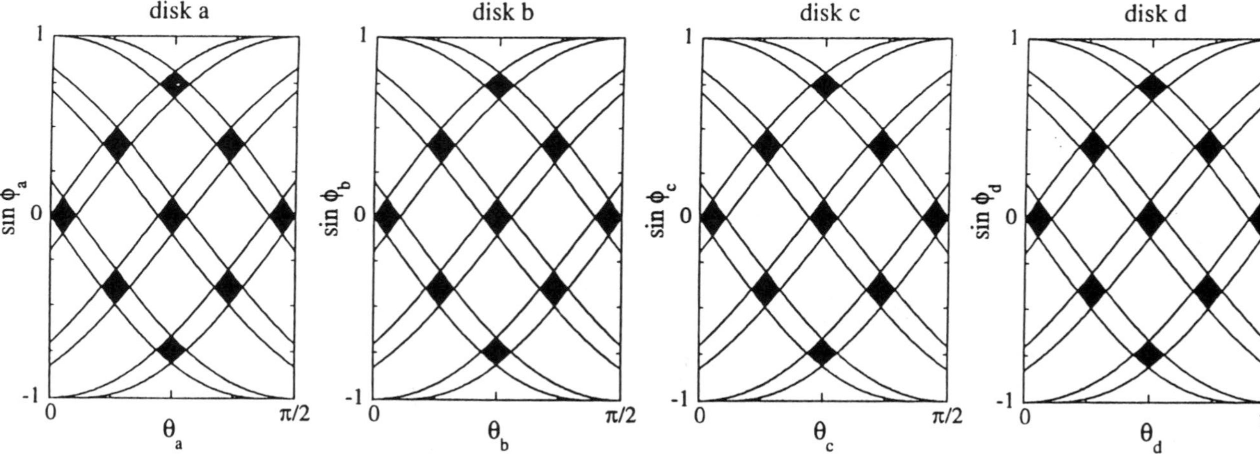

Figure 16. Same as in Fig. 15 for the four-disk scatterer. Here the number of cells multiplies by a factor 3 at each new collision.

where $A_{\omega_0\cdots\omega_{n-1}} = A(\mathbf{X})$ for some $\mathbf{X}$ belonging to the cell $C_{\omega_0\cdots\omega_{n-1}}$ (with a choice of the origin of time). In Eq. (3.46), the set of orbits entering the sum $\omega_0 \cdots \omega_{n-1}$ is composed of the orbits such that the time they take to visit the cells successively from ω_0 till ω_{n-1} is contained between T and $T + \Delta T$. For chaotic systems, the constant ΔT disappears after the limit is taken.

Among the different physical observables, the Lyapunov exponents measure the exponential separation between nearby trajectories, and they play a special role because they can be used to define the invariant measure. Indeed, the local instability of the dynamics in each cell of phase space determines the time spent on average in each of these cells. The more unstable the dynamics in the cell is, the more rapidly the particle escapes from the cell, and the smaller is the probability weight of the cell. The definition [Eq. (3.9)] of the Lyapunov exponents suggests the following definition for stretching factors associated with the action of the dynamics on each cell of the partition

$$\Lambda_{\omega_0\cdots\omega_{n-1}} = \operatorname*{Inf}_{\mathbf{X}_0 \in C_{\omega_0\cdots\omega_{n-1}}} \operatorname{Sup}_{\mathrm{e}} |\mathbf{M}(T; \mathbf{X}_0) \cdot \mathbf{e}| \tag{3.47}$$

The maximum average Lyapunov exponent can then be expressed by Eq. (3.46) in terms of the logarithms of the stretching factors.

In mechanical systems with two degrees of freedom, the maximum Lyapunov exponent coincides with the only positive Lyapunov exponent when it exists. In this case, it is possible to define invariant measures in terms of the stretching factors according to [122–124]

$$\mu_\beta(\omega_0 \cdots \omega_{n-1}) = \frac{(\Lambda_{\omega_0\cdots\omega_{n-1}})^{-\beta}}{\Sigma_{\omega_0\cdots\omega_{n-1}} (\Lambda_{\omega_0\cdots\omega_{n-1}})^{-\beta}} \tag{3.48}$$

with $\mu_\beta(\omega_0 \cdots \omega_{n-1})$ being non-zero if the transitions $\omega_0 \cdots \omega_{n-1}$ are allowed and otherwise zero. These invariant measures depend continuously on the parameter β. They are similar to the Gibbs measures of statistical mechanics if we assimilate the energies with the logarithms of the stretching factors. This analogy is at the origin of the development of the thermodynamic formalism [122–124]. In this way, Ruelle introduced the topological pressure, which we define here by

$$P(\beta) = \lim_{T\to\infty} \frac{1}{T} \ln \sum_{\substack{\omega_0\cdots\omega_{n-1} \\ T < T_{\omega_0\cdots\omega_{n-1}} < T+\Delta T}} (\Lambda_{\omega_0\cdots\omega_{n-1}})^{-\beta} \tag{3.49}$$

where the sum extends over the allowed strings $\omega_0 \cdots \omega_{n-1}$ [124]. This

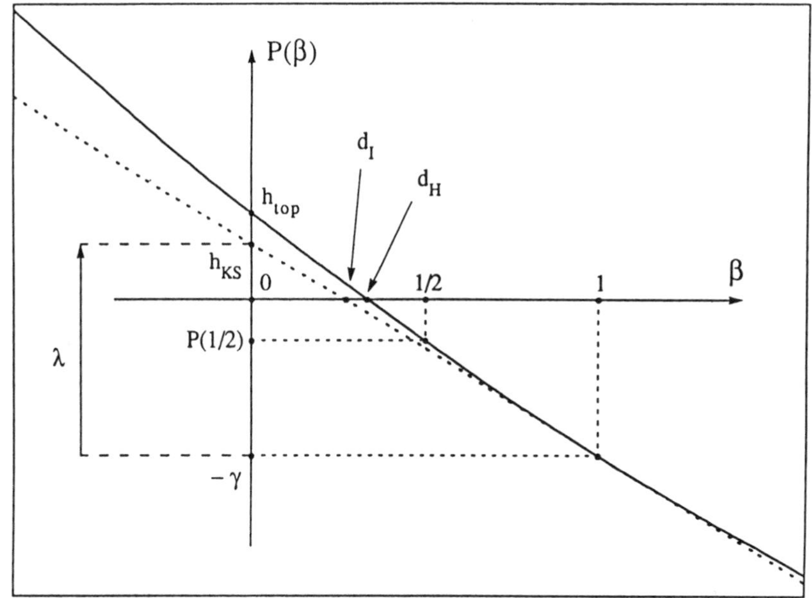

Figure 17. Schematic representation of the Ruelle pressure function and several of the characteristic quantities of chaos.

function is known to be convex and is schematically plotted in Fig. 17 for a typical hyperbolic system. The different characteristic quantities of chaos can be obtained by geometric construction around the pressure function. If we define the Kolmogorov–Sinai (KS) entropy per unit time and the average Lyapunov exponent associated with the invariant measure μ_β, we have the general properties

$$\lambda(\mu_\beta) = -P'(\beta) \tag{3.50}$$

$$h_{\mathrm{KS}}(\mu_\beta) = \beta\lambda(\mu_\beta) + P(\beta) \tag{3.51}$$

where the prime denotes the derivative with respect to β [4, 125, 126].

Among the different measures μ_β, a special role is played in classical mechanics by the natural measure with $\beta = 1$, where the phase-space cells have probability weights that are inversely proportional to the stretching factors. The standard KS entropy and average Lyapunov exponent are defined for this natural measure: $h_{\text{KS}} = h_{\text{KS}}(\mu_1)$, $\lambda = \lambda(\mu_1)$. Moreover, if the system is open and of the scattering type, the classical escape rate is given by

$$\gamma = -P(1) \tag{3.52}$$

According to Eq. (3.51), we obtain a generalization of the Pesin formula to open systems under the form [125]

$$h_{\text{KS}} = \lambda - \gamma \tag{3.53}$$

The KS entropy per unit time is the amount of information we need to record per unit time in order to reconstruct the trajectory of the system without ambiguity. In this regard, the KS entropy characterizes dynamical randomness. In closed hyperbolic systems, the exponential separation of nearby trajectories correlates entirely with randomness so that the KS entropy is then equal to the Lyapunov exponent. On the other hand, in open systems, the exponential separation contributes in part to the escape of particles so that dynamical randomness is reduced, as described by Eq. (3.53) [4].

The topological entropy is given by the pressure at $\beta = 0$, such that the sum in Eq. (3.49) becomes a simple counter of the number of periodic orbits [119]

$$h_{\text{top}} = P(0) \tag{3.54}$$

In general, the intersections of a line with the set of unstable manifolds of the repeller is a multifractal that can be characterized by generalized dimensions. These latter are calculated as the following root of the pressure function [126]

$$P[q + (1-q)d_q] = qP(1) \tag{3.55}$$

with values between 0 and 1. The partial Hausdorff dimension is given when $q = 0$, i.e., by the zero of the pressure

$$P(d_{\text{H}}) = 0 \tag{3.56}$$

while the partial information dimension $d_{\text{I}} = d_1$ is obtained by differentiation of Eq. (3.55) with respect to q according to Young's formula [126]

$$d_{\text{I}} = \frac{h_{\text{KS}}}{\lambda} \tag{3.57}$$

The topological pressure [Eq. (3.49)] is also known to be the leading pole $s = P(\beta)$ of the following Ruelle zeta function [124, 127]

$$\zeta_\beta(s) = \prod_p \left[1 - \frac{\exp(-sT_p)}{|\Lambda_p|^\beta}\right]^{-1} \tag{3.58}$$

where the product extends over the prime periodic orbits of period T_p and of stability eigenvalue $|\Lambda_p| > 1$. Equation (3.58) provides a very powerful method for accurate determination of the characteristic quantities of chaos. Moreover, let us mention here that a special role is played in semiclassical quantization by the pressure at the value $\beta = \frac{1}{2}$ [114].

After this review of classical chaos, we begin the presentation of the semiclassical theory of quantization.

IV. TIME DOMAIN: TRACE OF THE PROPAGATOR

Our objective in this section is to develop a semiclassical method to calculate the real or complex energy eigenvalues $\{E_b, E_{cr}\}$. Since the trace of the Green operator is related by Eq. (2.4) to the trace of the evolution operator, we first need to calculate tr $\hat{K}(T)$.

A. Regularizations

First, we consider the trace of the propagator in bounded systems with a discrete energy spectrum where

$$\text{tr}\,\hat{K}(T) = \text{tr}\exp\left(-\frac{i}{\hbar}\hat{\mathcal{H}}T\right) = \sum_{n=0}^{N}\exp\left(-\frac{i}{\hbar}E_n T\right) \tag{4.1}$$

If the Hilbert space has a finite dimension N, the trace is well defined.

However, the Hilbert space is infinite dimensional in many systems so that the sum may diverge, particularly along the real time axis. For this reason, we extend the definition of the trace to the complex plane of the time $T = \text{Re}T + i\text{Im}T$. In the lower half-plane $\text{Im}T < 0$, the above sum is convergent. This regularization is equivalent to multiplying the propagator by the heat kernel at an inverse temperature $\beta = -\text{Im}T/\hbar$. In particular, the trace of the heat kernel, $\text{tr}\exp(-\beta\hat{\mathcal{H}})$, is well defined for every positive temperature that corresponds to the lower part of the imaginary time axis.

The singularities that are met on the real time axis depend on the system but they may be everywhere dense on the real axis. To illustrate the problems caused by the ubiquity of those singularities, we consider a particle in a box with periodic boundary conditions for which $E_n = (\hbar^2/2m)(2\pi n/L)^2$ with $n = 0, 1, 2, \ldots$. After a rescaling of the time $\tau =$

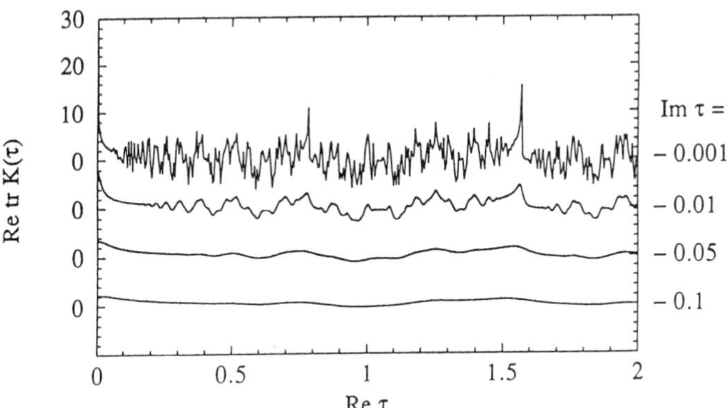

Figure 18. Trace of the propagator of a particle in a box with periodic boundary conditions versus Re τ for different values of Im $\tau = -0.001, -0.01, -0.05, -0.1$.

$T(\hbar/2m)(2\pi/L)^2$, the trace of the propagator becomes

$$\mathrm{tr}\ \hat{K}(\tau) = \sum_{n=0}^{\infty} \exp(-i\tau n^2) \tag{4.2}$$

which is of period 2π in the variable τ. Figure 18 shows this function versus Re τ for different values of Im τ. When Im $\tau \rightarrow 0$, singularities appear everywhere on the real time axis. In spite of this high irregularity, the function appears more regular in the vicinity of the origin, $\tau = 0$. The behavior at the origin can be obtained by replacing the sum by an integral,

$$\mathrm{tr}\ \hat{K}(\tau) \simeq \int_0^{+\infty} \exp(-i\tau n^2)\, dn = \frac{\sqrt{\pi}}{2} e^{-i\pi/4} \frac{1}{\sqrt{\tau}} \tag{4.3}$$

which shows that the function has a branch point at the origin where the function diverges like $\tau^{-1/2}$. The same divergence $\beta^{-1/2}$ appears in the trace of the heat kernel. We may expect many such branch points on the real time axis.

This example shows that the trace of the propagator may be a highly singular object to manipulate. Let us mention here that mathematicians have proposed to cure this problem by considering tr $\exp(-\alpha\hat{\mathscr{H}}^{\nu})$, which has a better behavior for appropriate exponents ν [128, 129]. In the present example the appropriate exponent would be $\nu = \frac{1}{2}$ so that the dense set of singularities may be transformed into a set of poles along the real axis of α. This procedure is called a homogenization of the spectrum, which is thereby transformed into a spectrum with a uniform distribution. A similar procedure is to consider the staircase function not only as a function of the energy but also of the inverse of the Planck constant $y = \hbar^{-1}$. The quantization in energy is then equivalent to a quantization of the inverse of the Planck constant [128]

$$N(E, y) = \sum_{n=0}^{\infty} \theta[E - E_n(y)] = \sum_{n=0}^{\infty} \theta[y - y_n(E)] \tag{4.4}$$

(This quantization of the y variable will be illustrated with the Morse oscillator in Section X.D.) A Laplace transform of the staircase function with respect to the variable y becomes a trace, i.e., $\Sigma_{n=0}^{\infty} \exp[-\alpha y_n(E)]$, which may be better behaved. Here we will not continue in this direction but we believe that such a procedure may turn out to be important.

Another method of regularization is to consider the trace of the propagator multiplied by a well-behaved observable $\hat{A}$, that is, $\text{tr}\,\hat{A}\hat{K}(T)$. Such a regularization is naturally carried out when physical observables, such as the photoabsorption cross-section [Eq. (2.11)], are considered. If the initial state is the ground state $|\phi_0\rangle$, the integrand of Eq. (2.11) becomes a time-dependent oscillation plus the trace of the propagator multiplied by $\hat{A} = \hat{\mathbf{d}}|\phi_0\rangle \cdot \langle\phi_0|\hat{\mathbf{d}}$. This trace turns out to be well defined because of the property of exponential damping of the ground state ϕ_0 at large distances, which introduces an effective cutoff at high quantum numbers n in Eq. (4.1). This regularization has an effect that is similar to the analytic continuation in the complex time plane.

In unbounded systems of the scattering type, there is another divergence that appears and that is due to the infinite volume of space. Indeed, the trace can be calculated in the position representation like

$$\text{tr}\,\hat{K}(T) = \int d\mathbf{q} K(\mathbf{q}, \mathbf{q}, T) \tag{4.5}$$

which is infinite, particularly for the propagator [Eq. (2.50)] of the free particle, as the integration volume is infinite. This divergence is cured by considering the difference between the propagator of the problem of interest and the propagator that describes the asymptotic dynamics of the particles at large separations [58]. Since both propagators have the same behavior at large distances their difference may have a well-defined trace

$$\text{tr}[\hat{K}(T) - \hat{K}_{\text{as}}(T)] \tag{4.6}$$

This regularization has already been used in Eqs. (2.10) and (2.14), where the problem has the same origin. Physically, it is natural to consider this difference since it amounts to comparing a property like a time delay between the system under study and a reference system [58, 64].

B. Short-Time Behavior

The behavior of the trace function around the origin $T = 0$ is of particular importance because it determines the Thomas–Fermi–Weyl–Wigner average level density. This behavior is derived in a systematic way thanks to the quasiclassical method where the heat kernel is expressed in the Weyl–Wigner representation [29, 67–69]. The heat kernel obeys an equation similar to Eq. (2.17) and can be expanded in powers of the Planck constant as in Eq. (2.18) if the potential is smooth. This expansion is standard and can be found elsewhere [130]. After the substitution $\beta = iT/\hbar$, we obtain

$$\begin{aligned} \operatorname{tr} \hat{K}(T) = \int \frac{d^f q d^f p}{(2\pi\hbar)^f} \exp\left(-\frac{i}{\hbar} \mathcal{H}_{\mathrm{cl}} T\right) \\ \times \left\{1 + \frac{T^2}{8} \partial_{\mathbf{q}}^2 V - i \frac{T^3}{24\hbar} [(\partial_{\mathbf{q}} V)^2 + (\mathbf{p} \cdot \partial_{\mathbf{q}})^2 V] + \cdots\right\} \end{aligned} \tag{4.7}$$

where the expansion appears as a series of powers of the time T rather than of the Planck constant. Equation (4.7) is convergent for $\operatorname{Im} T < 0$, as aforementioned. The short-time behavior is essentially determined by the classical Hamiltonian. If we assume that the potential increases in every direction as $V \sim |\mathbf{q}|^\alpha$ at large distances, the trace of the propagator will diverge at the origin as $1/T^{f(1/2+1/\alpha)}$.

C. Path Integral

The trace of the propagator can be expressed as a Feynman path integral by the same method as developed with Eqs. (2.51–2.56) as

$$\begin{aligned} \operatorname{tr} \hat{K}(T) &= \operatorname{tr} \hat{K}(T/N)^N \\ &= \int d\mathbf{q}_0 \langle \mathbf{q}_0 | \left[\exp\left(-\frac{i\hat{\mathcal{H}}T}{\hbar N}\right)\right]^N | \mathbf{q}_0 \rangle \\ &= \left(\frac{N}{2\pi i\hbar T}\right)^{Nf/2} \int d\mathbf{q}_0 \cdots d\mathbf{q}_{N-1} \exp\left[\frac{i}{\hbar} W(\mathbf{q}_0 \cdots \mathbf{q}_{N-1})\right] \end{aligned} \tag{4.8}$$

We remark that similar integrals are encountered in the evaluation of partition functions in statistical mechanics [79].

Contrary to the path integral of Section II, here the multiple integral extends over closed paths with $\mathbf{q}_0 = \mathbf{q}_N$, which is a condition imposed by the trace [19]. Therefore, the classical orbits that are selected by the stationary-phase condition are also closed: $\mathbf{q}(0) = \mathbf{q}(T)$. Two kinds of classical orbits satisfy this condition:

1. There are the equilibrium points for which $\mathbf{q}(t) = \mathbf{q}_s$ at all times $0 < t < T$ if $\partial_{\mathbf{q}} V(\mathbf{q}_s) = 0$. The fixed points occur at the critical energies $E_s = V(\mathbf{q}_s)$. In typical systems, they are isolated at the bottom of the potential well or at a saddle, but they may also be nonisolated in systems such as billiards, at $E_s = 0$.
2. There are the periodic orbits of period T for which $\mathbf{q}(t+T) = \mathbf{q}(t)$ at all times $0 < t < T$. The periodic orbits are multiples r of a prime periodic orbit p of period T_p so that $T = rT_p$. Since the prime period $T_p = T_p(E)$ generally varies with energy, we find a discrete set of periodic orbits at the different energies determined by $T = rT_p(E_{pr})$.

We already described these two kinds of orbits and their stability in Section III, where several examples were given, in particular in the period-energy diagrams. These diagrams will play an important role in the following considerations.

For both types of classical orbits, the same methods as developed in Section II for the semiclassical propagator will be applied in the following two sections.

D. Contribution of Equilibrium Points

The equilibrium points are characterized by their linear and nonlinear stabilities [71]. The linear stability is determined by the eigenvalues of the Hessian of the potential according to Eq. (3.26). Near an equilibrium point, the Feynman path integral [Eq. (4.8)] can be evaluated by the stationary-phase method of Section II. The action functional is expanded in a Taylor series in terms of the deviations with respect to the equilibrium point, $\xi^a = \{q_{in} - q_{is}\}_{n=0}^{N-1}$, as in Section II for the propagator, except that here we have an extra integral, since the path is closed due to the trace. We have a series of imaginary Gaussian integrals and their moments, as in Eq. (2.74), with $\int d^{Nf}\xi$ instead of $\int d^{(N-1)f}\xi$ and $W_{\mathrm{cl}} = -TV(\mathbf{q}_s) = -E_s T$. Moreover, here the matrix of the second derivatives of the action is the following $Nf \times Nf$ matrix

$$\mathbb{D} = \frac{1}{\Delta t}\begin{pmatrix} \mathbf{a} & -\mathbf{1} & 0 & 0 & 0 & \cdots & 0 & 0 & -\mathbf{1} \\ -\mathbf{1} & \mathbf{a} & -\mathbf{1} & 0 & 0 & \cdots & 0 & 0 & 0 \\ 0 & -\mathbf{1} & \mathbf{a} & -\mathbf{1} & 0 & \cdots & 0 & 0 & 0 \\ \vdots & \vdots & \vdots & \vdots & \vdots & \ddots & \vdots & \vdots & \vdots \\ 0 & 0 & 0 & 0 & 0 & \cdots & -\mathbf{1} & \mathbf{a} & -\mathbf{1} \\ -\mathbf{1} & 0 & 0 & 0 & 0 & \cdots & 0 & -\mathbf{1} & \mathbf{a} \end{pmatrix} + \mathcal{O}(\Delta t^2) \qquad (4.9)$$

where the $f \times f$ matrix $\mathbf{a}$ has the elements

$$a_{ij} = 2\delta_{ij} - \Delta t^2 \frac{\partial^2 V}{\partial q_i\, \partial q_j}(\mathbf{q}_s) \qquad (4.10)$$

The matrix [Eq. (4.9)] is similar to Eq. (2.75) except that we find elements in the corners of the matrix [Eq. (4.9)] and the diagonal submatrices are all identical because of the steadiness of the trajectory at the equilibrium point. The higher derivatives are given by Eq. (2.78) evaluated at the equilibrium point.

The imaginary Gaussian integrals and their moments can be evaluated by Eqs. (2.79) and (2.80) with $M = Nf$.

The determinant of the matrix $\mathbb{D}$ can be calculated as follows. Each submatrix [Eqs. (4.10)] can be diagonalized by the orthogonal transformation $\mathbf{O}$, which diagonalizes the Hessian of the potential. The elements of the transformed matrix $\tilde{\mathbb{D}}$ can then be gathered in N blocks $f \times f$, which have the same form as Eq. (4.9) but with scalars instead of matrices. The determinant of each block can then be calculated by recurrence to obtain

$$\det \mathbb{D} = \frac{1}{\Delta t^{Nf}} \prod_{k=1}^{f} (\Lambda_k + \Lambda_k^{-1} - 2)[1 + \mathcal{O}(\Delta t)] \qquad (4.11)$$

where $\Lambda_k = \exp(i\omega_k T)$ or $\exp(\lambda_k T)$ depending on whether the direction is stable or unstable. On the other hand, the inverse of the matrix $\mathbb{D}$ is given by the classical Green function solution of Eq. (2.73) along the trajectory $\mathbf{q}_{\text{cl}}(t) = \mathbf{q}_s$ and with periodic boundary conditions $\mathcal{G}_{ij}(t, t') = \mathcal{G}_{ij}(t + T, t')$.

Equation (2.71) can be solved in normal coordinates Q_i where the Hessian of the potential is diagonal. The Green function is then given by $\mathbf{G}(t, t') = \mathbf{O}^{\mathrm{T}}\mathbf{g}(t, t')\mathbf{O}$, where $\mathbf{O}$ is the orthogonal matrix that diagonalizes the Hessian and $\mathbf{g}$ is a diagonal matrix with the diagonal elements

$$g_k(t,t') = -\frac{\cos[\omega_k(t-t'+T/2)]}{2\omega_k \sin(\omega_k T/2)} \qquad (t<t')$$
$$g_k(t,t') = -\frac{\cos[\omega_k(t'-t+T/2)]}{2\omega_k \sin(\omega_k T/2)} \qquad (t>t') \tag{4.12}$$

The frequency ω_k has to be replaced by $\pm i\lambda_k$ for an unstable direction.

The contribution of the equilibrium points (e.p.) to the trace Eq. (4.8) can then be written as

$$\operatorname{tr} \hat{K}(T)\Big|_{\text{e.p.}} = \sum_s \exp\left(-\frac{i}{\hbar} E_s T\right) \prod_{j=1}^{f} \frac{1}{2i \sin(\omega_j T/2)} \exp \sum_{n=1}^{\infty} (i\hbar)^n C_n \tag{4.13}$$

where the first coefficient is

$$C_1(T) = \frac{1}{8} f_{kkll} \int_0^T dt\, g_k(t,t) g_l(t,t)$$
$$+ \frac{1}{8} f_{kkl} f_{lmm} \int_0^T dt\, dt' g_k(t,t) g_l(t,t') g_m(t',t')$$
$$+ \frac{1}{12} f^2_{klm} \int_0^T dt\, dt' g_k(t,t') g_l(t,t') g_m(t,t') \tag{4.14}$$

with the notation $f_{i_1\cdots i_n} = \partial^n V/\partial Q_{i_1}\cdots \partial Q_{i_n}$ for the derivatives of the potential in the normal coordinates. The coefficient [Eq. (4.14)] as well as the following ones can be represented by diagrams where a vertex with n legs is associated with each factor $-f_{i_1\cdots i_n}$ and a line with each Green function $g_k(t,t')$. Integrals are then performed over the times of the different vertices (see Fig. 4). Note the similarity with the case of the propagator in Section II. In Eq. (4.13), we used the important property that the result of the Taylor series appearing in Eq. (2.74) can be transformed into the exponential of a series involving only connected diagrams.

By using the Taylor expansion in a series of $x = \exp(-i\omega_j T)$, the leading term of Eq. (4.13) corresponding to the equilibrium point s can be rewritten into the following form

$$\operatorname{tr} \hat{K}(T)\Big|_{s,\text{leading}} = \sum_{n_1 \cdots n_f = 0}^{\infty} \exp\left\{-\frac{i}{\hbar} T\left[E_s + \sum_{j=1}^{f} \hbar\omega_j(n_j + \tfrac{1}{2})\right]\right\} \tag{4.15}$$

which is a sum of oscillators with frequencies equal to the energy eigenvalues at the harmonic approximation, including the zero-point energy.

If we consider the term corresponding to the first term of the correction C_1 in Eq. (4.14), we have

$$\operatorname{tr} \hat{K}(T)\Big|_{kkll} = \frac{i\hbar}{8} f_{kkll} \exp\left(-\frac{i}{\hbar} E_s T\right) \prod_{j=1}^{f} \frac{e^{-i\omega_j T/2}}{1 - e^{-i\omega_j T}}$$

$$\times \frac{T}{2i\omega_k 2i\omega_l} \frac{1 + e^{-i\omega_k T}}{1 - e^{-i\omega_k T}} \frac{1 + e^{-i\omega_l T}}{1 - e^{-i\omega_l T}} \tag{4.16}$$

We see that the imaginary exponentials appear in functions of the following types, which can also be expanded in Taylor series as done for the leading terms

$$\frac{1}{1-x} = \sum_{n=0}^{\infty} x^n \tag{4.17}$$

$$\frac{1}{(1-x)^2} = \sum_{n=0}^{\infty} (n+1)x^n \tag{4.18}$$

$$\frac{x}{(1-x)^2} = \sum_{n=1}^{\infty} n x^n \tag{4.19}$$

for $|x| < 1$. Using this property, we get for $k \neq l$

$$\operatorname{tr} \hat{K}(T)\Big|_{kkll} = -\frac{i\hbar T f_{kkll}}{8\omega_k \omega_l} \sum_{n_1 \cdots n_f = 0}^{\infty} \left(n_k + \frac{1}{2}\right)\left(n_l + \frac{1}{2}\right)$$

$$\times \exp\left\{-\frac{i}{\hbar} T\left[E_s + \sum_{j=1}^{f} \hbar\omega_j\left(n_j + \frac{1}{2}\right)\right]\right\} \tag{4.20}$$

We observe that this correction is proportional to T, which turns out to be a general property. Therefore, if the corrections are again brought into the exponential form, $\exp(-iTE/\hbar)$, corrections are added to the harmonic energy eigenvalues, $E = E_s + \Sigma_j \hbar\omega_j(n_j + \frac{1}{2})$, and we obtain the anharmonic corrections as will be shown explicitly in the following Section V.

E. Contribution of Periodic Orbits

Different kinds of periodic orbits are encountered whether they are real or complex, isolated or nonisolated, stable or unstable, bifurcating or not. Moreover, they may proliferate exponentially, as in the chaotic case, or subexponentially in more regular systems such as the circle billiard. As illustrated with the examples of Section III the spectrum of the lengths is of special importance because it contains information on the discrete or continuous degeneracy of the periodic orbits [111].

The integration is similar to the case of the equilibrium points. The integral [Eq. (4.8)] is evaluated around each periodic orbit by the stationary-phase method, which transforms the integral into an imaginary Gaussian integral and a series of its moments, as in Eq. (2.74). The type of the periodic orbit turns out to be essential when we expand the path action in Taylor series around the orbit. Indeed, the result of the Gaussian integrals are given in terms of the determinant and the inverse of the matrix $(\mathbb{D})_{ab} = W_{,ab}$ of the second variations of the path action according to Eqs. (2.79) and (2.80). The integration has a meaning under the condition that those quantities exist so that we must require $\det \mathbb{D} \neq 0$.

1. *Real, Unstable, Isolated Periodic Orbit*

Let us suppose that the periodic orbit is real, unstable, isolated, and far from any bifurcation. Since it is a one-dimensional solution of Newton's equations, one of the positions takes an arbitrary value, that is, q_{10}, while all the other positions are then determined by Newton's equations and the fixed value of T. Therefore, the separation of the path with respect to the classical solution is a $(Nf-1)$-component vector $\{\xi^a\} = \{q_{in} - q_i^{(\text{cl})}(n\Delta t)\}$ with the double index $a = (i, n)$ running over the values $n = 0, \ldots, N-1$ and $i = 1, \ldots, f$. The component with $i = 1$ and $t = n\Delta t = 0$ is now vanishing.

The trace Eq. (4.8) can then be expanded as in Eq. (2.74) except that the integral is carried out on [19]

$$\int d^{Nf}q \cdots = \sum_{p,r} \int dq_{10}\, d^{Nf-1}\xi \cdots \tag{4.21}$$

instead of $\int d^{(N-1)f}\xi$. In (4.21), the sum extends over all the prime periodic orbits p and their repetitions $r = 1, 2, \ldots$. The classical action is now evaluated for the periodic orbit $T = rT_p$

$$W_{\rm cl} = \oint_{p,r} L(\mathbf{q}, \dot{\mathbf{q}})\, dt \tag{4.22}$$

Here the second derivatives of the action form the $(Nf-1) \times (Nf-1)$ matrix

$$\mathbb{D} = \frac{1}{\Delta t} \begin{pmatrix} \tilde{\mathbf{A}}_0 & -\tilde{\mathbf{1}} & 0 & 0 & 0 & \cdots & 0 & 0 & -\tilde{\mathbf{1}} \\ -\tilde{\mathbf{1}}^{\rm T} & \mathbf{A}_1 & -\mathbf{1} & 0 & 0 & \cdots & 0 & 0 & 0 \\ 0 & -\mathbf{1} & \mathbf{A}_2 & -\mathbf{1} & 0 & \cdots & 0 & 0 & 0 \\ 0 & 0 & -\mathbf{1} & \mathbf{A}_3 & -\mathbf{1} & \cdots & 0 & 0 & 0 \\ \vdots & \vdots & \vdots & \vdots & \vdots & \ddots & \vdots & \vdots & \vdots \\ 0 & 0 & 0 & 0 & 0 & \cdots & -\mathbf{1} & \mathbf{A}_{N-2} & -\mathbf{1} \\ -\tilde{\mathbf{1}}^{\rm T} & 0 & 0 & 0 & 0 & \cdots & 0 & -\mathbf{1} & \mathbf{A}_{N-1} \end{pmatrix} + \mathcal{O}(\Delta t^2) \tag{4.23}$$

where the $f \times f$ matrices $\mathbf{A}_n$ have the elements

$$A_{ij}(n\Delta t) = 2\delta_{ij} - \Delta t^2 \frac{\partial^2 V}{\partial q_i\, \partial q_j} [\mathbf{q}_{\rm cl}(n\Delta t)] \tag{4.24}$$

with $i, j = 1, \ldots, f$. The matrix $\tilde{\mathbf{A}}_0$ is the same matrix as Eq. (4.24) but for $i, j = 2, \ldots, f$ and at $t = 0$. Here $\mathbf{1}$ denotes the $f \times f$ identity matrix, while $\tilde{\mathbf{1}}$ is the $(f-1) \times f$ matrix formed by removing the first line from the $f \times f$ identity matrix. The symbol T denotes the transpose. The corrections in Δt^2 are negligible since we will take the limit $\Delta t \to 0$ at the end of the calculation and keep only the expressions forming standard integrals over $T = N\Delta t$.

The higher derivatives [Eq. (2.78)] are now evaluated along the periodic orbit $\mathbf{q}_{\rm cl}(n\Delta t)$. The imaginary Gaussian integrals and their moments are calculated from Eqs. (2.79) and (2.80) with $M = Nf - 1$.

The determinant of Eq. (4.23) is nonvanishing and given by [131]

$$\det \mathbb{D} = -\frac{1}{\Delta t^{Nf}} \dot{q}_{01}^2\, \partial_E T \det[\mathbf{m}(T) - \mathbf{1}][1 + \mathcal{O}(\Delta t)] \tag{4.25}$$

where $\mathbf{m}(T)$ is the monodromy matrix obtained by linearizing the Poincaré symplectic return map over the full period T at the prime periodic orbit p as explained in Section III. Introducing the stability eigenvalues $\{\Lambda^{(k)}(T)\}$ of the periodic orbit p over time T we have

$$\det[\mathbf{m}(T) - \mathbf{1}] = \prod_{k=1}^{f-1} [\Lambda^{(k)}(T) + \Lambda^{(k)}(T)^{-1} - 2] \tag{4.26}$$

These stability eigenvalues refer to perturbations that are transverse to the flow and are related to the Lyapunov exponents by $|\Lambda^{(k)}(T)| = \exp(\lambda_k T)$.

By recovering the continuous time limit, the inverse of the second derivative matrix [Eq. (4.23)] can be expressed in terms of the classical Green function, which is a solution of Eq. (2.71). The fact that a line and a column are missing in Eq. (4.23) implies that the Green function must satisfy the following particular boundary conditions [19]

$$\begin{gathered}\mathcal{G}_{1j}(t_0, t') = \mathcal{G}_{1j}(t_0 + T, t') = 0 \\ \mathcal{G}_{ij}(t_0, t') = \mathcal{G}_{ij}(t_0 + T, t') \quad \text{for} \quad i = 2, \ldots, f \quad j = 1, \ldots, f\end{gathered} \tag{4.27}$$

where t_0 is the initial time corresponding to the arbitrary coordinate q_{10}.

When the results are gathered, the contribution of the periodic orbits to the trace of the evolution operator is given by [19]

$$\operatorname{tr} \hat{K}(T)\Big|_{\text{p.o.}} = \sum_{\substack{p,r \\ (T=rT_p)}} \frac{T_p \exp\left[\frac{i}{\hbar} rW_p - i\frac{\pi}{2} r\mu_p + i\frac{\pi}{4} \operatorname{sgn} \partial_E T_p + \sum_{n=1}^{\infty} (i\hbar)^n C_n(rT_p)\right]}{|2\pi\hbar r(\partial_E T_p) \det(\mathbf{m}_p^r - \mathbf{1})|^{1/2}} \tag{4.28}$$

where $W(T) = rW_p$. Together with the other phase coming from $\operatorname{sgn}(\partial_E T_p)$, the Maslov index μ_p is determined as in Section II by the number of negative eigenvalues of the second derivative matrix [Eq. (4.23)]. The index entering in Eq. (4.28) can be related to the Morse index ν_ℓ of the propagator Eq. (2.84). When we take the trace of the propagator Eq. (2.84), we identify the initial with the final positions, $\mathbf{q}_0 = \mathbf{q}$, and we perform an extra integral over $\mathbf{q}$ that can be evaluated with the stationary-phase method. This last integral selects the periodic orbits of classical dynamics. The extra stationary-phase integral introduces an extra phase that has to be added to the phase $\exp(-i\pi\nu_\ell/2)$ to give the phase in Eq. (4.28), namely, $\exp[-i\pi\mu_p/2 + i\pi \operatorname{sgn}(\partial_E T_p)/4]$ of the periodic orbit. The extra index is determined as before by counting the number of negative eigenvalues of the Hessian appearing in the last stationary-phase integral. The Maslov index will be discussed in detail in one of the following sections.

The corrections are given as a series in powers of $\hbar$ with coefficients C_n, which can be calculated in terms of the classical Green functions and the high derivatives of the potential along the periodic orbit. We have explicitly evaluated the first of these coefficients, which is given by [19]

$$\begin{aligned} C_1(T) = &\frac{1}{8}\frac{1}{T}\int_0^T dt_0 \int_{t_0}^{t_0+T} dt \frac{\partial^4 V(t)}{\partial q_i\, \partial q_j\, \partial q_k\, \partial q_l} \mathcal{G}_{ij}(t,t)\mathcal{G}_{kl}(t,t) \\ &+ \frac{1}{24}\frac{1}{T}\int_0^T dt_0 \int_{t_0}^{t_0+T} dt\, dt' \frac{\partial^3 V(t)}{\partial q_i\, \partial q_j\, \partial q_k} \frac{\partial^3 V(t')}{\partial q_l\, \partial q_m\, \partial q_n} \\ &\times [3\mathcal{G}_{ij}(t,t)\mathcal{G}_{kl}(t,t')\mathcal{G}_{mn}(t',t') + 2\mathcal{G}_{il}(t,t')\mathcal{G}_{jm}(t,t')\mathcal{G}_{kn}(t,t')] \end{aligned} \tag{4.29}$$

Diagrams are associated with the integrals defining the coefficients, as in Fig. 4 of Section II, except that here there is an extra time average $T^{-1}\int_0^T dt_0$ over the initial time t_0, which occurs in the boundary condition [Eq. (4.27)].

2. *Other Types of Periodic Orbits*

If the periodic orbit is nonisolated and forms a continuous family of dimension ν in the energy shell, the integration of tr $\hat{K}(T)$ must be carried out by leaving out ν among the integration variables:

$$\int d^{Nf}q \cdots = \sum_{p,r} \int \prod_{i=1}^{\nu} dq_{i0}\, d^{Nf-\nu}\xi \cdots \tag{4.30}$$

This is the case for systems with continuous symmetries, as in the circle billiard, as well as in systems without continuous symmetry, as the Sinai billiard or the Bunimovich stadium, where there appear continuous families of marginally unstable periodic orbits called bouncing ball orbits [132–133].

Considering the order of magnitude of their contribution to the trace of the propagator, the amplitudes of the nonisolated periodic orbits are of the order $\hbar^{-\nu/2}$ with $\nu > 1$, whereas the isolated periodic orbits have amplitudes with $\nu = 1$. We see that the nonisolated periodic orbits have larger contributions than the isolated ones.

At this point, let us also mention the complex periodic orbits. Those orbits may also contribute to the path integral evaluated in the limit $\hbar \to 0$. Their contribution may be necessary to guarantee the consistency of the semiclassical asymptotic series as shown for the quartic oscillator [129]. From the physical point of view, they are essential to explain the metastability of quantized states in several situations dominated by tunneling phenomena, as we will see in a following section.

F. Simple Examples

To illustrate the preceding technique, we consider two simple examples, which are the harmonic and the Morse oscillators described in Section III.

1. *Harmonic Oscillator*

The harmonic oscillator is a singular case in several respects. The trace of its propagator can be directly evaluated from Eq. (4.1)

$$\mathrm{tr}\,\hat{K}(T) = \frac{1}{2i\,\sin(\omega T/2)} \tag{4.31}$$

for $T \neq 2\pi r/\omega$. The same result can be obtained with the path integral Eq. (4.8), which is immediately an imaginary Gaussian integral without taking a Taylor expansion around its equilibrium point at $q=0$.

The trace [Eq. (4.31)] presents poles on the real time axis at $\mathrm{Re}\,T = 2\pi r/\omega$ with $r \in \mathbb{Z}$ where the trace diverges as $-i(-)^r/(T-2\pi r/\omega)$. The trace has an analytic continuation in the upper half-plane of the complex time T.

2. *Morse Oscillator*

The situation is more complicated for the Morse oscillator [Eq. (3.31)] because of the anharmonicity, as seen in Fig. 7. The period increases near the dissociation energy according to Eq. (3.33). The quantum eigenenergies are given by [98]

$$E_n = \hbar\omega_0\left(n+\frac{1}{2}\right) - \frac{(\hbar\omega_0)^2}{4D}\left(n+\frac{1}{2}\right)^2 \tag{4.32}$$

where $\omega_0 = \alpha\sqrt{2D}$ is the frequency of the harmonic oscillations near the equilibrium point and where the quantum number n takes the values

$$n = 0, 1, 2, \ldots, n_{\max} \qquad \text{with} \qquad n_{\max} = \mathrm{Int}\left(\frac{2D}{\hbar\omega_0} - \frac{1}{2}\right) \tag{4.33}$$

$\mathrm{Int}(\cdot)$ denoting the integer part.

At the semiclassical level, the trace of the propagator has two contributions: the first one from the equilibrium point $q=0$ and the second one from the periodic orbits. The contribution of the equilibrium point is of the form

$$\operatorname{tr} \hat{K}(T)\Big|_{\text{e.p.}} \simeq \frac{1}{2i \sin(\omega_0 T/2)} \tag{4.34}$$

up to higher order corrections due to anharmonicities near the equilibrium point [see Eqs. (4.16–4.20)].

On the other hand, the periodic orbits contribute by Eq. (4.28), but we have $\det(\mathbf{m}_p^r - \mathbf{1}) = 1$ for this one-degree-of-freedom system. The action W for the single periodic orbit p repeated r times can be obtained by $W = S - ET$ from the reduced action $S = \oint p dq$ at the energy E

$$S(E) = \frac{4\pi r}{\omega_0} D(1 - \sqrt{1 - E/D}) \tag{4.35}$$

The period [Eq. (3.33)] is obtained as $T = \partial_E S$. Accordingly, we have

$$W = \frac{4\pi r}{\omega_0} D - \left(\frac{2\pi r}{\omega_0}\right)^2 \frac{D}{T} - DT \tag{4.36}$$

There exist thresholds where higher repetitions r occur at times $T = 2\pi r/\omega_0$. A new term then enters the periodic-orbit (p.o.) contribution. In principle, the behavior of the trace should be replaced by uniform approximations at these thresholds. We obtain

$$\operatorname{tr} \hat{K}(T)\Big|_{\text{p.o.}} \simeq \left(\frac{4\pi D}{\hbar \omega_0^2 T}\right)^{1/2} \sum_{r=1}^{r_{\max} = \operatorname{Int}(\omega_0 T/2\pi)} \times \exp\left\{\frac{i}{\hbar}\left[\frac{4\pi r D}{\omega_0} - \left(\frac{2\pi r}{\omega_0}\right)^2 \frac{D}{T} - DT\right] - i\pi r + i\frac{\pi}{4}\right\} \tag{4.37}$$

The semiclassical approximation to the trace of the propagator is therefore given by

$$\operatorname{tr} \hat{K}(T) = \operatorname{tr} \hat{K}(T)\Big|_{\text{e.p.}} + \operatorname{tr} \hat{K}(T)\Big|_{\text{p.o.}} \tag{4.38}$$

The preceding formulas reproduce the oscillations in the trace of the propagator very well, as seen in Fig. 19 on the real time axis. The short-time behavior is taken into account by the first pole of the equilibrium point contribution at $T = 0$. Moreover, the equilibrium point contribution is at the origin of the alternating peaks in the trace but this

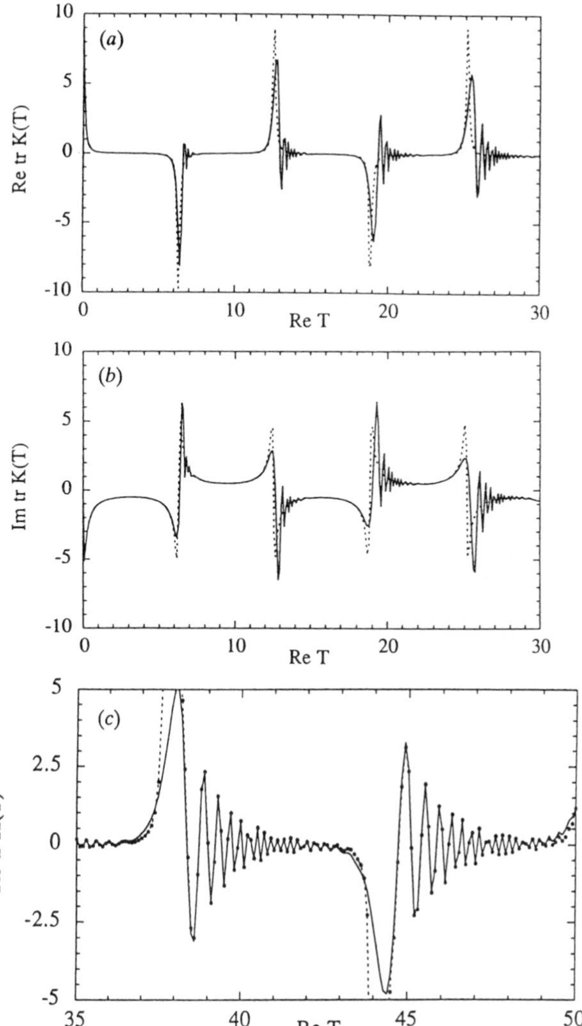

Figure 19. Trace of the propagator of the Morse oscillator [Eq. (3.31)] with $\hbar = 0.25$, $\omega_0 = 1$, and $D = 50$ versus Re T for Im $T = -0.1$, and comparison with its semiclassical approximation [Eqs. (4.34–4.38)]. (*a*) and (*b*) show the real and imaginary parts of the exact trace (solid line) compared with the equilibrium-point contribution [Eq. (4.34)] to the trace (dashed line), which indicates that the peak at the origin as well as the following peaks are due to the equilibrium-point contribution although the extra high-frequency oscillations are not. (*c*) Exact trace (solid line) compared with the semiclassical trace [Eq. (4.38)] including both the contributions of the equilibrium point and of the periodic orbits (dashed line with dots), which shows that the extra oscillations are due to the periodic orbits. There is a remarkable agreement between the successive thresholds $T_r = 2\pi r/\omega_0$. In the vicinity of the thresholds, the semiclassical trace could be improved by a uniform approximation.

contribution cannot explain the oscillations on top of these peaks. These oscillations become more and more important as T increases due to anharmonicities in the oscillations. These anharmonicities are taken into account by the periodic-orbit contribution [Eq. (4.37)]. We find matching problems between the different contributions at the level of the thresholds. The equilibrium point produces a pole at each threshold $T = 2\pi r/\omega_0$ while the rth repetition of the real periodic orbit gives a discontinuity that, in principle, should be smoothed by uniform approximation. Nevertheless, we have neglected the anharmonic corrections entering in the equilibrium point contribution [Eq. (4.13)]. These corrections contribute to multiple poles at each threshold with a multiplicity increasing with the power of $\hbar$ of the correction. Therefore, we conclude that there must exist a cancelation mechanism between these singularities. To our knowledge, this problem is unsolved today in spite of the simplicity of the Morse oscillator. A similar behavior is expected around the equilibrium points in systems with more degrees of freedom. This example illustrates the fact that both equilibrium points and periodic orbits contribute to the trace of the propagator in the time domain.

We will now turn to the energy domain since our purpose is the calculation of the energy eigenvalues.

V. ENERGY DOMAIN: EQUILIBRIUM-POINT QUANTIZATION

The trace of the Green operator is given by the semisided Fourier transform [Eq. (2.4)] of the trace of the evolution operator, which we have obtained in Section IV. First, we will evaluate the semisided Fourier transform according to the principles of the theory of distributions in time and energy. The integration of a kernel like $\exp[-i(E_n - E)T/\hbar]$ then gives a distribution of the form $1/(E - E_n + i0)$. According to this method, the periodic orbits turn out to have a negligible contribution. In this case of resummation around the equilibrium points, we will speak about equilibrium point quantization.

A. Evaluation of the Integral

The semisided Fourier transform [Eq. (2.4)] is evaluated keeping the terms that oscillate linearly in time. Such terms appear in the contributions of the equilibrium points to the trace of the propagator. We observed for the example of the Morse oscillator in Section IV.F.2 that the short-time behavior at $T = 0$ is taken into account by the equilibrium point contribution [Eq. (4.13)]. It is true in general that the equilibrium point quantization includes the short-time behavior of the propagator, which can be obtained alternatively by the Weyl–Wigner method of Section IV.B.

Our purpose will be to derive the contribution of the Green operator to the trace by keeping all the $\hbar$-corrections in Eq. (4.13), which are due to the anharmonicities around the equilibrium points.

After obtaining the corrections in the form of Eq. (4.20), followed by a regrouping of these corrections into an exponential form, $\exp(-iTE/\hbar)$, we turn to the energy domain and calculate the trace of the Green operator [Eq. (2.4)]. The leading term [Eq. (4.15)] of Eq. (4.13) contributes by the familiar expression

$$\operatorname{tr} \hat{G}(E)\bigg|_{s,\text{leading}} = \sum_{n_1 \cdots n_f = 0}^{\infty} \frac{1}{E - E_s - \Sigma_{j=1}^{f} \hbar\omega_j(n_j + \frac{1}{2})} \tag{5.1}$$

where we recognize the usual energy levels of a set of harmonic oscillators. Note that ω_j in Eq. (5.1) must be replaced by $-i\lambda_j$ for an unstable direction. The other possibility, $+i\lambda_j$, is excluded in the Green operator of the forward semigroup where the integral [Eq. (2.4)] extends over $T > 0$.

The term of Eq. (4.13) next to the leading one, which corresponds to Eq. (4.20), brings in anharmonic corrections that can be evaluated by noting that the factor T in Eq. (4.20) is equivalent to a derivative $(\hbar/i)(\partial/\partial E)$ in front of the integral [Eq. (2.4)] defining the Green operator. Since the corrections can be transformed into an exponential, we conclude that the contribution of the equilibrium point $\mathbf{q}_s$ is of the form

$$\operatorname{tr} \hat{G}(E)\bigg|_{s} = \sum_{n_1 \cdots n_f = 0}^{\infty} \exp\left(\gamma_{n_1 \cdots n_f} \frac{\partial}{\partial E}\right) \frac{1}{E - E_s - \Sigma_{j=1}^{f} \hbar\omega_j(n_j + \frac{1}{2})} \tag{5.2}$$

so that the energy levels are shifted by the quantities $\gamma_{n_1 \cdots n_f}$ with respect to the eigenenergies of the harmonic oscillators. For instance, we have

$$\gamma_{n_1 \cdots n_f} = -\frac{\hbar^2 f_{kkll}}{8\omega_k \omega_l}\left(n_k + \frac{1}{2}\right)\left(n_l + \frac{1}{2}\right) \tag{5.3}$$

for the term [Eq. (4.20)].

Assuming that there is no degeneracy between the frequencies $\{\omega_k\}$, we obtain the eigenenergies as a series of the form

$$E = E_s + \Delta E_2 + \sum_{k=1}^{f} \hbar\omega_k(n_k + \tfrac{1}{2}) + \sum_{k \le l = 1}^{f} x_{kl}(n_k + \tfrac{1}{2})(n_l + \tfrac{1}{2}) + \mathcal{O}(\hbar^3) \tag{5.4}$$

where the coefficients x_{kl} and the constant ΔE_2 are functions of the frequencies and the derivatives f_{klm} and f_{kkll} of the potential,

$$x_{kk} = \frac{\hbar^2}{16\omega_k^2}\left[f_{kkkk} - \sum_{m=1}^{f} \frac{f_{kkm}^2(8\omega_k^2 - 3\omega_m^2)}{\omega_m^2(4\omega_k^2 - \omega_m^2)}\right] \tag{5.5}$$

$$x_{kl} = \frac{\hbar^2}{4\omega_k\omega_l} \times \left\{f_{kkll} - \sum_{m=1}^{f} \frac{f_{kkm}f_{llm}}{\omega_m^2} + \sum_{m=1}^{f} \frac{2f_{klm}^2(\omega_k^2 + \omega_l^2 - \omega_m^2)}{[(\omega_k + \omega_l)^2 - \omega_m^2][(\omega_k - \omega_l)^2 - \omega_m^2]}\right\} \tag{5.6}$$

$$\Delta E_2 = \sum_{k=1}^{f} \frac{\hbar^2}{64\omega_k^2}\left(f_{kkkk} - \frac{7f_{kkk}^2}{9\omega_k^2}\right) + \sum_{k\neq l=1}^{f} \frac{3\hbar^2 f_{kkl}^2}{64\omega_k^2(4\omega_k^2 - \omega_l^2)} - \sum_{k\neq l\neq m=1}^{f} \frac{\hbar^2 f_{klm}^2}{24[(\omega_k + \omega_l)^2 - \omega_m^2][(\omega_k - \omega_l)^2 - \omega_m^2]} \tag{5.7}$$

A similar result can be calculated by conventional perturbation theory or by Van Vleck contact transformation [134–136]. The same result with $\Delta E_2 = 0$ can be derived from the classical Birkhoff normal form of the Hamiltonian with the semiclassical substitution $(P_j^2 + \omega_j^2 Q_j^2)/2 \rightarrow \hbar\omega_j(n_j + \frac{1}{2})$. The term ΔE_2 may be interpreted as the contribution of the anharmonicities of the potential to the zero-point energy.

The result [Eqs. (5.4–5.7)] is valid for the bottom of a potential as well as for a saddle point with $(\pm i\omega, \pm\lambda)$. In the latter case, Eq. (5.4) gives the complex eigenenergies of the metastable states associated with the saddle point [137, 138]. A calculation shows that $\operatorname{Im} \Delta E_2 = 0$, $\operatorname{Im} x_{11} = 0$, $\operatorname{Im} x_{22} = 0$, and $\operatorname{Re} x_{12} = 0$. Accordingly, the real energies and the half-widths of the resonant states are given by

$$\varepsilon_{nk} = E_s + \text{Re}\,\Delta E_2 + \hbar\omega(n+\tfrac{1}{2}) + \text{Re}\,x_{11}(n+\tfrac{1}{2})^2 + \text{Re}\,x_{22}(k+\tfrac{1}{2})^2 + \mathcal{O}(\hbar^3) \tag{5.8}$$

$$\Gamma_{nk} = \hbar\lambda(1+2k) - 2\text{Im}\,x_{12}(n+\tfrac{1}{2})(k+\tfrac{1}{2}) + \mathcal{O}(\hbar^3) \tag{5.9}$$

The resonances that dominate the dynamics at long time have the smallest half-width, that is, $k=0$. We see that all these dominant resonances have the same lifetimes in the harmonic approximation. The anharmonicities modify this result. The lifetimes may then take different values but the important point is that the resonances remain labeled by the same integers as in the harmonic approximation. In this sense, the assignment of quantum numbers remains possible, which we associate with the notion of effective separability.

B. Discussion

The effective separability exists as long as the Dunham expansion [Eq. (5.4)] is valid [139]. However, this expansion breaks down for different reasons.

Classically, the expansion [Eq. (5.4)] is equivalent to the existence of the Birkhoff normal form of the Hamiltonian [71, 140–141]. The normal form is established by using a nonlinear canonical transformation from the original coordinates **p**, **q** to a new set of coordinates **P**, **Q**. This nonlinear transformation can be expressed as a series that is known to diverge in the presence of homoclinic orbits [141]. Conversely, homoclinic orbits are at the origin of classical chaos so that there is a close relationship between the breakdown of the reduction to a Birkhoff normal form and the emergence of chaos. The reduction to the previous normal form often becomes impossible when the frequencies are close to rational ratios like 1/2, 1/3, 1/4, where the coefficients x_{kl}, y_{klm}, ... of the Dunham expansion may become large due to small denominators. It is under the same conditions of near commensurability between the frequencies that the KAM theorem no longer applies [2, 3, 71].

In quantum mechanics, the breakdown of the validity of the Dunham expansion may be understood in similar terms. When the frequencies are close to rationals, the denominators that define the coefficients [Eqs. (5.5) and (5.6)] are small and the perturbation series badly approximates the eigenenergies. This situation is known in molecular systems by the names of Fermi or Darling–Dennison interactions [134]. In these circumstances, off-diagonal terms of the Hamiltonian start to play a role that can no longer be neglected. Due to those off-diagonal terms, the energy levels are affected by Wigner repulsion so that the discrete spectrum acquires the properties of the eigenvalue spectra of random matrices [142].

In this section, we showed that the resummation on an equilibrium point is essentially equivalent to perturbation theory using as basis functions the eigenfunctions of the normal mode harmonic oscillators or, equivalently, to a reduction by quantum Van Vleck contact transformations. If we disregard the constant ΔE_2, this method is equivalent to the reduction to a Birkhoff normal form by a classical contact transformation followed by a semiclassical quantization of the normal mode action variables. These procedures extend the WKB method from 1F systems to systems with more than one degree of freedom. The method is successful even in the presence of mild classical chaos if the quasiperiodic motions remain dominant in phase space [143–146]. This success is based on the results of the KAM theorem, which guarantees that the motion remains partially regular in the vicinity of the equilibrium points where an effective separability holds.

In this chapter, we would like to emphasize the role of the periodic orbits. According to our previous discussion in Section IV, the equilibrium points are not the only classical trajectories that contribute to the trace formulas. The periodic orbits also contribute. Near an equilibrium point, we may understand that their role is negligible because a line of constant T rarely crosses a period, as a function of energy, near the energy of an equilibrium point in Figs. 6 and 7. From this point of view, the anharmonicities enhance the role of the periodic orbits. However, the anharmonicities do not necessarily preclude the existence of an effective separability [138].

The periodic orbits proliferate exponentially in the highly excited regimes where the classical motion becomes chaotic, which signals the transition to a regime where the equilibrium point quantization fails. The effective separability breaks down and the role of the periodic orbits becomes dominant in these regimes where a particular equilibrium point around which the phase space structure is ordered does not exist. It is the purpose of the following section to describe the periodic-orbit quantization.

VI. ENERGY DOMAIN: PERIODIC-ORBIT QUANTIZATION

The semisided Fourier transform (or Laplace transform) in Eq. (2.4) can also be evaluated by the stationary-phase method. According to this method, we find several critical points: the first one is the origin of time,

$T = 0$, which contributes to the Fermi–Thomas–Weyl–Wigner average level density. Next, we find the contributions from the periodic orbits where the phase $S = W(T) - ET$ is stationary. The equilibrium points do not contribute in the stationary-phase method because the phases of their contributions increase linearly with time. Accordingly, no critical point is selected if the stationary-phase condition is applied to terms of the form (4.20). Therefore, the equilibrium points do not contribute to the trace of the Green operator if $\hat{G}(E)$ is derived from $\hat{K}(T)$ by the stationary-phase method. Thus, we will refer to periodic-orbit quantization when resummation is carried out around the periodic orbits.

A. Contribution of the Paths of Zero Length

In the stationary-phase integration of Eq. (2.4), $T = 0$ is a critical point that produces a very important contribution to the trace of the Green operator. This contribution can be attributed to paths of zero length since the propagation is infinitesimal if $T = 0$. The real part of $\operatorname{tr} \hat{G}(E)$ is in general divergent but its imaginary part is finite for bounded systems.

According to Eq. (2.9), the imaginary part of $\operatorname{tr} \hat{G}(E)$ gives the average level density considered by Fermi, Thomas, Weyl, Wigner, and others. The level density is the derivative of the staircase function, defined by

$$N(E) = \sum_{n=0}^{\infty} \theta(E - E_n) \tag{6.1}$$

where $\theta(\cdot)$ is the Heaviside function. The smooth function obtained by averaging over the stairs of this function is the average staircase function $N_{\text{av}}(E)$, which can be obtained by integrating the average level density.

As for the equilibrium points and the periodic orbits, this contribution can be expanded in powers of the Planck constant [130]. For smooth potentials, the first terms of the average staircase function are given by

$$\begin{aligned} N_{\text{av}}(E) = & \int \frac{d^f q\, d^f p}{(2\pi\hbar)^f} \bigg\{ \theta(E - \mathcal{H}_{\text{cl}}) - \frac{\hbar^2}{8m} \partial_{\mathbf{q}}^2 V\, \partial_E^2 \theta(E - \mathcal{H}_{\text{cl}}) \\ & + \frac{\hbar^2}{24m} \Big[(\partial_{\mathbf{q}} V)^2 + m^{-1} (\mathbf{p} \cdot \partial_{\mathbf{q}})^2 V \Big]\, \partial_E^3 \theta(E - \mathcal{H}_{\text{cl}}) + \mathcal{O}(\hbar^4) \bigg\} \\ = & \frac{1}{\Gamma(1 + f/2)} \left(\frac{m}{2\pi\hbar^2} \right)^{f/2} \int_{V(\mathbf{q}) < E} [E - V(\mathbf{q})]^{f/2}\, d^f q + \mathcal{O}(\hbar^{-f+2}) \end{aligned} \tag{6.2}$$

This expression is obtained using the Weyl–Wigner representation of operators as functions over the $(\mathbf{q}, \mathbf{p})$ space (see Section II.D). In systems with Dirichlet, Neumann, or more complicated boundary conditions, further corrections appear that are in odd powers of $\hbar$ with respect to the leading term [147].

In two-dimensional billiards, the average staircase function of the Helmholtz equation, $(\Delta + \kappa^2)\psi = 0$, is given by a formula the leading term of which was derived by Weyl while the next-to-leading terms were obtained thereafter by several authors [148]

$$N_{\rm av}(\kappa) = \frac{A\kappa^2}{4\pi} \mp \frac{L\kappa}{4\pi} + \frac{1}{12\pi} \oint \frac{d\ell}{R(\ell)} + \sum_{\rm corners} \frac{\pi^2 - \theta_i^2}{24\pi\theta_i} + \frac{1-h}{6} + \mathcal{O}(\kappa^{-\eta} \ln \kappa) \tag{6.3}$$

with $0 < \eta \leq 1$. The quantity κ is the wavenumber related to the energy by $\kappa = \sqrt{2mE}/\hbar$ so that here the wavenumber plays the role of the inverse of the Planck constant, $\kappa \sim \hbar^{-1}$. The parameter A is the area of the billiard, L is the length of its perimeter, the parameters θ_i are the interior corner angles, $R(\ell)$ is the local radius of curvature (measured positive when concave), $\oint d\ell$ is an integral along the arc of perimeter, and h is the number of smooth holes in the domain that characterizes the connectivity of the billiard. The different signs correspond, respectively, to the Dirichlet and Neumann boundary conditions. We see that the hard walls introduce terms in the average staircase function that have odd powers of the Planck constant with respect to the leading term and that are sensitive to the boundary conditions. These new terms are not present in Eq. (6.2) for smooth potentials. We also remark that the corner term converges to the curvature term if we approximate the smooth billiard by a polygonal billiard and take the continuous limit.

Similarly, in three-dimensional billiards we have [12, 148]

$$N_{\rm av}(\kappa) = \frac{V\kappa^3}{6\pi^2} \mp \frac{S\kappa^2}{16\pi} + \frac{\kappa}{12\pi^2} \oint d\sigma \left(\frac{1}{R_1} + \frac{1}{R_2}\right) + \frac{\kappa}{24\pi} \sum_{\rm edges} \int \frac{\pi^2 - \theta_i(\ell)^2}{\pi\theta_i(\ell)} d\ell + \mathcal{O}(\kappa^0) \tag{6.4}$$

where V is the volume of the billiard, S is the area of its surface, R_1 and R_2 are the local two main radii of curvatures of the surface (measured positive when concave), $\oint d\sigma$ is a surface integral, $\theta_i(\ell)$ are the angles formed by the walls at the edges of the billiard, and $\int d\ell$ are integrals along the arcs of length of the edges. Balian and Bloch [12] also derived a formula for general boundary conditions where the wave function is proportional to its gradient on the surface of the billiard.

To illustrate the power of these formulas, Fig. 20 shows a nice agreement between the exact and average staircase functions for the circle billiard.

B. Contribution of Unstable and Isolated Periodic Orbits

According to the discussion in Section V, periodic orbits become important in nonseparable systems, especially when the system is classically chaotic. We will now continue the calculations of Section IV.E and obtain the contribution of the periodic orbits to the trace of the Green operator.

1. *Stationary-Phase Integration*

As in Section IV.E.1, we suppose that the periodic orbit is unstable and isolated. To go from the time to the energy domain, we have to carry out the semisided Fourier transform [Eq. (2.4)] to obtain the periodic orbit

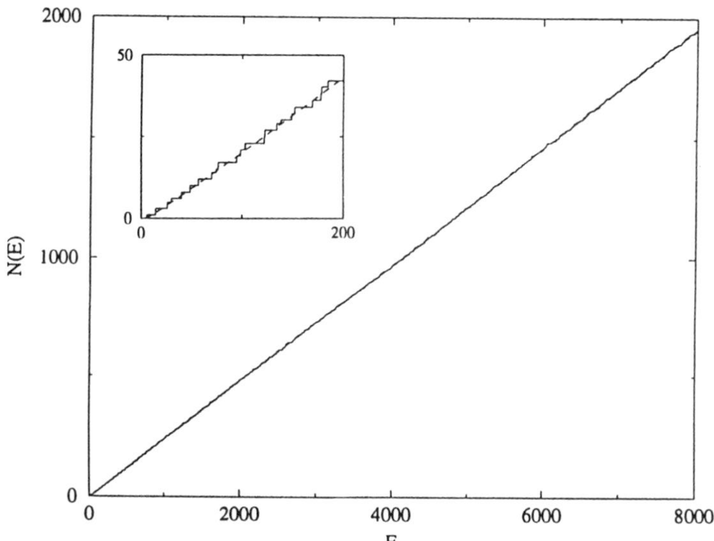

Figure 20. Staircase function of the circle billiard: comparison between the exact (solid line), and the average staircase function [Eq. (6.3)] (dashed line). Insert: initial part of the staircase function.

contribution

$$\operatorname{tr} \hat{G}(E)\Big|_{\text{p.o.}} = \frac{1}{i\hbar} \int_{\varepsilon}^{\infty} dT \exp\left(\frac{i}{\hbar} ET\right) \operatorname{tr} \hat{K}(T)\Big|_{\text{p.o.}} \tag{6.5}$$

where the lower limit $\varepsilon > 0$ is to remind us that the contribution from the critical point $T = 0$ has already been taken into account, as in Section VI.A.

By introducing the trace of the propagator given by Eq. (4.28), we have a new integral over time T with the oscillatory kernel $\exp[(i/\hbar)(W + TE)]$, which we evaluate once again by the stationary-phase method. The phase is stationary when $\partial_T W + E = 0$, that is, at the periods $T = rT_p$ [75]. In going from the time to the energy domain, it is convenient to introduce the reduced action

$$S(E) = \oint \mathbf{p} \cdot d\mathbf{q} = W(T) + ET \tag{6.6}$$

where the time T is eliminated for the energy E by the period-energy relations $T = rT_p(E)$ for each periodic orbit. Equation (6.6) is a Legendre transform from the action function $W(T)$ to the reduced action $S(E)$. From the previous stationary-phase condition $\partial_T W + E = 0$, we obtain $T = \partial_E S$ for the full period $T = rT_p$ and $T_p = \partial_E S_p$ for the fundamental period (with $S = rS_p$).

By Taylor expansion around the stationary periods, the integral is reduced to a sum of Gaussian integrals and its moments. Performing the integrals as before with Eq. (2.57), we obtain [19]

$$\operatorname{tr} \frac{1}{E - \mathscr{H}}\Big|_{\text{p.o.}} = \sum_{\substack{p,r \\ (T=rT_p)}} \frac{T_p \exp[\frac{i}{\hbar} rS_p - i\frac{\pi}{2} r\mu_p + \Sigma_{n=1}^{\infty} (i\hbar)^n \tilde{C}_n(rT_p)]}{i\hbar |\det(\mathbf{m}_p^r - \mathbf{1})|^{1/2}} \tag{6.7}$$

where the coefficients $\tilde{C}_n$ depend on the energy E, since they are evaluated at $T = rT_p(E)$. For the first of them, we get [19]

$$\tilde{C}_1(rT_p) = C_1(rT_p) - \frac{\partial_E^2 B_{pr}}{2r(\partial_E^2 S_p) B_{pr}} + \frac{(\partial_E^3 S_p)^2}{6r(\partial_E^2 S_p)^3} - \frac{\partial_E^4 S_p}{8r(\partial_E^2 S_p)^2} \tag{6.8}$$

where C_1 is given by Eq. (4.29) and

$$B_{pr}(E) = \frac{T_p}{|(\partial_E T_p) \det(\mathbf{m}_p^r - \mathbf{1})|^{1/2}} \tag{6.9}$$

The next coefficients can be obtained systematically.

2. *The Maslov Index*

It is time to come back to the Maslov index μ_p as it appears in Eq. (6.7). We have already discussed the Morse or Maslov index in Section II on the semiclassical formula for the propagator from $\mathbf{q}_0$ to $\mathbf{q}$ in a time T. The Morse index ν_ℓ associated with the classical orbit ℓ in Eq. (2.84) is given by the number of negative eigenvalues of the second derivative matrix [Eq. (2.75)]. The works of Morse have shown that it is equal to the number of conjugate points on the classical orbit ℓ as we explained in Section II [72].

We have not mentioned an alternative possibility to go from the time to the energy domain, which is to derive a semiclassical approximation of the Green function by taking the semisided Fourier transform [Eq. (2.4)] of the semiclassical propagator [Eq. (2.84)] [11, 149]. Each new stationary-phase integral adds a new term to the index that is given by the number of negative eigenvalues of the quadratic form appearing in the corresponding imaginary Gaussian. This principle is very general and we can see an illustration of this mechanism with the addition of the phase $-(\pi/4)\,\mathrm{sgn}\,\partial_E T_p$ when going from the time domain equation (4.28) to its energy domain correspondent [Eq. (6.7)]. Indeed, we recognize the sign of the quadratic form entering the imaginary Gaussian in the last time integral since $\mathrm{sgn}\,\partial_E T_p = \mathrm{sgn}\,\partial_E^2 S_p$. Accordingly, the semiclassical Green function would have a Morse index given by $\kappa_\ell = \nu_\ell + \sigma_\ell$, where σ_ℓ is the contribution to the index from the time integral.

In this context, we can take the trace of the semiclassical Green function, $G_{sc}(\mathbf{q}, \mathbf{q}_0, E)$, to obtain the trace formula [Eq. (6.7)] by identifying the initial with the final positions, $\mathbf{q}_0 = \mathbf{q}$, and performing an extra integral over $\mathbf{q}$, which can be evaluated with the stationary-phase method. This last integral selects the periodic orbits of the classical dynamics. The extra stationary-phase integral introduces an additional phase that has to be added to the Morse index κ_ℓ to give the Maslov index μ_p of the periodic orbit. In this way, the works by Gutzwiller [149] have shown that the Maslov index μ_p is composed of two contributions, which emerge from the derivation of the trace formula. The first contribution is associated with the energy-domain Green function, while the second contribution is related to the formation of the trace with respect to $G_{sc}(\mathbf{q}, \mathbf{q}_0, E)$ as we just mentioned. According to Robbins [150], the Maslov index may be evaluated as

$$\mu_p = \kappa_p - \tfrac{1}{2}\,\mathrm{sgn}(\mathbf{F}) \tag{6.10}$$

The first contribution counts the zero-crossings of the off-diagonal matrix element $(\partial \mathbf{q}/\partial \mathbf{p})$ of the stability matrix in the surface-of-section coordinate system $(\mathbf{q}, \mathbf{p})$. In the second contribution, $\text{sgn}(\mathbf{F})$ refers to the difference between the number of positive and negative eigenvalues of the $2(f-1) \times 2(f-1)$ matrix $\mathbf{F}$,

$$\mathbf{F} = \begin{pmatrix} \mathbf{c}\mathbf{a}^{-1} & (\mathbf{a}^{\mathrm{T}})^{-1} - \mathbf{1} \\ \mathbf{a}^{-1} - \mathbf{1} & -\mathbf{a}^{-1}\mathbf{b} \end{pmatrix} \tag{6.11}$$

where $\mathbf{a}$, $\mathbf{b}$, $\mathbf{c}$, and $\mathbf{d}$ are $(f-1) \times (f-1)$ matrices entering the definition of the $2(f-1) \times 2(f-1)$ monodromy matrix

$$\mathbf{m} = \begin{pmatrix} \mathbf{a} & \mathbf{b} \\ \mathbf{c} & \mathbf{d} \end{pmatrix} \tag{6.12}$$

The matrix $\mathbf{F}$ appears in the evaluation of the stationary-phase integral of the trace of the semiclassical Green operator, $\int d\mathbf{q} G_{\text{sc}}(\mathbf{q}, \mathbf{q}; E)$, which contains the phase $\exp[iS_\ell(\mathbf{q}, \mathbf{q}; E)/\hbar]$. The quadratic form of the stationary-phase integral involves the expression of the Poincaré mapping as an interlaced transformation between the variables $(\mathbf{q}_{n-1}, \mathbf{q}_n)$ and $(\mathbf{p}_{n-1}, \mathbf{p}_n)$ instead of the forward transformation from the variables $(\mathbf{q}_{n-1}, \mathbf{p}_{n-1})$ to $(\mathbf{q}_n, \mathbf{p}_n)$. This interlacing of the variables explains the complicated form of the matrix [Eq. (6.11)] [150].

Inspired by the works of Maslov, Creagh et al. [151] emphasized that the Maslov index is a canonical invariant of phase space (like the other quantities appearing in the trace formula), and thus is independent of the coordinate representation. In fact, the Maslov index can be defined topologically as the winding number of the invariant manifolds around a periodic orbit. The invariant manifolds are known to be Lagrangian manifolds, a property that applies to a torus in the case of an integrable system, as well as to the stable and unstable manifolds of an unstable orbit in a non-integrable system (see Section II.E.4). Caustics appear in coordinate space on the fold lines formed by the projection of a Lagrangian manifold from phase space to the coordinate space. The fold causes a singularity in the inverse mapping.

According to the concept of μ_p being a winding number associated with a Lagrangian manifold [151], one may explicitly identify the invariant manifolds and follow them in the surface of section transverse to the periodic orbit. For Hamiltonian flows, the surface of section can be constructed in a local coordinate system as proposed by Eckhardt and Wintgen (Section III.A) [92]. The number of intersections of the invariant manifolds with the momentum axis of the surface of section then corresponds to the Maslov index [150, 151]. This number will be even if the periodic orbit is hyperbolic, and odd if the periodic orbit is hyperbolic with reflection, as in the latter case, the orbit will not return to its initial position, but to a position related by inversion symmetry to the initial one.

C. Dynamical Zeta Functions

1. Level Density and Staircase Function

Let us now carry out the summation over all the unstable periodic orbits of the system that is assumed to be hyperbolic and possibly chaotic.

A direct consequence of Eqs. (6.2) and (6.7) is Gutzwiller's semiclassical formula for the level density [Eq. (2.9)] [11]

$$\begin{aligned} n(E) &= -\frac{1}{\pi} \operatorname{Im} \operatorname{tr} \frac{1}{E - \hat{\mathcal{H}} + i0} \\ &= n_{\text{av}}(E) + n_{\text{osc}}(E) \\ &= \int \frac{d^f q\, d^f p}{(2\pi\hbar)^f} \delta[E - \mathcal{H}_{\text{cl}}(\mathbf{q}, \mathbf{p})] + \mathcal{O}(\hbar^{-f+1}) \\ &\quad + \frac{1}{\pi\hbar} \sum_p \sum_{r=1}^{\infty} T_p \frac{\cos[\frac{r}{\hbar} S_p(E) - r\frac{\pi}{2}\mu_p]}{|\det(\mathbf{m}_p^r - \mathbf{1})|^{1/2}} + \mathcal{O}(\hbar^0) \end{aligned} \tag{6.13}$$

The first term is the aforementioned average level density while the second term is an oscillatory part due to the periodic orbits. Let us remember that the level density is a distribution in energy rather than a smooth function. Truncating the sum over the periodic orbits gives a smooth function. Therefore, we expect that the sum over the periodic orbits is not convergent to a function but could converge to a distribution in energy. This is the case for one-degree-of-freedom systems like the harmonic or the Morse oscillators. Integrating Eq. (6.13) with respect to energy gives the staircase function that is a discontinuous function of energy and, therefore, more regular

$$\begin{aligned} N(E) &= N_{\text{av}}(E) + N_{\text{osc}}(E) \\ &= \int \frac{d^f q\, d^f p}{(2\pi\hbar)^f} \theta[E - \mathcal{H}_{\text{cl}}(\mathbf{q}, \mathbf{p})] + \mathcal{O}(\hbar^{-f+1}) \\ &\quad + \sum_p \sum_{r=1}^{\infty} \frac{1}{\pi r} \frac{\sin[\frac{r}{\hbar} S_p(E) - r\frac{\pi}{2}\mu_p]}{|\det(\mathbf{m}_p^r - \mathbf{1})|^{1/2}} + \mathcal{O}(\hbar) \end{aligned} \tag{6.14}$$

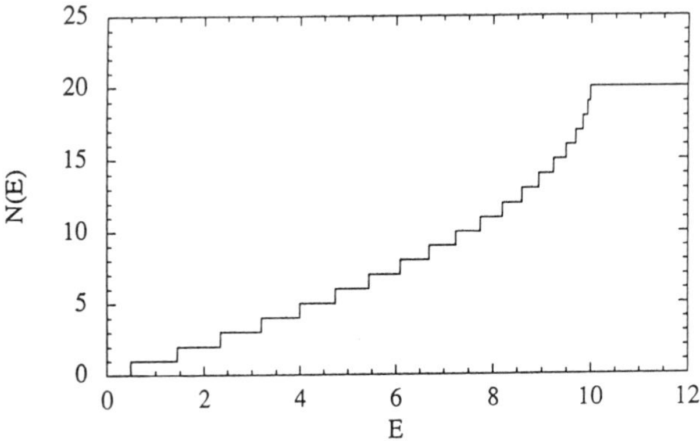

Figure 21. Staircase function of the Morse oscillator with $\hbar = 1$, $\omega_0 = 1$, and $D = 10$: comparison between the exact (solid line), and the semiclassical $\hbar^0$ approximation [Eq. (6.15)] (dashed line), which is so good that it is hidden by the curve of the exact staircase function.

For one-degree-of-freedom systems with one well ($f = 1$), we have

$$N(E) = \frac{S_p(E)}{2\pi\hbar} + \sum_{r=1}^{\infty} \frac{1}{\pi r} \sin\left[\frac{r}{\hbar} S_p(E) - r\frac{\pi}{2}\mu_p\right] + \mathcal{O}(\hbar) \tag{6.15}$$

for the single periodic orbit p. Figure 21 shows an excellent agreement of this formula with the exact quantum mechanical staircase function in the case of the Morse oscillator where $S_p(E) = (4\pi/\omega_0)[D - \sqrt{D(D-E)}]$.

2. *Selberg and Ruelle Zeta Functions*

In classically chaotic systems, the periodic orbits proliferate exponentially so that the sum over the periodic orbits is not guaranteed to converge everywhere for complex energies. In order to study these convergence properties and to calculate the energy levels or the resonances, we introduce the so-called Selberg and Ruelle [127, 152] zeta functions. With this approach, the corrections in $(i\hbar)^n$ are neglected, which is an approximation for general hyperbolic systems.

For simplicity, we assume that all the periodic orbits are unstable with $2(f-1)$ transverse stability eigenvalues over the fundamental period T_p: $\{\Lambda_p^{(k)}, 1/\Lambda_p^{(k)}\}$ $(k = 1, \ldots, f-1)$. The determinant involving the monodromy matrix can be calculated in the coordinates where the monodromy matrix is diagonal. In this case, we find the diagonal elements $\Lambda_p^{(k)} - 1$ and $\Lambda_p^{(k)-1} - 1$ with $k = 1, \ldots, f-1$. Therefore, we get

$$\det(\mathbf{m}_p^r - \mathbf{1}) = \prod_{k=1}^{f-1} (\Lambda_p^{(k)r} - 1)\left(\frac{1}{\Lambda_p^{(k)r}} - 1\right) \tag{6.16}$$

Assuming that none of the stability eigenvalues is on the unit circle, $|\Lambda_p^{(k)}| > 1$, we can expand in Taylor series [153] and obtain

$$\begin{aligned} |\det(\mathbf{m}_p^r - \mathbf{1})|^{-1/2} &= \prod_{k=1}^{f-1} \frac{1}{|\Lambda_p^{(k)}|^{r/2}(1 - \frac{1}{\Lambda_p^{(k)r}})} \\ &= \sum_{m_1,\ldots,m_{f-1}=0}^{\infty} \frac{1}{|\Lambda_p^{(1)}|^{r/2}\Lambda_p^{(1)\,m_1 r}|\Lambda_p^{(2)}|^{r/2}\Lambda_p^{(2)\,m_2 r}\cdots|\Lambda_p^{(f-1)}|^{r/2}\Lambda_p^{(f-1)\,m_{f-1} r}} \end{aligned} \tag{6.17}$$

We remark that all the terms inside the sum [Eq. (6.17)] are of the form of a certain quantity to the power r, which suggests the possibility of performing the summation over the repetition number r. Moreover, if we neglect the $\hbar$-corrections in Eq. (4.28), we have the property

$$\frac{T_p}{i\hbar} \exp\left(\frac{i}{\hbar} rS_p\right) = -\frac{1}{r}\frac{\partial}{\partial E} \exp\left(\frac{i}{\hbar} rS_p\right) \tag{6.18}$$

Accordingly, we obtain a series over the repetition number r of the form

$$-\sum_{r=1}^{\infty} \frac{x^r}{r} = \ln(1-x) \tag{6.19}$$

By gathering the previous results, we get

$$\operatorname{tr} \frac{1}{E - \hat{\mathcal{H}}}\bigg|_{\text{p.o.}} = \frac{\partial}{\partial E} \ln Z(E) + \mathcal{O}(\hbar^0) \tag{6.20}$$

where we introduced the Selberg zeta function [154] defined by

$$Z(E) = \prod_{m_1,\ldots,m_{f-1}=0}^{\infty} \prod_p \left[1 - \frac{\exp(\frac{i}{\hbar}S_p - i\frac{\pi}{2}\mu_p)}{\prod_{k=1}^{f-1} |\Lambda_p^{(k)}|^{1/2}\Lambda_p^{(k)\,m_k}}\right] \tag{6.21}$$

In Eq. (6.20), we note that the logarithmic derivative of $Z(E)$ is in $\hbar^{-1}$.

Thanks to the zeta function, we can rewrite the level density in the form

$$n(E) = n_{\rm av}(E) - \frac{1}{\pi} \operatorname{Im} \frac{\partial}{\partial E} \ln Z(E) + \mathcal{O}(\hbar^0) \qquad (6.22)$$

Since the energy levels or the scattering resonances are poles of the trace of the resolvent in Eq. (6.20) these poles should appear as zeros of the Selberg zeta function [Eq. (6.21)]. Indeed, if the zeta function has a zero at E_0 so that $Z(E) = A(E) \times (E - E_0)$, where $A(E_0) \neq 0$, then its logarithmic derivative has a pole at E_0 since

$$\frac{\partial}{\partial E} \ln Z(E) = \frac{\partial}{\partial E} [\ln(E - E_0) + \ln A(E)] = \frac{1}{E - E_0} + \frac{\partial_E A(E)}{A(E)} \qquad (6.23)$$

Therefore, we reach the periodic-orbit quantization condition, $Z(E) = 0$, which generalizes the Bohr–Sommerfeld and WKB conditions to systems with many isolated periodic orbits.

We may also consider systems where periodic orbits are isolated and unstable but have several stable directions for which the stability eigenvalues are $\Lambda_p^{(k)\pm 1}$ for $k = 1, \ldots, u$ and $\exp(\pm i2\pi\rho_l)$ for $l = 1, \ldots, s$ with $s + u = f - 1$. In these systems, the zeta function appearing in Eq. (6.20) is given by

$$Z(E) = \prod_{\substack{m_1,\ldots,m_u=0 \\ n_1,\ldots,n_s=0}}^{\infty} \prod_p \left\{ 1 - \frac{\exp[\frac{i}{\hbar} S_p - i\frac{\pi}{2}\tilde{\mu}_p - i2\pi \sum_{l=1}^{s} \rho_l(n_l + \frac{1}{2})]}{\prod_{k=1}^{u} |\Lambda_p^{(k)}|^{1/2} \Lambda_p^{(k)\, m_k}} \right\} \qquad (6.24)$$

The zeta function for one-degree-of-freedom systems is given by

$$Z(E) = 1 - \exp\left[\frac{i}{\hbar} S_p(E) - i\frac{\pi}{2}\mu_p\right] \qquad (6.25)$$

so that we recover the standard Bohr–Sommerfeld quantization condition $S_p(E) = \oint p\,dq = 2\pi\hbar(n + \mu_p/4)$.

In scattering systems, it may be interesting to consider the factorization of the Selberg zeta function into semiclassical Ruelle's zeta functions according to [114],

$$Z(E) = \prod_{m_1,\ldots,m_{f-1}=0}^{\infty} [\zeta_{m_1,\ldots,m_{f-1}}(E)]^{-1} \qquad (6.26)$$

In systems where the factorization [Eq. (6.26)] is valid, the poles of the Ruelle zeta functions give the scattering resonances.

3. *Convergence and the Topological Pressure*

In systems with two degrees of freedom, the semiclassical Ruelle zeta functions can be compared with the classical Ruelle zeta function [Eq. (3.58)] we considered in Section III.D on the thermodynamic formalism of dynamical systems [97]. In this section we mentioned the result that the leading pole of the zeta function [Eq. (3.58)] is equal to the topological pressure $s = P(\beta)$ defined by Eq. (3.49). This zeta function is analytic in the upper half-plane $\mathrm{Re}\, s \geq P(\beta)$ but has poles in the other half-plane. The same property holds if we multiply the weight given to each orbit by a phase as in

$$\zeta_\beta(s) = \prod_p \left[1 - \frac{\exp(-sT_p + i\varphi_p)}{|\Lambda_p|^\beta}\right]^{-1} \tag{6.27}$$

for arbitrary φ_p. In this complex case, $s = P(\beta)$ is no longer a pole of Eq. (6.27) but $\mathrm{Re}\, s = P(\beta)$ is still the threshold of analyticity. This general result can be applied to the semiclassical Ruelle zeta function

$$\zeta_m(E) = \prod_p \left[1 - \frac{\exp(\frac{i}{\hbar}S_p - i\frac{\pi}{2}\mu_p)}{|\Lambda_p|^{1/2}\Lambda_p^m}\right]^{-1} \tag{6.28}$$

of two-degree-of-freedom systems, or to systems with similar zeta functions. The existence of a threshold at a finite value of the classical rate s implies that the threshold is proportional to the Planck constant in the complex energy surface. Indeed, we know that the imaginary part of a complex energy $E = \varepsilon - i\Gamma/2$ is related to the lifetime by $\Gamma = \hbar/\tau$. Accordingly, we have

$$S_p(E) = S_p\left(\varepsilon - i\frac{\hbar}{2\tau}\right) = S_p(\varepsilon) - i\frac{\hbar}{2\tau}\,T_p(\varepsilon) + \mathcal{O}(\hbar^2) \tag{6.29}$$

Inserting in Eq. (6.28) and neglecting terms in $\hbar$ [consistently with the leading semiclassical approximation, Eq. (6.14)], we obtain the Ruelle zeta function with $s = -1/2\tau$, a parameter $\beta = 1/2 + m$, and some particular phase φ_p. According to the previous result on the poles of the Ruelle zeta function, the scattering resonances of the quantum system are localized in the complex energy surface such that [97, 114]

$$\frac{1}{\hbar}\,\mathrm{Im}\,E_n \le P\left(\frac{1}{2}+m\right) \qquad \text{with} \qquad m = 0, 1, 2, \ldots \tag{6.30}$$

in terms of the topological pressure $P(\beta)$. The pressure is a nonincreasing function of β so that the scattering resonances with the highest imaginary parts are given by the poles of the first semiclassical Ruelle zeta function [Eq. (6.28)] with $m = 0$. As a consequence, the pressure gives us a bound on the lifetimes of the scattering resonances $\{E_n = \varepsilon_n - i\Gamma_n/2\}$ in the semiclassical limit ($\hbar \to 0$)

$$\begin{aligned}\frac{\Gamma_n}{\hbar} = \frac{1}{\tau_n} \ge -2P\left(\frac{1}{2}\right) &= \lambda(\mu_{1/2}) - 2h_{\mathrm{KS}}(\mu_{1/2})\\ &\ge \lambda(\mu_1) - 2h_{\mathrm{top}}\end{aligned} \tag{6.31}$$

at energy ε_n. For complex energies above this threshold, the zeta function is absolutely convergent. Equation (3.51) shows how to express the pressure in terms of the KS entropy and the mean Lyapunov exponent but of the unusual invariant measure with $\beta = \frac{1}{2}$ [114]. The last bound follows from the geometry of the pressure function in Fig. 17 [155]. The bound [Eq. (6.31)] is effective as long as the partial Hausdorff dimension of the repeller is smaller that one half. Otherwise, the pressure becomes positive and the bound [Eq. (6.31)] is overwhelmed by the known result that $\Gamma_n \ge 0$. Related results have been proved in [156, 157].

Other results above the convergence of the zeta functions have been obtained recently by several groups [158–160].

4. *Approximating the Zeta Functions*

Different schemes have been proposed to approximate the zeta functions.

One approach is the cycle expansion of Cvitanović [161]. Let us assume that the system is topologically equivalent to a Smale horseshoe so that its symbolic dynamics has two symbols. The idea of the cycle expansion is to develop the product over prime periodic orbits defining the zeta functions as follows. For the case of a Ruelle zeta function,

$$\begin{aligned}\zeta_m^{-1} = \prod_p (1 - x t_p) &= 1 - x t_0 - x t_1 - x^2(t_{01} - t_0 t_1)\\ &\quad - x^3(t_{001} - t_0 t_{01} + t_{011} - t_{01} t_1) + \mathcal{O}(x^4)\end{aligned} \tag{6.32}$$

with $t_p = \exp(iS_p/\hbar - i\pi\mu_p/2)/(|\Lambda_p|^{1/2}\Lambda_p^m)$. The role of the fictitious parameter x is to help counting the period in terms of numbers of symbols, and truncating the series accordingly. It is set to $x=1$ afterwards. Besides the first three terms, the other terms appear as differences between the amplitude corresponding to a longer period and products of amplitudes corresponding to the different orbits of shorter periods that are topologically composing the longer period. If we suppose that the longer periods are exact combinations of the fundamental periods 0 and 1, then we have exact factorizations such as $t_{01}=t_0 t_1$ so that all the higher order terms vanish and the zeta function reduces to $\zeta_m^{-1}=1-t_0-t_1$. In general, we should not expect that this cancelation mechanism is exact, but rather that it is approximate such that the higher order terms are of small amplitudes and can be neglected at a large enough order.

A different approach by Gaspard and Rice [114] is based on the transfer operator. If we have a generating partition in a Poincaré surface of section, we may assume that the dynamics is piecewise linear in each cell of the partition if the cells are small enough. The amplitudes of transitions between these cells can be approximated semiclassically and gathered in a semiclassical transfer matrix

$$Q_{ab}^{(m)}=\frac{\exp(\frac{i}{\hbar}S_{ab}-i\frac{\pi}{2}\mu_{ab})}{|\Lambda_{ab}|^{1/2}\Lambda_{ab}^{m}}\,\Xi_{ab} \tag{6.33}$$

where a and b are strings of symbols labeling cells of the partition. The parameters S_{ab}, μ_{ab}, and Λ_{ab} are the contributions to the action, the Maslov index, and the stability eigenvalue coming from the segments of the classical orbits that go from cell a to cell b. The matrix element Ξ_{ab} is equal to 1 if the transition between the cells a and b is allowed by the phase-space geometry and 0 otherwise. The matrix $\{\Xi_{ab}\}$ is the transition matrix of the topological Markov chain induced by the dynamics on the adopted partition, and it defines a symbolic dynamics in the sense of Section III.C [119]. The topological Markov chain can be represented by a graph that is weighted by the elements of the transfer matrix [Eq. (6.33)]. The chain is characterized by the topological entropy per symbol, which is the growth rate of the set of periodic orbits with respect to their period, and which is given by the logarithm of the largest eigenvalue of the matrix $\{\Xi_{ab}\}$.

This method essentially corresponds to the truncation of the cycle expansion at a period p approximately equal to the length of the strings a of the partition. The approximate Ruelle zeta function can be written as

$$\zeta_m^{-1}=\det(\mathbf{1}-x\mathbf{Q}^{(m)}) \tag{6.34}$$

where x is the same fictitious parameter as in Eq. (6.32), which is set to

$x = 1$ at the end of the calculation. We observe that the Fredholm determinant Eq. (6.34) of the transfer matrix can be expanded in the form of the cycle expansion [Eq. (6.32)], where a cancelation mechanism automatically occurs, which truncates the expansion. Finally, the approximate Selberg zeta function of a two-degree-of-freedom system becomes

$$Z(E) = \prod_{m=0}^{\infty} \det[\mathbf{1} - \mathbf{Q}^{(m)}(E)] \tag{6.35}$$

where the integer m is the power of stability eigenvalues that appears in the denominator of Eq. (6.33). The presence of the product is related to the number of degrees of freedom. The proliferation of factors increases with the number of degrees of freedom, as shown in Eq. (6.21).

Let us suppose that the system is equivalent to a Smale horseshoe, which has a topological entropy equal to $\ln 2$ per symbol. Different approximations of the form of Eq. (6.33) can be constructed depending on the size of the cells. We can choose the cells as the 2^n curvilinear rectangles labeled by n-uples of symbols 0 and 1. In this case, the transfer matrix is a $2^n \times 2^n$ matrix. Each line of the transfer matrix has only two nonvanishing elements. For the line labeled by $a = \omega_{-m-1}\omega_{-m} \cdots \omega_{-2} \cdot \omega_{-1}\omega_0 \cdots \omega_{n-m-2}$, the two nonvanishing columns are $b = \omega_{-m} \cdots \omega_{-1} \cdot \omega_0\omega_1 \cdots \omega_{n-m-2}\omega_{n-m-1}$ with $\omega_{n-m-1} = 0$ or 1. For $n = 1$, we have the transfer matrix

$$\mathbf{Q}^{(m)} = \begin{matrix} & \begin{matrix} 0 & 1 \end{matrix} \\ \begin{matrix} 0 \\ 1 \end{matrix} & \begin{pmatrix} q_{00} & q_{01} \\ q_{10} & q_{11} \end{pmatrix} \end{matrix} \tag{6.36}$$

Here the approximate zeta function is equal to

$$\zeta_m^{-1} = 1 - xq_{00} - xq_{11} - x^2(q_{01}q_{10} - q_{00}q_{11}) \tag{6.37}$$

Now we have a shadowing mechanism that is very similar to the one working in the cycle expansion. Thus, if the stretching and folding dynamics is quasilinear, we would have $q_{00}q_{11} \simeq q_{01}q_{10}$. The diagonal matrix elements can be approximated by the values of the factors t_p associated with the periodic orbits going through the corresponding cells: $q_{00} \simeq t_0$ and $q_{11} \simeq t_1$. From this, we obtain the same approximate zeta function as given by truncating the cycle expansion [Eq. (6.32)] at the level of the fundamental orbits of unit period.

For a finer partition, $n=2$, we have the transfer matrix

$$\mathbf{Q}^{(m)} = \begin{array}{c} \\ 0\cdot 0 \\ 0\cdot 1 \\ 1\cdot 0 \\ 1\cdot 1 \end{array} \begin{array}{c} \begin{array}{cccc} 0\cdot 0 & 0\cdot 1 & 1\cdot 0 & 1\cdot 1 \end{array} \\ \begin{pmatrix} q_{000} & q_{001} & 0 & 0 \\ 0 & 0 & q_{010} & q_{011} \\ q_{100} & q_{101} & 0 & 0 \\ 0 & 0 & q_{110} & q_{111} \end{pmatrix} \end{array} \tag{6.38}$$

and the approximate zeta function

$$\zeta_m^{-1} = 1 - xq_{000} - xq_{111} - x^2(q_{010}q_{101} - q_{000}q_{111}) + \mathcal{O}(x^3) \tag{6.39}$$

Here, the correspondence with the cycle expansion is obtained with the identifications $q_{000} \simeq t_0$, $q_{111} \simeq t_1$, and $q_{010}, q_{101} \simeq \sqrt{t_{01}}$, and by observing that approximate cancelations occur at the orders $x^3, x^4, \ldots$. In this case, the approximate zeta function is equivalent to the truncation of the cycle expansion (6.32) between the orders x^2 and x^3. In this way, successive approximations can be obtained in the transfer matrix approach and shown to be equivalent to those of the cycle expansion approach. A transfer matrix built on a partition with 2^n cells contains 2^{n+1} nonvanishing elements. Such a transfer matrix corresponds to the inclusion of all the fundamental periodic orbits with a period less than n in the cycle expansion, which represent a set of about $2^n/(n \ln 2)$ periodic orbits. The advantage of the cycle expansion is the high precision that is available in favorable systems, but the search for periodic orbits may be tedious. The advantage of the transfer matrix approach is that we can take an arbitrary partition suitable for numerical purposes and automatize the construction of the transfer matrix.

At each level of approximation, the transfer matrix can be represented by a weighted graph, as shown in Fig. 22. The vertices represent the classical states (i.e., the partition cells), while the lines correspond to the transitions and are weighted by the transition amplitudes given by the elements of Eq. (6.33).

Since the work by Gaspard and Rice, the transfer matrix approach has been applied to a number of systems. In particular, Bogomolny developed an extension of the transfer matrix method to bounded systems [162, 163].

5. *Bounded Systems*

Instability is a natural property of scattering systems. The large Lyapunov exponents of the classical repeller introduce a damping corresponding to the finite lifetimes of the scattering resonances. Convergence is therefore

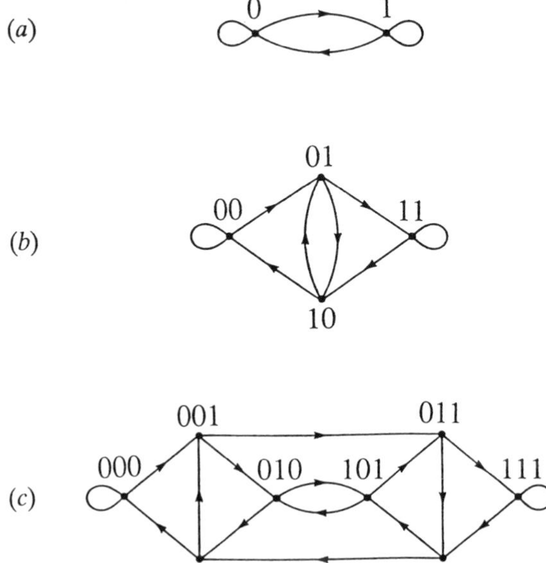

Figure 22. Graphs of the topological Markov chains corresponding to the transfer matrices associated with a binary symbolic dynamics $\{0, 1\}$. (*a*) Graph of the transfer matrix [Eq. (6.36)] based on single-symbol strings. (*b*) Graph of the transfer matrix [Eq. (6.38)] based on two-symbol strings. (*c*) Graph of the transfer matrix based on three-symbol strings.

favored in scattering systems. On the other hand, the pressure function of classically chaotic bounded systems is always positive, which introduces a barrier that prevents us from reaching the real energy eigenvalues. Several schemes have been proposed to remedy this difficulty.

One proposition is to perform a convolution of the level density or the staircase function with a Gaussian to achieve regularization [164, 165]. This Gaussian smoothing justifies the use of the trace formula (6.14) if the sum only includes periodic orbits up to a period determined by the average level density [164]. In the same context, a related method was proposed where the eigenvalues are the zeros of the functional equation [166]

$$\mathrm{Re}[e^{-i\pi N_{\mathrm{av}}(E)} Z(E)] = 0 \tag{6.40}$$

$N_{\mathrm{av}}(E)$ being the average staircase function. This method has been applied to the anisotropic Kepler problem as well as to the closed three-disk billiard [167, 168].

6. *Inclusion of the $\hbar$-Corrections*

For the level density or the staircase function, we can always take into account the $\hbar$-corrections by calculating the coefficients $\tilde{C}_n(rT_p)$ for each periodic orbit and its repetitions [19].

However, methods using the zeta function in Selberg's form [Eq. (6.21)] are neglecting the higher corrections in the Planck constant. The inclusion of these corrections will in general modify the form of the function $Z(E)$ according to the dependency of the corrections C_n on the repetition number r. In hyperbolic systems, the classical Green functions $\mathcal{G}_{ij}(t, t')$ feature the clustering property, that is, they decrease to zero when $|t - t'| \to \infty$ together with $T \to \infty$. The decrease is often exponential and controlled by the Lyapunov exponents of the periodic orbit. According to this property, the coefficients $C_n(rT_p)$ will in general depend on the repetition number r as $r^\gamma \exp(-\lambda_k rT_p)$. The consequence of this behavior is that the corrections in powers of $\hbar$ remain of a simple and controllable form at the level of the first Ruelle zeta function in the product [Eq. (6.26)]. With the corrections in $\hbar^n$, this becomes [20, 169]

$$\zeta^{\mathrm{c}}_{0\ldots0}(E) = \prod_p \left\{1 - \frac{\exp[(i/\hbar)S_p(E) - i(\pi/2)\mu_p + \Sigma_{n=1}^{\infty} (i\hbar)^n \tilde{c}_{np}(E)]}{\Pi_{k=1}^{f-1} |\Lambda_p^{(k)}(E)|^{1/2}}\right\}^{-1} \tag{6.41}$$

with

$$\tilde{c}_{np}(E) = \lim_{r\to\infty} \frac{1}{r} \tilde{C}_n[rT_p(E)] \tag{6.42}$$

where $\tilde{C}_n$ are the coefficients of the series in $(i\hbar)^n$ appearing in Eq. (6.7). The first of them is given explicitly by Eq. (6.8) with Eq. (4.29).

However, the next Ruelle zeta functions are not modified in a simple way by the corrections in $\hbar^n$ so that we do not know the corrected expression for the Selberg zeta function. Nevertheless, in a number of scattering systems, the scattering resonances with the longest lifetimes are given by the poles of the first Ruelle zeta function as explained above. For these scattering systems, which appear in many applications, Eq. (6.41) may be sufficient to obtain the resonances.

An example of resummation in the two- and three-disk scatterers will be presented in Section IX, where we illustrate the above theoretical methods in the case of billiards [19, 20].

D. Complex Periodic Orbits

1. *Tunneling in 1F Systems*

Complex periodic orbits arise in different contexts where a process is possible according to quantum mechanics but is classically forbidden [170, 171]. The purely wave mechanical process can then be taken into account by a complex classical orbit, which may join two regions of phase space separated by a barrier. Semiclassically, the classically forbidden processes are described in terms of complex periodic orbits.

The best known example is tunneling through a potential energy barrier, for instance in one-degree-of-freedom systems, of the type

$$\mathcal{H} = \frac{p^2}{2m} + V(q) \tag{6.43}$$

The potential may be bounded with several wells separated by barriers where tunneling leads to a splitting of degenerate energy levels. On the other hand, the potential may be bounded on one side with a well and unbounded beyond a barrier on the other side. Such a potential is of the scattering type where tunneling may lend a finite lifetime to quasibound states of the well.

Recent works by Strunz [172] have shown that tunneling in bounded 1F systems with N wells can be taken into account by an appropriate sum over complex periodic orbits so that the level density can be written in the form of Eq. (6.22), but here with the following zeta function

$$Z(E) = \det(\mathbf{1} - \mathbf{Q}) \tag{6.44}$$

The transfer operator $\mathbf{Q}$ contains the amplitudes of transitions between the semiclassical states with positive and negative momenta in each well. This transfer operator has an interpretation that is similar to the transfer operator [Eq. (6.33)], where the motion is described by coherent transitions between different semiclassical states. As shown in Fig. 23, the semiclassical state is associated with a point q_j^0 located in each well ($j = 1, 2, \ldots, N$) such that the reduced actions from this point to each one of the left and right turning points, $q_{<j}$ and $q_{j>}$, are equal,

$$\int_{q_j^0}^{q_{j>}} p\, dq = \int_{q_{<j}}^{q_j^0} p\, dq = \frac{S_j}{2} \qquad \text{with} \qquad p = \sqrt{2m[E - V(q)]} \tag{6.45}$$

where S_j is the reduced action from the left turning point to the right one. The states with positive and negative momenta are labeled by the symbols $(j, +)$ and $(j, -)$ with $j = 1, 2, \ldots, N$. Figure 23 shows the graph that

(a)

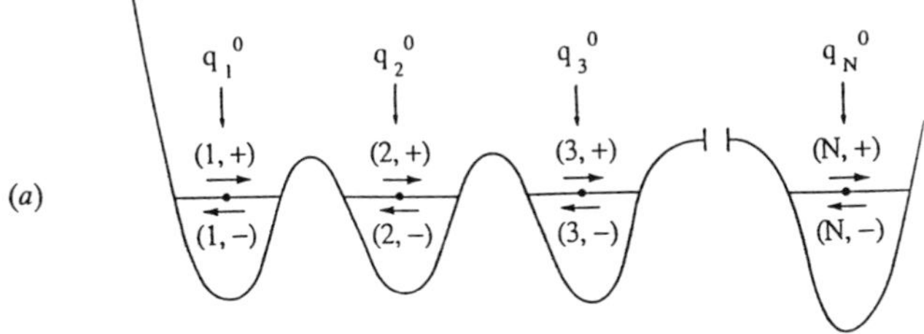

(b)

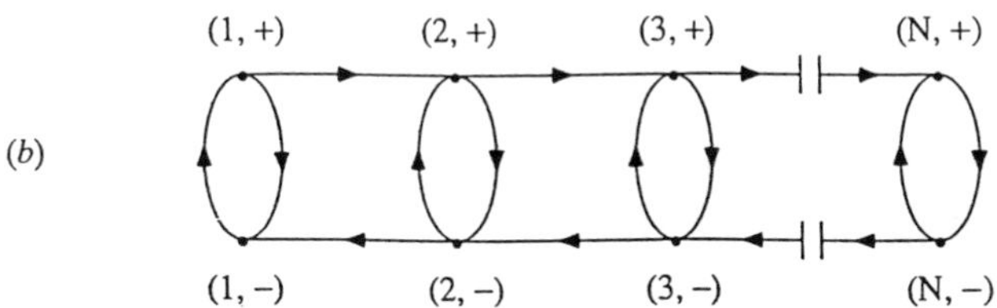

Figure 23. Multiple-well potential (*a*) and the corresponding graph (*b*) of the associated transfer operator Eqs. (6.46)–(6.47) describing tunneling. The symbols $(j, \pm)$ correspond to semiclassical states with positive or negative momentum at the mid-points q_j^0 in each well.

describes the possible transitions between the different vertices associated with the semiclassical states.

The amplitudes of the transitions are calculated by the uniform approximation technique by matching WKB wave functions between pairs of wells. Accordingly, the transfer operator is a $2N \times 2N$ matrix of the form

$$\mathbf{Q} = \begin{pmatrix} \mathbf{Q}_{++} & \mathbf{Q}_{+-} \\ \mathbf{Q}_{-+} & \mathbf{Q}_{--} \end{pmatrix} \tag{6.46}$$

with the matrix elements [172]

$$\begin{aligned}
(\mathbf{Q}_{++})_{jk} &= \sqrt{1-\rho_j^2}\, e^{-i\phi_j}\, e^{i(S_j+S_{j+1})/2}\delta_{j+1,k} \\
(\mathbf{Q}_{+-})_{jk} &= -i\rho_j\, e^{-i\phi_j}\, e^{iS_j}\delta_{j,k} \\
(\mathbf{Q}_{-+})_{jk} &= -i\rho_{j-1}\, e^{-i\phi_{j-1}}\, e^{iS_j}\delta_{j,k} \\
(\mathbf{Q}_{--})_{jk} &= \sqrt{1-\rho_{j-1}^2}\, e^{-i\phi_{j-1}}\, e^{i(S_{j-1}+S_j)/2}\delta_{j-1,k}
\end{aligned} \tag{6.47}$$

with $\hbar = 1$ to simplify the notation. In each well, the amplitude to go from one turning point to the other is given by $\exp(iS_j)$, where S_j is the reduced action of the path [Eq. (6.45)]. At each barrier, part of the amplitude is transmitted by tunneling and another part is reflected. The parameter ρ_j denotes the reflection amplitude from the jth to the $(j+1)$th well and ϕ_j the corresponding phase shift. The semiclassical reflection amplitude and phase shift are given by [172]

$$\rho = \frac{\exp(-\pi\theta)}{\sqrt{1+\exp(-2\pi\theta)}} \tag{6.48}$$

$$\phi = \arg[\Gamma(1/2 + i\theta)] - \theta \ln|\theta| + \theta \tag{6.49}$$

in terms of the tunneling integral

$$\begin{aligned} \theta(E) &= -\frac{1}{\pi\hbar}\int_{q_>}^{q_<} dq\sqrt{2m[V(q)-E]} \qquad \text{for} \qquad E \le V_{\max} \\ \theta(E) &= -\frac{i}{\pi\hbar}\int_{q_+}^{q_-} dq\sqrt{2m[E-V(q)]} \qquad \text{for} \qquad E > V_{\max} \end{aligned} \tag{6.50}$$

In the case of tunneling through the barrier, the energy is below the maximum $V_{\max}$ of the barrier and there are two real turning points, $q_>$ and $q_<$. Far below the barrier, the phase shift goes to zero while the reflection amplitude tends to the value 1 because $\theta \to -\infty$.

On the other hand, for energies above the barrier, there are two complex-conjugate turning points, q_+ and q_-. When the energy is far above a maximum of the potential, the two adjacent wells merge and form a larger well, because the reflection amplitude tends to zero (since now $\theta \to +\infty$) while the phase shift vanishes in this limit. The behavior of the reflection amplitude and of the phase shift as a function of the tunneling integral θ are shown in Fig. 24.

On the left and right sides of the potential, the potential increases without limit so that these reflection amplitudes are always equal to 1 with zero phase shifts. Moreover, a change of sign of the momentum implies a multiplication by a phase $(-i)$. For bound potentials, the transfer matrix [Eqs. (6.46) and (6.47)] is unitary

$$\mathbf{Q}\mathbf{Q}^{\dagger} = \mathbf{Q}^{\dagger}\mathbf{Q} = \mathbf{1} \tag{6.51}$$

As a first example, we consider a double-well potential, as shown in

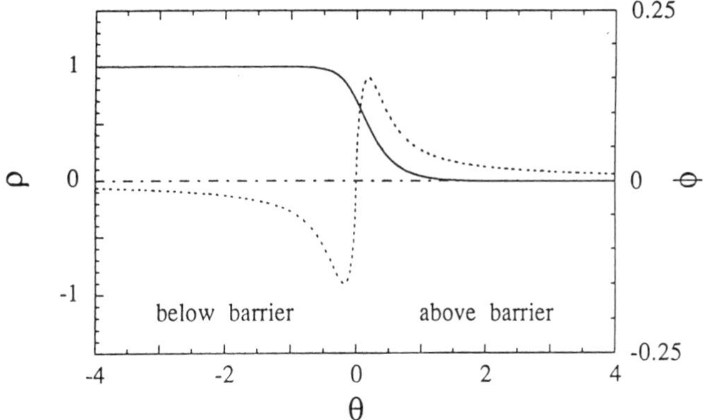

Figure 24. Reflection amplitude ρ [Eq. (6.48)] (solid line and left-hand axis) and phase shift ϕ [Eq. (6.49)] (dashed line and right-hand axis) versus the tunneling integral θ [Eq. (6.50)].

Fig. 25, with the corresponding graph. The transfer matrix is given by

$$
\mathbf{Q} = \begin{array}{c} \\ (1,+) \\ (1,-) \\ (2,+) \\ (2,-) \end{array}
\begin{array}{c}
\begin{array}{cccc} (1,+) & (1,-) & (2,+) & (2,-) \end{array} \\
\begin{pmatrix}
0 & -i\rho\, e^{iS_1 - i\phi} & \sqrt{1-\rho^2}\, e^{i(S_1+S_2)/2 - i\phi} & 0 \\
-i\, e^{iS_1} & 0 & 0 & 0 \\
0 & 0 & 0 & -i\, e^{iS_2} \\
0 & \sqrt{1-\rho^2}\, e^{i(S_1+S_2)/2 - i\phi} & -i\rho\, e^{iS_2 - i\phi} & 0
\end{pmatrix}
\end{array}
\tag{6.52}
$$

from which we obtain the quantization condition in terms of the zeta function

$$
0 = Z(E) = 2e^{(i/\hbar)(S_1+S_2)}\left[\cos\left(\frac{S_1+S_2}{\hbar} - \phi\right) + \rho\cos\left(\frac{S_1-S_2}{\hbar}\right)\right] \tag{6.53}
$$

Deep in the wells, the reflection amplitude is nearly equal to 1, while the phase shift vanishes so that the right-hand side of Eq. (6.53) nearly factorizes into $\cos(S_1/\hbar)\cos(S_2/\hbar)$, and each well sustains levels that are quasiindependent of each other. Nevertheless, tunneling couples both wells through the transmission amplitude, which causes a well-known splitting in the case where a level of the first well is nearly degenerate with another level of the second well.

Another example is given by the potential of Fig. 26 where the quantum states of the well acquire a finite lifetime by tunneling through

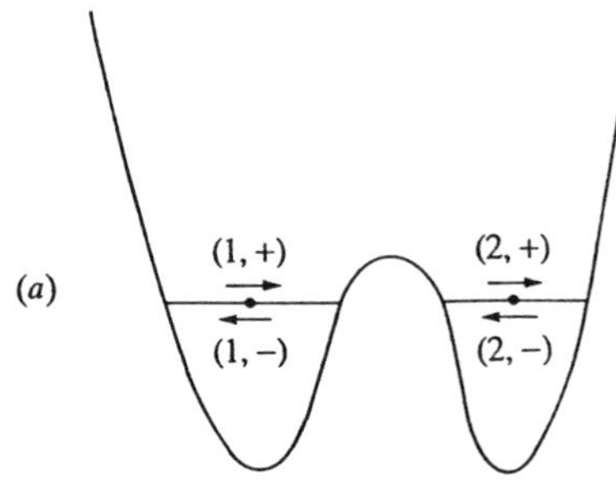

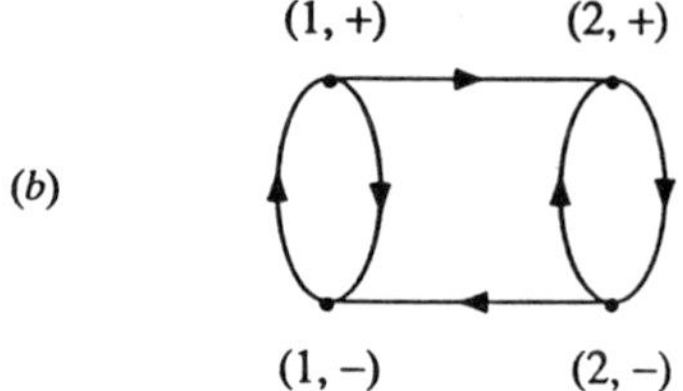

Figure 25. Double-well potential (*a*) and the corresponding graph (*b*) of the associated transfer operator [Eq. (6.52)]. In the double-well potential, tunneling induces a splitting of the low-lying energy eigenvalues.

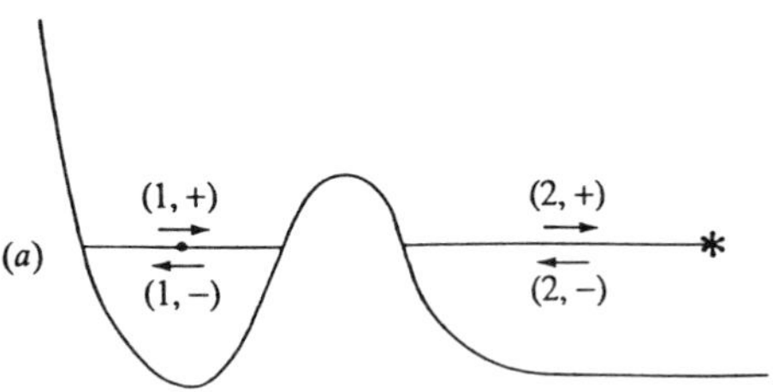

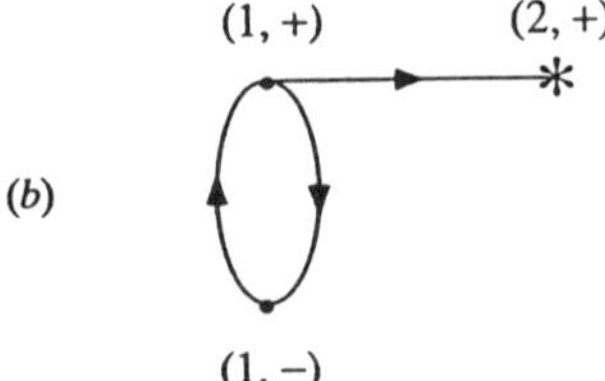

Figure 26. Single-well potential with barrier (*a*) and the corresponding graph (*b*) of the associated transfer operator [Eq. (6.54)]. In this potential, the energy eigenvalues correspond to metastable states that acquire a finite lifetime because of tunneling through the barrier.

the barrier. This potential, which is of the scattering type, may be considered as the limit of the double-well potential of Fig. 25 when the right-hand wall is moved a large distance. The reflection at this wall takes so long that we replace it by a boundary with infinite absorption. The graph is reduced to a two-state graph and the transfer matrix is given by

$$\mathbf{Q} = \begin{matrix} & (1,+) & (1,-) \\ \begin{matrix}(1,+)\\(1,-)\end{matrix} & \multicolumn{2}{c}{\begin{pmatrix} 0 & -i\rho\, e^{iS-i\phi} \\ -i\, e^{iS} & 0 \end{pmatrix}} \end{matrix} \tag{6.54}$$

For this open potential with the scattering boundary conditions we adopted, the transfer matrix is no longer unitary as it was for a bounded potential. The scattering resonances are obtained as the complex zeros of the following zeta function

$$0 = Z(E) = 1 + \rho \exp\left(\frac{i}{\hbar} S - i\phi\right) \tag{6.55}$$

Deep inside the well, the scattering resonances are given by $E_n = \varepsilon_n - i\Gamma_n/2$ with

$$S(\varepsilon_n) \simeq 2\pi\hbar(n + \tfrac{1}{2}) \qquad \Gamma_n \simeq \frac{\hbar}{T(\varepsilon_n)} \exp[2\pi\theta(\varepsilon_n)] \tag{6.56}$$

so that the half-width Γ_n vanishes for a large enough barrier where $\theta \to -\infty$. We observe that the half-width is exponentially small with a nonanalytic behavior, of the form $\exp(-C/\hbar)$ $(C > 0)$, which is typical of tunneling phenomena.

We end this section with several observations. We see that the one-freedom zeta function [Eq. (6.44)] has a form that is similar to the two-freedom zeta function [Eq. (6.35)]. The difference comes from the extra product over the integer m, which characterizes two-dimensional systems. In systems with more degrees of freedom, the product would extend over several integers $m_1, \ldots, m_f$ as in Eq. (6.21). This remark suggests that a 1F system acquires a kind of topological chaos if we consider its complex dynamics rather than its real dynamics [173]. Indeed, the existence of a nontrivial transfer matrix associated with the graph of Fig. 23 shows that here the complex periodic orbits also proliferate exponentially as in real chaos. It is possible to characterize this topological chaos by a topological entropy per symbol given by the logarithm of the largest eigenvalue of the adjacency matrix of the graph of Fig. 23 [i.e., the skeleton of the transfer matrix in Eq. (6.46) with elements equal to 1 if two vertices are joined and 0 otherwise]. We obtain

$$h_{\text{top}} = \ln\left(2\cos\frac{\pi}{N+2}\right) \text{ per symbol} \tag{6.57}$$

where N is the number of wells. For one well, this quantity vanishes while it tends to $\ln 2$ in the limit $N \to \infty$.

2. *Tunneling in 2F Systems*

In systems with two degrees of freedom, the situation becomes more complicated. We find two types of tunneling: (1) tunneling through a barrier and (2) dynamical tunneling out of quasiperiodic islands.

The first type of tunneling is a generalization of tunneling in 1F systems. In systems like the Hénon–Heiles Hamiltonian, at energies below the saddle equilibrium points, we may find potential energy barriers that cannot be crossed by real classical orbits although the quantum mechanical waves may tunnel through them. As in 1F systems, complex classical orbits may be invoked to estimate the tunneling amplitude in these situations. Among the complex classical orbits, the simplest ones are the periodic orbits of the inverted potential. The tunneling amplitude can be calculated in terms of the action of such complex periodic orbits.

On the other hand, there is no potential energy barrier in the phenomenon of dynamical tunneling. In 2F systems, we may find stable periodic orbits surrounded by a main quasiperiodic island at energies higher than any barrier. Trajectories in the vicinity of the stable periodic orbit are forever trapped by the concentric tori of the quasiperiodic island. Therefore, the lifetime is infinite for such orbits according to classical dynamics. Nevertheless, the quantum mechanical wave functions can extend from inside the island to the outside continuum, allowing the quantum states supported by the island to have a finite lifetime. We remark that the island can support such metastable states only if its area A is larger than $2\pi\hbar$, the number of supported metastable states being given approximately by $A/2\pi\hbar$. The states deep inside the island have a long lifetime while the outer states have shorter lifetimes. Dynamical tunneling is due to complex classical orbits, although very few results are known for the moment [174]. Progress in this question may be expected for Hamiltonian mappings and their quantum extensions. Adachi [175], as well as Shudo et al. [176], devoted several works to tunneling in the time domain at the level of the propagator. In the energy domain, the real parts of the eigenenergies of the states supported by a quasiperiodic island can be obtained from the work on quantum maps by Balazs et al.

[104] but the calculation of the imaginary parts remains an open problem. In the integrable limit, the dynamics in a Poincaré surface of section transverse to the orbits can be approximated by a 1F flow so that 2F dynamical tunneling can be reduced to barrier tunneling in a 1F effective system in such a limit. However, such an approach is not systematic. Important hints for a better understanding of dynamical tunneling in area-preserving maps are provided by several recent works on the complex classical dynamics of maps. In particular, Percival and coworkers [177] showed that the invariant tori of the quasiperiodic island have an analytic continuation in complex coordinates up to an uncrossable frontier where singularities are dense. Moreover, the structure of the Julia set, that is, the set of trapped complex orbits, has been recently studied by Gelfreich et al. [178].

In bifurcations, new stable periodic orbits surrounded by their quasiperiodic island may appear. Therefore, we may expect that complex periodic orbits play a special role in the quantization of the dynamics near bifurcations. This problem is the object of a revived interest stimulated by the experimental observation of bifurcations in the periodic orbits emerging from the wave dynamics in the Rydberg states of hydrogen-like atoms in a magnetic field [34, 35]. We may expect important advances in this exciting problem.

3. *Partition Functions of Statistical Mechanics and Thermal Reaction Rates*

The calculation of partition functions and of thermal reaction rates also involves the use of complex periodic orbits with respect to the real classical dynamics or real periodic orbits in the inverted potential [66, 179]. Indeed, if we follow a calculation of the partition function, which is analogous to Section II.F, we obtain

$$\mathcal{Q}(\beta) = \operatorname{tr} \exp(-\beta \hat{\mathcal{H}}) = \left(\frac{N}{2\pi\hbar^2\beta}\right)^{Nf/2} \int d\mathbf{q}_1\, d\mathbf{q}_2 \cdots d\mathbf{q}_{N-1}$$
$$\times \exp\left\{ -\sum_{n=0}^{N-1} \left[\frac{N}{2\hbar^2\beta} (\mathbf{q}_{n+1} - \mathbf{q}_n)^2 + \frac{\beta}{N} V(\mathbf{q}_n) \right] \right\} \tag{6.58}$$

where $\beta = 1/k_{\mathrm{B}}T$ is the inverse temperature. Equation (6.58) can be evaluated by the steepest-descent method. The points $\{\mathbf{q}_n\}$, where the argument of the exponential is at a maximum, are determined. These points correspond to the solutions of Newton's equation in the inverted potential

$$\frac{d^2\mathbf{q}}{d\tau^2} = \frac{\partial V}{\partial \mathbf{q}} \qquad \text{with} \qquad \mathbf{q}(\tau = 0) = \mathbf{q}(\tau = \hbar\beta) \tag{6.59}$$

in terms of the fictitious time

$$\tau = it(=\hbar\beta) \tag{6.60}$$

Since the fictitious time τ is the complex-rotated real time t, we may speak about complex periodic orbits. Contrary to the stationary-phase method, the steepest descent method only selects the maxima of the exponential in (6.58) so that we do not have to consider every solution of Eq. (6.59), but only those corresponding to a maximum.

Similar integrals also appear in the calculation of thermal reaction rates, as shown elsewhere [66, 179].

E. Nonisolated Periodic Orbits

Periodic orbits may also be nonisolated, for instance, in integrable systems or, equivalently, in systems with continuous symmetries. In such systems, the periodic orbits are found in continuous families, but these families proliferate at most algebraically as T^η with their period. Since the families of periodic orbits do not proliferate exponentially, as in the chaotic case, the topological entropy vanishes in these systems ($h_{\text{top}} = 0$).

When the periodic orbits are nonisolated, the integration of the trace formula requires a different procedure, as mentioned in Section IV.E.2. Indeed, the presence of a continuous family implies that the matrix of second derivatives [Eq. (4.23)] of the action is degenerate due to one or more extra neutral directions besides the direction of the flow. As a consequence, the integration must be carried out on a smaller set of variables, as shown in Eq. (4.30). The sum over periodic orbits should be replaced by a sum over different continuous families of periodic orbits, and the corresponding amplitudes will necessarily differ.

In this context, Berry and Tabor [180] derived a trace formula for integrable systems. In this case, there exist action-angle variables such that the Hamiltonian turns out to depend only on the actions, $\mathcal{H} = \mathcal{H}(J_1, \ldots, J_f)$. According to the EBK quantization condition, the energy levels are then obtained by the substitution

$$J_i = \frac{1}{2\pi} \oint_{C_i} \mathbf{p} \cdot d\mathbf{q} = \hbar\left(n_i + \frac{\mu_i}{4}\right) \tag{6.61}$$

where μ_i is the Maslov index corresponding to the loop C_i around the

Lagrangian manifold labeled by $J_1, \ldots, J_f$. The energy eigenvalues are then given by

$$E_{n_1 \ldots n_f} = \mathscr{H}[\hbar(n_1 + \mu_1/4), \ldots, \hbar(n_f + \mu_f/4)] \tag{6.62}$$

Inserting this result into the level density and converting the sums over the quantum numbers into integrals and sums over Fourier components by means of the Poisson summation formula, Berry and Tabor [180] obtained the following trace formula for the level density

$$n(E) = n_{\text{av}}(E) + \frac{2}{\hbar^{(f+1)/2}} \sum_{\mathbf{m}} \sum_{r=1}^{\infty} \frac{1}{r^{(f-1)/2}} \times \frac{\cos[(r/\hbar)S_{\mathbf{m}}(E) - r(\pi/2)\mathbf{m}\cdot\boldsymbol{\mu} - (\pi/4)\alpha_{\mathbf{m}}]}{|\mathbf{m}|^{(f-1)/2}|\boldsymbol{\omega}(\mathbf{J}_{\mathbf{m}})||K_{\mathbf{m}}|^{1/2}} + \mathscr{O}(\hbar^{-(f-1)/2}) \tag{6.63}$$

The quantity $S_{\mathbf{m}}(E)$ is the total reduced action of one of the prime periodic orbits of the continuous family labeled by the set of positive integers $\mathbf{m}$. These integers are relatively prime and count the rotations around the different circuits C_i of the torus over one prime period. The frequencies of rotation of the angle variables are given by $\boldsymbol{\omega} = \partial_{\mathbf{J}}\mathscr{H}$. The quantity $K_{\mathbf{m}}$ is the scalar Gaussian curvature of the action surface $E = \mathscr{H}(\mathbf{J})$ defined by

$$K_{\mathbf{m}} = \frac{1}{|\mathbf{m}|^{f-1}} \det\left[\frac{\partial^2(\mathbf{m}\cdot\mathbf{J})}{\partial\xi_i\,\partial\xi_j}\right]_{i,j=1,\ldots,f-1} \tag{6.64}$$

for a set of intrinsic coordinates $\{\xi_i\}_{i=1}^{f-1}$ of the action surface $E = \mathscr{H}(\mathbf{J})$. Equation (6.64) is evaluated at the point $\mathbf{J}_{\mathbf{m}}$. The parameter $\alpha_{\mathbf{m}}$ is the excess of positive over negative eigenvalues of the matrix inside the determinant [Eq. (6.64)].

The Berry–Tabor formula shows that continuous families of periodic orbits have a much larger contribution, namely in $\hbar^{-(f+1)/2}$, to the level density than the isolated unstable periodic orbits, the contribution of which is in $\hbar^{-1}$ according to the Gutzwiller trace formula [Eq. (6.13)]. Recently, Creagh and Littlejohn have shown in detail how the presence of continuous symmetries converts the Gutzwiller-type contribution into the Berry–Tabor one [181]. The $\hbar$-corrections can be calculated for Eq. (6.63) by methods similar to those developed in the previous sections.

It is important to remark that continuous families of periodic orbits may exist in systems without continuous symmetries, such as the stadium, the Sinai billiards, and the hard sphere gas [133, 182]. In these systems, a different procedure is necessary to obtain the contributions from the families of periodic orbits.

We should also mention that several periodic orbits may coalesce at bifurcations. As a consequence, the assumption of isolation of the periodic orbits is no longer satisfied at a bifurcation, where we may also expect an enhancement of the amplitude. The quantization of bifurcations is the subject of recent works.

F. Periodic Orbits Close to Bifurcations

Thus, close to bifurcations, we may encounter another situation where the trace formula for real unstable periodic orbits does not provide an accurate description. Complex orbits (which have been called "ghost-orbits" [183]) may be invoked to improve the analysis, which thus bears some similarity to the analysis of tunneling phenomena. The limitation of the trace formula, Eq. (6.13), in this context lies in the fact that the amplitude factor in Eq. (6.13) depends entirely on the *linear* stability of the orbit, and will thus diverge at a bifurcation point where the eigenvalues of the linear stability matrix pass through the value one. This divergence may be prevented by a modified amplitude factor, which takes into account terms beyond the linear stability, as shown by Ozorio de Almeida and Hannay [7, 184]. We quote the amplitude factor given in [184]:

$$A_p^{(r)} = (2\pi\hbar i)^{-1} \int dP\, dQ \left| \frac{\partial^2 S_p^{(r)}}{\partial P\, \partial Q} \right|^{1/2} \exp\left\{ -\frac{i}{\hbar} [S_p^{(r)}(P, Q) - PQ] \right\} \tag{6.65}$$

where $S_p^{(r)}(P, Q)$ is the generating function for the Poincaré map in terms of the canonical conjugate variables (Q, P) transverse to the periodic orbit p, and the index r refers to the rth repetition of the orbit. If only the quadratic part of $S_p^{(r)}$ is taken into account, the integral reduces to a Gaussian form, and can be related to the linear stability matrix $\mathbf{m}_p$,

$$A_p^{(r)} = e^{-i(\pi/2)\sigma_p^{(r)}} |\det(\mathbf{m}_p^r - \mathbf{1})|^{-1/2} \tag{6.66}$$

which is the factor found in Eq. (6.7). The index $\sigma_p^{(r)}$ is equal to 0 or 1 depending on the sign of the quadratic form in the Gaussian integral.

In the neighborhood of a bifurcation, the generating function takes on a particular normal form, characteristic of the given type of bifurcation. The normal form is determined by the topology of the bifurcation, and contains all information on the generation or annihilation of satellite orbits around the central orbit. In particular, in dependence of a control parameter ε, which indicates the deviation from the critical energy of the bifurcation point, the normal form features a different number of extrema, which correspond to the fixed points of the Poincaré map. On one side of the bifurcation, complex-valued extrema of the normal form may occur, which correspond to the "ghost orbits" mentioned above [183].

In fact, the normal forms for the generating function $S_p^{(r)}$ with $r = 1$ correspond to the so-called "elementary catastrophes", or structurally stable caustics [185, 186]. For example, a tangent bifurcation corresponds to the fold catastrophe, with its characteristic normal form [7]

$$S_p^{(1)}(P, Q) - PQ = P^2 + Q^3 + \varepsilon Q^2 \tag{6.67}$$

Alternatively, in the presence of the symmetry $Q \to -Q$, we may find pitchfork bifurcations with the characteristic normal form [7]

$$S_p^{(1)}(P, Q) - PQ = P^2 \pm Q^4 + \varepsilon Q^2 \tag{6.68}$$

Examples of such bifurcations will be given in the context of the dynamics of the molecular transition complex (see Section XII). The integrals [Eq. (6.65)] involving the normal forms are standard diffraction catastrophe integrals [185, 186]. If evaluated by the stationary-phase method, they will yield contributions at the extrema of the normal form, which may include "ghost orbits", as shown in [183]. The general result is that the amplitude $A_p^{(r)}$ increases at a bifurcation in a way controlled by the nonlinear stability of the critical periodic orbit. As a consequence, there is an effect of nonlinear saturation of the amplitude in contrast to the prediction resulting from linear stability analysis, that the amplitude diverges.

The contributions involving higher than quadratic terms of the generating function $S_p^{(r)}$ correspond to higher $\hbar$ orders, which were obtained in Section VI.B. The contributions of these $\hbar$-corrections may be small when the periodic orbit is far from a bifurcation, but they become non-negligible and essential at bifurcations in the joint limit $\varepsilon \to 0$ and $\hbar \to 0$ [7, 184].

VII. SEMICLASSICAL AVERAGES OF QUANTUM OBSERVABLES

A. Diagonal Matrix Elements

In many problems of interest, we would like to evaluate, with a semiclassical method, not only the eigenenergies but also matrix elements of some observable $\hat{A}$ between eigenstates of the Hamiltonian $\hat{\mathscr{H}}_0$.

We found it most enlightening to formulate this problem in terms of perturbation theory [101]. We consider a new Hamiltonian including the operator $\hat{A}$, which is assumed to be Hermitian and acts as a perturbation with respect to the Hamiltonian $\hat{\mathcal{H}}_0$

$$\hat{\mathcal{H}} = \hat{\mathcal{H}}_0 + \lambda\hat{A} \tag{7.1}$$

There is a result of perturbation theory stating that the first-order perturbation of the energy levels is given by the average of $\hat{A}$ over the corresponding eigenstates of $\hat{\mathcal{H}}_0$ [187]

$$E_n^{(1)} = \frac{dE_n}{d\lambda}\bigg|_{\lambda=0} = \langle \phi_n(0)|\hat{A}|\phi_n(0)\rangle \tag{7.2}$$

for

$$\hat{\mathcal{H}}(\lambda)|\phi_n(\lambda)\rangle = E_n(\lambda)|\phi_n(\lambda)\rangle \tag{7.3}$$

where

$$E_n(\lambda) = E_n^{(0)} + \lambda E_n^{(1)} + \mathcal{O}(\lambda^2) \tag{7.4}$$

If the eigenvalue problem [Eq. (7.3)] has been solved, the matrix element of $\hat{A}$ may be calculated as the derivative of the eigenvalues $E_n(\lambda)$ with respect to the perturbation parameter λ. The eigenenergies of Eq. (7.1) can be obtained be periodic-orbit quantization. With these results, we can calculate the averages of Eq. (7.2) in the limit $\lambda \to 0$ in terms of the periodic orbits of the classical Hamiltonian corresponding to $\hat{\mathcal{H}}$. The classical Hamiltonian can be defined as the Wigner transform $\mathcal{H}_{\rm cl} = \mathcal{H}_{\rm W}(\mathbf{q}, \mathbf{p})$ of $\hat{\mathcal{H}}$.

The perturbed eigenenergies are given by the zeros of the Selberg zeta function [Eq. (6.21)] for the Hamiltonian [Eq. (7.1)]. The zeta function can be expanded into a series according to the method of the cycle expansion

$$Z(E;\lambda) = \prod_{p\mathbf{m}} [1 - t_{p,\mathbf{m}}(E;\lambda)] = \sum_{n=0}^{\infty} \sum_{p_1\mathbf{m}_1\cdots p_n\mathbf{m}_n} (-)^n t_{p_1\mathbf{m}_1} \cdots t_{p_n\mathbf{m}_n} \tag{7.5}$$

with

$$t_{p\mathbf{m}}(E;\lambda) = \exp\left[\frac{i}{\hbar} S_p(E;\lambda) + C_{p\mathbf{m}}(E;\lambda) + \mathcal{O}(\hbar)\right] \tag{7.6}$$

where $C_{p\mathbf{m}}$ contains the contributions of the stability eigenvalues and of the Morse indexes of the periodic orbits. The periodic orbits are labeled by p, and $\mathbf{m}$ denotes the vector of integers appearing in Eq. (6.21).

The derivative of the zeros with respect to the parameter λ is given by taking the total derivative $dZ[E_n(\lambda); \lambda]/d\lambda = 0$. Setting $\lambda = 0$, we get

$$\begin{aligned}
\left.\frac{dE_n}{d\lambda}\right|_{\lambda=0} &= -\left.\frac{\partial_\lambda Z}{\partial_E Z}\right|_{E_n;\lambda=0} \\
&= -\left.\frac{\sum_{\{p_i\mathbf{m}_i\}} x_{\{p_i\mathbf{m}_i\}}[\partial_\lambda S_{p_1} + \cdots + \partial_\lambda S_{p_n} + \mathcal{O}(\hbar)]}{\sum_{\{p_i\mathbf{m}_i\}} x_{\{p_i\mathbf{m}_i\}}[T_{p_1} + \cdots + T_{p_n} + \mathcal{O}(\hbar)]}\right|_{E_n;\,\lambda=0} \\
&= \langle E_n^{(0)}|\hat{A}|E_n^{(0)}\rangle
\end{aligned} \tag{7.7}$$

where we used the notation $x_{\{p_i\mathbf{m}_i\}} = (-)^n t_{p_1\mathbf{m}_1} \cdots t_{p_n\mathbf{m}_n}$ as well as the result $T_p = \partial_E S_p$. We observe that the leading semiclassical approximation to the diagonal matrix elements of $\hat{A}$ only involves $\partial_\lambda S_p$, while the derivatives $\partial_\lambda C_{p\mathbf{m}}$ appear in the next correction of order $\hbar$ in the numerator. At the leading semiclassical approximation, those corrections $\mathcal{O}(\hbar)$ are neglected. Accordingly, we only need to calculate $\partial_\lambda S_p$ for each periodic orbit, which is given by the following formula of classical mechanics [188]

$$\frac{\partial S}{\partial \lambda} = -\oint \frac{\partial \mathcal{H}_{\mathrm{W}}}{\partial \lambda}\, dt = -\oint A_{\mathrm{W}}\, dt \tag{7.8}$$

where S is the reduced action of the periodic orbit p and A_{W} is the classical function obtained from the operator $\hat{A}$ by the Wigner–Weyl correspondence. The circular integrals are evaluated over the periodic orbit.

In the same context, it is interesting to mention the formulas

$$\mathrm{tr}\,\frac{1}{E - \hat{\mathcal{H}}} = \mathrm{tr}\,\frac{\partial}{\partial E}\ln(E - \hat{\mathcal{H}}) \tag{7.9}$$

$$\mathrm{tr}\,\frac{\hat{A}}{E - \hat{\mathcal{H}}_0} = -\mathrm{tr}\,\frac{\partial}{\partial \lambda}\ln(E - \hat{\mathcal{H}}_0 - \lambda\hat{A})|_{\lambda=0} \tag{7.10}$$

for complex values of E and with the operations ordered as shown. Commuting the operations may require subtractions and/or regularizations of infinities. These formulas show the similarity with the logarithmic derivative appearing in Eq. (6.20). The derivative is taken, in one case, with respect to energy and, in the other case, with respect to the perturbation parameter.

B. Semiclassical Photoabsorption Cross-Section

Using Eq. (7.10), a formula can be obtained, which gives the photoabsorption cross-section for atoms or molecules [189]. In Section II.C, we mentioned that the photoabsorption cross-section is given by the Fourier transform of the autocorrelation function of the electric dipole operator. At zero temperature, the atom or the molecule is initially in its ground state ϕ_0 of eigenenergy E_0 so that Eq. (2.11) becomes

$$\kappa_\omega = \frac{\omega}{6\hbar c\varepsilon_0} \int_{-\infty}^{+\infty} dt\, e^{-i\omega t} e^{-iE_0 t/\hbar} \langle \phi_0 | \hat{\mathbf{d}} \cdot \exp\left(\frac{i}{\hbar} \hat{\mathcal{H}} t\right) \hat{\mathbf{d}} | \phi_0 \rangle \tag{7.11}$$

which may be rewritten as follows:

$$\kappa_\omega = \mathrm{tr}[\hat{A}\delta(E_\omega - \hat{\mathcal{H}}_0)] = -\frac{1}{\pi} \operatorname{Im} \mathrm{tr} \frac{\hat{A}}{E_\omega - \hat{\mathcal{H}}_0 + i0} \tag{7.12}$$

where $E_\omega = E_0 + \hbar\omega$ and

$$\hat{A} = \frac{\pi\omega}{3c\varepsilon_0} \hat{\mathbf{d}} | \phi_0 \rangle \cdot \langle \phi_0 | \hat{\mathbf{d}} \tag{7.13}$$

and where we used the property of the Green operator that

$$\hat{G}^+(E) = \frac{1}{E - \hat{\mathcal{H}} + i0} = \mathcal{P} \frac{1}{E - \hat{\mathcal{H}}} - i\pi\, \delta(E - \hat{\mathcal{H}}) \tag{7.14}$$

with $\mathcal{P}$ denoting Cauchy's principal part.

A semiclassical expression can be obtained for the cross-section using Eq. (7.10) applied to Eq. (6.14) giving the staircase function of the perturbed Hamiltonian $\hat{\mathcal{H}} = \hat{\mathcal{H}}_0 + \lambda\hat{A}$ with the operator in Eq. (7.13). After setting $\lambda = 0$, we get the following semiclassical formula for the photoabsorption cross-section [189]

$$\begin{aligned}
\kappa_\omega &= -\frac{1}{\pi} \operatorname{Im} \mathrm{tr} \frac{\hat{A}}{E - \hat{\mathcal{H}}_0 + i0} \\
&= \int \frac{d^f q\, d^f p}{(2\pi\hbar)^f} A_{\mathrm{W}}(\mathbf{q}, \mathbf{p})\, \delta[E - \mathcal{H}_{\mathrm{cl}}(\mathbf{q}, \mathbf{p})] + \mathcal{O}(\hbar^{-f+1})
\end{aligned}$$

$$+\frac{1}{\pi\hbar}\sum_{p}\sum_{r=1}^{\infty}\left(\oint_{p} A_{\mathrm{W}}\, dt\right)\frac{\cos[\frac{r}{\hbar}S_p(E) - r\frac{\pi}{2}\mu_p]}{|\det(\mathbf{m}_p^r - \mathbf{1})|^{1/2}} + \mathcal{O}(\hbar^0) \tag{7.15}$$

with $E = E_\omega$. The semiclassical level density [Eq. (6.13)] is recovered in the special case where $\hat{A} = \hat{I}$. The quasiclassical term contributes to a background in the photoabsorption cross-section, while the periodic-orbit terms contribute to oscillations on top of the quasiclassical background [190]. This formula applies to either discrete or continuous spectra.

For nonzero temperatures, however, the preceding semiclassical method does not seem to remain applicable and we need the quasiclassical method as we will explain below (see Section XI.A).

VIII. QUANTUM BILLIARDS

A. Wave Equation and Generalities

Billiards occupy a special position among the dynamical systems because of the relative simplicity of their dynamics where hard particles fly in free motion between elastic collisions. The Schrödinger equation is therefore identical to the free particle equation

$$(\Delta_f + \kappa^2)\psi = 0 \tag{8.1}$$

where Δ_f is the f-dimensional Laplacian and $\kappa = \hbar^{-1}(2mE)^{1/2}$ is the wavenumber. The specificity of the system is expressed by the boundary condition such as the Dirichlet boundary condition $\psi = 0$, which is taken on the impenetrable border of the billiard.

The border of the billiard is simply the walls of the cavity for billiards, such as the Bunimovich stadium, the Lorentz, or the Sinai billiards. For the hard sphere or hard disk gases, the border is formed by the walls of the container and by the hypersurface of the locations of the collisions between the particles [182].

Effective realizations of billiards have been carried out in microwave cavities as well as in electronic nanometric semiconductor devices in order to investigate how the transition from regular to chaotic ray motion influences the wave properties [41, 48].

As we mentioned earlier, billiards may be bounded or open. In the first case, the eigenstates are solution of the Helmholtz equation (8.1) with Dirichlet boundary conditions that select eigenvalues of the wavenumber κ so that the energy spectrum is discrete. In the second case, the billiard is a scattering system. The quantum dynamics is then given by the scattering S-matrix and by the scattering resonances that are the complex poles of the S-matrix, as well as of the Green function, in the second Riemann sheet of the complex energy surface. In this regard, the situation is very similar to the case of smooth Hamiltonian systems described in Section II.

However, the fact that the particle is in free flight between its elastic collisions on the walls already introduces significant simplifications into the classical dynamics in the sense that the classical action has the same form of Eq. (3.40) for each billiard. Similarly, the simplicity of the Schrödinger equation (8.1) inside the billiard considerably reduces the quantum mechanical treatment of billiards. In particular, explicit expressions can be given for the total Green function, as shown elsewhere [97], while the Feynman path integral reduces to multiple but ordinary integrals so that the quantization can be expressed as series of usual integrals. For these reasons, billiards are often used as solvable models in theoretical argumentation as well as to test numerical procedures and, especially, semiclassical methods. In this way, it is our purpose in this chapter to illustrate our theory of $\hbar$-expansion with the two- and the three-disk scatterers.

B. Multiple Scattering Expansion

In the following, we will focus on two-dimensional billiards ($f = 2$). The solution of the Helmholtz equation (8.1) requires the Green function satisfying

$$(\Delta_f + \kappa^2)G = \delta(\mathbf{q} - \mathbf{q}') \tag{8.2}$$

together with the Dirichlet boundary condition that $G(\mathbf{q}, \mathbf{q}'; \kappa) = 0$ for $\mathbf{q}$ belonging to the border of the billiard. For $f = 2$, the free Green function, which is solution of Eq. (8.2) without any boundary condition on the border of the billiard, is given by [132]

$$G_0(\mathbf{q}, \mathbf{q}'; \kappa) = -\frac{i}{4} H_0^{(1)}(\kappa|\mathbf{q} - \mathbf{q}'|) \tag{8.3}$$

where $H_0^{(1)}(z)$ is the first Hankel function.

According to Green's theorem, a solution ψ of Eq. (8.1) must satisfy the relation [191, 192]

$$\oint_{\partial D} ds' \left[G_0(\mathbf{q}, \mathbf{q}') \frac{\partial}{\partial n'} \psi(\mathbf{q}') - \psi(\mathbf{q}') \frac{\partial}{\partial n'} G_0(\mathbf{q}, \mathbf{q}') \right]$$
$$= \begin{cases} \psi(\mathbf{q}), & \mathbf{q} \in D \\ 0, & \mathbf{q} \notin D \end{cases} \tag{8.4}$$

where $\partial/\partial n$ denotes the derivative along a direction normal to the border and interior to the domain D of the billiard, that is, exterior to the obstacles, such as disks or hard walls. In Eq. (8.4), the integral extends over the arc of perimeter of the border. Because of the Dirichlet boundary condition, the second term vanishes and we get an equation giving the wave function in the domain in terms of its normal derivative on the border. We would like to have a closed equation for the function

$$u(s) \equiv \frac{\partial}{\partial n} \psi[\mathbf{q}(s)] \tag{8.5}$$

which is the normal component of the gradient of ψ on the border ∂D. This gradient is a function of the arc of perimeter s. The gradient of ψ is discontinuous on the border because of the Dirichlet boundary condition. It is known that integral representations, of the form Eq. (8.4), give the value of the function at the middle of the jump at the discontinuities just as for Fourier series of discontinuous functions. By differentiating Eq. (8.4), we get the closed equation [191, 192]

$$u(s) = 2 \oint_{\partial \mathrm{D}} ds' \frac{\partial}{\partial n} G_0[\mathbf{q}(s), \mathbf{q}'(s')] u(s') \equiv (\hat{Q} u)(s) \tag{8.6}$$

which defines the integral operator $\hat{Q}(\kappa)$ acting on functions defined on the perimeter parametrized by the variable s.

The eigenvalues or the scattering resonances are therefore the zeros of the Fredholm determinant of the operator $\hat{I} - \hat{Q}(\kappa)$ in the complex plane of the wavenumber $\kappa = \sqrt{2mE}/\hbar$, which is the condition for the existence of a nontrivial solution $u(s)$ for Eq. (8.6). The Fredholm determinant is expanded as a series of its traces according to

$$0 = \det[\hat{I} - \hat{Q}(\kappa)] = \exp - \sum_{N=1}^{\infty} \frac{1}{N} \operatorname{tr} \hat{Q}^N \tag{8.7}$$

with

$$\operatorname{tr} \hat{Q}^N = 2^N \oint ds_1 \cdots ds_N \frac{\partial G_0}{\partial n_1}(\mathbf{q}_1, \mathbf{q}_N) \frac{\partial G_0}{\partial n_N}(\mathbf{q}_N, \mathbf{q}_{N-1}) \cdots$$
$$\frac{\partial G_0}{\partial n_3}(\mathbf{q}_3, \mathbf{q}_2) \frac{\partial G_0}{\partial n_2}(\mathbf{q}_2, \mathbf{q}_1) \tag{8.8}$$

where $\mathbf{q}_j = \mathbf{q}(s_j)$ belongs to the border of the billiard. Using the definition of Eq. (8.3) for the free Green function, we get

$$\frac{\partial G_0}{\partial n_{j+1}}(\mathbf{q}_{j+1}, \mathbf{q}_j) = -\frac{i\kappa}{4}\cos\varphi_{jj+1}H_1^{(1)}(\kappa\ell_{jj+1}) \tag{8.9}$$

where φ_{jj+1} is the angle between the path from the point $\mathbf{q}_j$ to the point $\mathbf{q}_{j+1}$ and $-\mathbf{n}_{j+1}$, which is minus the unit vector normal to the border and exterior to the obstacles. Here $H_1^{(1)}(z)$ is the first Hankel function of first order and $\ell_{jj+1} = |\mathbf{q}_{j+1} - \mathbf{q}_j|$ is the length between those points. Introducing Eq. (8.9) in Eq. (8.8), we obtain

$$\operatorname{tr}\hat{Q}^N = \left(\frac{-i\kappa}{2}\right)^N \oint ds_1 \cdots ds_N \prod_{j=1}^{N} [\cos\varphi_{jj+1}H_1^{(1)}(\kappa\ell_{jj+1})] \tag{8.10}$$

with the cyclic identification $N+1=1$.

C. Stationary Phase Method and Periodic Orbits

The integrals [Eq. (8.10)] are evaluated by the stationary phase method [81]. In the short wavelength limit, the wavenumber κ is large, and κ^{-1} is a small parameter. In the domain where all the lengths ℓ_{jj+1} remain nonvanishing during the integration of Eq. (8.10), the Hankel functions may be replaced by their asymptotic expansion for large z. In this case, Eq. (8.10) may be rewritten as [20]

$$\operatorname{tr}\hat{Q}^N = \left(-e^{-i\pi/4}\sqrt{\frac{\kappa}{2\pi}}\right)^N \oint ds_1 \cdots ds_N \exp\left[i\kappa\mathcal{L} + \mathcal{M} + \frac{i}{\kappa}\mathcal{N} + \mathcal{O}\left(\frac{1}{\kappa^2}\right)\right] \tag{8.11}$$

where we introduced the definitions

$$\mathcal{L} \equiv \sum_{j=1}^{N} \ell_{jj+1} \tag{8.12}$$

$$\mathcal{M} \equiv \sum_{j=1}^{N} \ln\frac{\cos\varphi_{jj+1}}{\sqrt{\ell_{jj+1}}} \tag{8.13}$$

$$\mathcal{N} \equiv \frac{3}{8}\sum_{j=1}^{N} \ell_{jj+1}^{-1} \tag{8.14}$$

$\mathcal{L}$ is the total length of the path joining the N border points in the integral [Eq. (8.11)]. We will work in a system of units where the lengths

(ℓ), the arcs of perimeter (s), and the wavenumber (κ) are dimensionless, for instance, taking the radius of the disks as unit length ($a = 1$).

According to the stationary phase method, the multiple integrals [Eq. (8.11)] are evaluated locally around the critical points of the integration domain $\{s_1, \ldots, s_N\}$, where the total length $\mathcal{L}$ of the path is extremal. A set of critical points are given by the trajectories of classical motion that are solutions of

$$\mathcal{L}_{,j} = \frac{\partial \mathcal{L}}{\partial s_j} = 0 \tag{8.15}$$

The stationary solutions are periodic orbits of several types [20]:

1. *Regular Periodic Orbits*. For those stationary solutions, all the N lengths ℓ_{jj+1} are real, nonvanishing, and exterior to the obstacles. At each elastic bounce, the incident angle equals the reflection angle, which is the law of geometric optics. A periodic orbit may be a multiple r of a prime periodic orbit $\mathcal{P}$ with P bounces if $N = rP$. We will denote this multiple by $r\mathcal{P}$. The integral [Eq. (8.10)] has P distinct stationary points ${}^0\mathcal{P}_i$ with $i = 1, \ldots, P$, which are associated with this prime periodic orbit $\mathcal{P}$. If $\{s_1^0 s_2^0 \cdots s_P^0\}$ are the coordinates of the impact points of the prime periodic orbit $\mathcal{P}$ on the border of the billiard, these P stationary points are located at the distinct cyclic permutations of

$$\{s_1 s_2 \cdots s_N\} = \underbrace{s_1^0 s_2^0 \cdots s_P^0}_{1} \underbrace{s_1^0 s_2^0 \cdots s_P^0}_{2} \cdots \underbrace{s_1^0 s_2^0 \cdots s_P^0}_{r}\} \tag{8.16}$$

   which are only P in number. They all contribute by the same amplitude because they lead to the same integrals.

   Regular periodic orbits may be of different stabilities. Here we will assume that they are unstable of saddle type and isolated.

2. *Degenerate Periodic Orbits*. For those periodic orbits, one or several of the lengths ℓ_{jj+1} are going to zero at the critical point [12]. Consequently, a degenerate periodic orbit contains consecutive bounces that are located at the same point: $s_j^0 = s_{j+1}^0$ so that $\ell_{jj+1}^0 = 0$.

   First, we consider the case where one length goes to zero. Let $\mathcal{P}$ be a regular prime periodic orbit whose impact points are $\{s_1^0 s_2^0 \cdots s_P^0\}$. We want to calculate the number of stationary points corresponding to $\mathcal{P}$ in the integral [Eq. (8.10)] with $N = rP + 1$. Any one of the P impact points s_j^0 with $j = 1, \ldots, P$ may be repeated so that we obtain P different orbits

$$
\begin{aligned}
\{s_1 s_2 \cdots s_N\} = &\{s_1^0 s_1^0 s_2^0 \cdots s_P^0 \cdots s_1^0 s_2^0 \cdots s_P^0\} \\
&\{s_1^0 s_2^0 s_2^0 \cdots s_P^0 \cdots s_1^0 s_2^0 \cdots s_P^0\} \\
&\qquad\vdots \\
&\{s_1^0 s_2^0 \cdots s_P^0 s_P^0 \cdots s_1^0 s_2^0 \cdots s_P^0\}
\end{aligned}
\tag{8.17}
$$

which cannot be mapped onto each other by cyclic permutations. Contrary to Eq. (8.16), each one of these P sequences now has $N = rP + 1$ distinct cyclic permutations. Hence, we get $PN = P(rP + 1)$ stationary points that we denote by ${}^1\mathcal{P}_i^j$, with $j = 1, \ldots, P$ and $i = 1, \ldots, N$. These $P(rP + 1)$ stationary points only contribute by P different amplitudes corresponding to the P sequences [Eq. (8.17)], which differ by the repeated impact point. We will see that these amplitudes contribute to corrections in κ^{-1}.

When two lengths go to zero, we find two cases where both lengths vanish either consecutively, as in ${}^2\mathcal{P}_i^j = \{\cdots s_{j-1}^0 s_j^0 s_j^0 s_j^0 s_{j+1}^0 \cdots\}$, or separately, as in ${}^{(1,1)}\mathcal{P}_i^{(j,k)} = \{\cdots s_{j-1}^0 s_j^0 s_j^0 s_{j+1}^0 \cdots s_{k-1}^0 s_k^0 s_k^0 s_{k+1}^0 \cdots\}$. The contributions of these periodic orbits appear in the κ^{-2}-corrections.

Since we will restrict ourselves to the κ^{-1}-corrections, we will focus on the degenerate periodic orbits where one length goes to zero.

Besides the regular and the degenerate periodic orbits, we also find ghost and complex periodic orbits that play no essential role in our present considerations. The ghost periodic orbits are special orbits that cross hard walls [12, 132]. Although unphysical the ghost orbits are solution of Eq. (8.15) but it turns out that their contributions vanish by cancelation [12]. On the other hand, complex periodic orbits may be important in particular billiards with nearly bifurcating periodic orbits. In certain billiards, such as the open stadium, sharp edges may introduce special critical points for the integration of Eq. (8.10) by the stationary-phase method. The edges are known to produce important diffraction effects so that the corresponding amplitudes may in general be large. Smooth edges, where the curvature of the border of the billiard has a discontinuity, are also encountered in bounded billiards such as the Bunimovich stadium. Alonso and Gaspard [133] showed elsewhere that such smooth edges may have contributions as important as the regular periodic orbits of the Gutzwiller type. Nevertheless, the regular and degenerate orbits give the dominant contributions in smooth billiards on which we will focus in the following sections.

D. Contribution of the Regular Periodic Orbits

Here, we assume that no length goes to zero. The functions in Eqs. (8.12–8.14) may be expanded in Taylor series around the stationary solutions in terms of the deviations $\{\xi_i = s_i - s_i^0\}$. We obtain

$$\mathcal{L} = \mathcal{L}^0 + \underbrace{\mathcal{L}^0_{,i}\xi_i}_{=0} + \tfrac{1}{2}\mathcal{L}^0_{,ij}\xi_i\xi_j + \tfrac{1}{6}\mathcal{L}^0_{,ijk}\xi_i\xi_j\xi_k + \tfrac{1}{24}\mathcal{L}^0_{,ijkl}\xi_i\xi_j\xi_k\xi_l + \mathcal{O}(\xi^5) \tag{8.18}$$

$$\mathcal{M} = \mathcal{M}^0 + \mathcal{M}^0_{,i}\xi_i + \tfrac{1}{2}\mathcal{M}^0_{,ij}\xi_i\xi_j + \mathcal{O}(\xi^3) \tag{8.19}$$

$$\mathcal{N} = \mathcal{N}^0 + \mathcal{O}(\xi) \tag{8.20}$$

where we adopted the convention of summation over repeated indexes.

The stationary phase method in its simplest form is based on the matrix of the second derivatives of the total length. This matrix controls the linear stability of the periodic orbit and contains the Lyapunov exponents. The inverse of the second derivative matrix,

$$\mathcal{D}_{ij} \equiv \mathcal{L}^0_{,ij} \qquad \mathcal{G}_{ij} \equiv (\mathcal{D}^{-1})_{ij} \tag{8.21}$$

propagates linear perturbations along the periodic orbit and is therefore playing the role of a classical Green function. Because the total length is of the form

$$\mathcal{L} = \sum_{j=1}^{N} \ell(s_j, s_{j+1}) \tag{8.22}$$

the matrix $\mathcal{D}$ is symmetric and of the cyclic type with only two nonvanishing bands beyond the diagonal

$$\mathcal{D}_{ii} = \frac{\partial^2 \ell(s_{i-1}, s_i)}{\partial s_i^2} + \frac{\partial^2 \ell(s_i, s_{i+1})}{\partial s_i^2} \tag{8.23}$$

$$\mathcal{D}_{ii+1} = \frac{\partial^2 \ell(s_i, s_{i+1})}{\partial s_i \, \partial s_{i+1}} \tag{8.24}$$

$$\mathcal{D}_{ij} = 0 \qquad \text{for} \qquad |i - j| \geq 2 \tag{8.25}$$

with periodic boundary conditions, such that $i = N$ is identified with $i = 0$

and $i = N+1$ with $i=1$. Consequently, the corners $\mathcal{D}_{1N}$ and $\mathcal{D}_{N1}$ of the matrix $\mathcal{D}$ are nonvanishing. Similarly, the tensors $\mathcal{M}^0_{,ij}$, $\mathcal{L}^0_{,ijk}$, and $\mathcal{L}^0_{,ijkl}$ are nonvanishing only when $|i-j|$, $|i-k|$, $|i-l| \leq 1$.

By introducing the above Taylor expansions in Eq. (8.11), the integrals can be carried out by methods analogous to those of the preceding sections. Now we have $\xi \sim \kappa^{-1/2}$. Finally, the contribution of the regular periodic orbit $r\mathcal{P}$ becomes

$$\operatorname{tr} \hat{Q}^N \Big|_{0\mathcal{P}} = \sum_{i=1}^{P} \operatorname{tr} \hat{Q}^N \Big|_{0\mathcal{P}_i} = P\mathcal{J}(r\mathcal{P}) \tag{8.26}$$

with

$$\mathcal{J}(r\mathcal{P}) \equiv (-)^{rP} \frac{\exp[ir\kappa\mathcal{L}_{\mathcal{P}} - ir\frac{\pi}{2}\mu_{\mathcal{P}} + \frac{ir}{\kappa}\mathcal{N}_{\mathcal{P}} + \frac{i}{\kappa}\mathcal{C}_1(r\mathcal{P}) + \mathcal{O}(\kappa^{-2})]}{|\Lambda^r_{\mathcal{P}} + \Lambda^{-r}_{\mathcal{P}} - 2|^{1/2}} \tag{8.27}$$

where we use the quantities defined by the prime periodic orbit $\mathcal{P}$: $N = rP$, $\mathcal{L}^0 = r\mathcal{L}_{\mathcal{P}}$, $\mathcal{N}^0 = r\mathcal{N}_{\mathcal{P}}$, $\mu(N) = r\mu_{\mathcal{P}}$, and $\Lambda(N) = \Lambda^r_{\mathcal{P}}$. The quantity $\Lambda_{\mathcal{P}}$ is the stability eigenvalue (whose absolute value is larger than unity) of the symplectic Poincaré map linearized near the periodic orbit $\mathcal{P}$. The parameter $\mu_{\mathcal{P}}$ is the Maslov index, which is the number of negative eigenvalues of the $P \times P$ matrix $\mathcal{D}$. In the case of defocusing billiards, such as the disk scatterers, these indexes are vanishing, $\mu_{\mathcal{P}} = 0$.

The second term of the corrections in κ^{-1} is given by [20]

$$\begin{aligned}\mathcal{C}_1(r\mathcal{P}) \equiv &-\tfrac{1}{8}\mathcal{L}^0_{,ijkl}\mathcal{G}_{ij}\mathcal{G}_{kl} \\ &+\tfrac{1}{8}\mathcal{L}^0_{,ijk}\mathcal{L}^0_{,lmn}\mathcal{G}_{ij}\mathcal{G}_{kl}\mathcal{G}_{mn} + \tfrac{1}{12}\mathcal{L}^0_{,ijk}\mathcal{L}^0_{,lmn}\mathcal{G}_{il}\mathcal{G}_{jm}\mathcal{G}_{kn} \\ &-\tfrac{1}{2}\mathcal{L}^0_{,ijk}\mathcal{M}^0_{,l}\mathcal{G}_{ij}\mathcal{G}_{kl} + \tfrac{1}{2}(\mathcal{M}^0_{,ij} + \mathcal{M}^0_{,i}\mathcal{M}^0_{,j})\mathcal{G}_{ij}\end{aligned} \tag{8.28}$$

This coefficient can be written in terms of connected diagrams, as shown in Fig. 27, where the filled vertices represent derivatives of the total length $\mathcal{L}$ and the open vertices represent derivatives of the logarithm of the amplitude $\mathcal{M}$. Each line corresponds to the classical Green function $\mathcal{G}$ and summation is carried out over the repeated indexes.

The next κ^{-2}-correction can also be calculated and, moreover, involves derivatives of $\mathcal{N}$. However, we point out that there are extra corrections in κ^{-1} as we will see in Section VIII.E.

E. Contribution of the Degenerate Periodic Orbits

In Section VIII.C, we observed that degenerate periodic orbits contribute $P(rP+1)$ stationary points ${}^1\mathcal{P}^j_i$ $(i = 1, \ldots, rP+1;\ j = 1, \ldots, P)$ to the

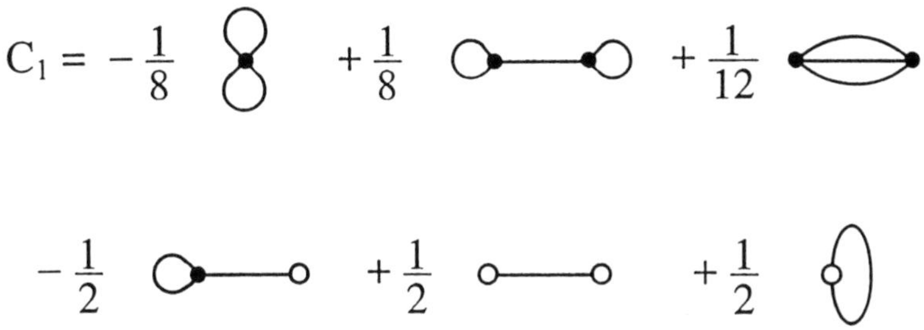

$$C_1 = -\frac{1}{8} \quad + \frac{1}{8} \quad + \frac{1}{12}$$

$$-\frac{1}{2} \quad + \frac{1}{2} \quad + \frac{1}{2}$$

Figure 27. Feynman diagrams contributing to the coefficient $\mathscr{C}_1$ [Eq. (8.28)] of the $\hbar$-correction in the case of billiards.

integrals tr $\hat{Q}^{N+1}$ with $N = rP$; among those, there are only P distinct amplitudes that we will label by ${}^1\mathscr{P}^j$, dropping the index i but keeping j. To fix the ideas, we will consider the case where $\ell_{N,N+1}$ is the length that goes to zero ($j = N$).

Here we are not allowed to use the asymptotic expansion of the Hankel function for the particular factor where $\ell_{N,N+1} \rightarrow 0$ at the stationary point, but we may use it for all the other factors since $\ell_{jj+1} \neq 0$ for $j \neq N$.

We will restrict ourselves to the case of defocusing billiards for which the path from s_N to s_{N+1} goes inside the obstacle (see Fig. 28). Locally,

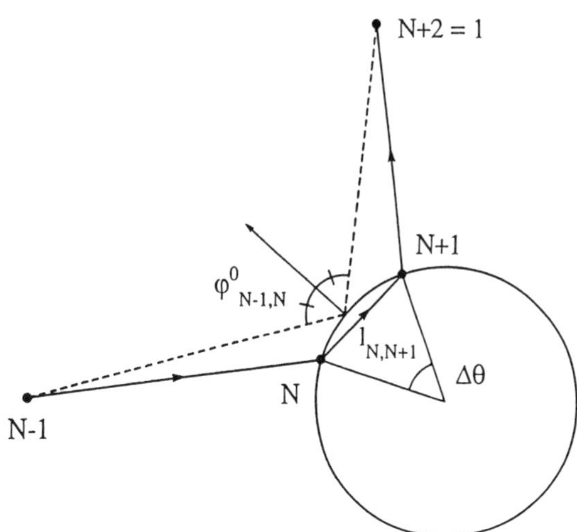

Figure 28. Geometry of a path where one length $\ell_{N,N+1}$ goes to zero between the successive collisions N and $N+1$. The dashed lines represent the stationary phase for which the incident angle $\varphi^0_{N-1,N}$ equals the reflection angle.

we can approximate the wall by an arc of circle of radius a, which is the inverse of the curvature of the border. With this assumption, the angle between the points s_N and s_{N+1} on the circle of radius a is given by

$$\Delta\theta = \frac{s_{N+1} - s_N}{a} \tag{8.29}$$

while the length of the path from N to $N+1$ is

$$\ell_{N,N+1} = 2a\left|\sin\frac{s_{N+1} - s_N}{2a}\right| \tag{8.30}$$

The cosine of the angle $\varphi_{N,N+1}$ between the direction of the path $(N, N+1)$ and the normal interior to the obstacle at the point $N+1$ is

$$\cos\varphi_{N,N+1} = -\left|\sin\frac{s_{N+1} - s_N}{2a}\right| = -\frac{\ell_{N,N+1}}{2a} \tag{8.31}$$

The integral [Eq. (8.10)] becomes

$$\operatorname{tr}\hat{Q}^{N+1}\Big|_{1\mathscr{P}N} = \left(\frac{-i\kappa}{2}\right)^{N+1}\oint ds_1\cdots ds_N ds_{N+1}$$
$$\times\left(-\frac{\ell_{N,N+1}}{2a}\right)H_1^{(1)}(\kappa\ell_{N,N+1})\prod_{j=0}^{N-1}[\cos\varphi_{jj+1}H_1^{(1)}(\kappa\ell_{jj+1})] \tag{8.32}$$

with $j = 0$ identified with $j = N+1$.

This integral has been evaluated elsewhere [20] and the result was

$$\operatorname{tr}\hat{Q}^{N+1}\Big|_{1\mathscr{P}} = \sum_{i=1}^{N+1}\sum_{j=1}^{P}\operatorname{tr}\hat{Q}^{N+1}\Big|_{1\mathscr{P}_i^j}$$
$$= (N+1)\left[\sum_{j=1}^{P}\frac{i}{2\kappa a(\cos\overset{0}{\varphi}_{jj+1})^3} + \mathcal{O}\left(\frac{1}{\kappa^2}\right)\right]\mathscr{I}(r\mathscr{P}) \tag{8.33}$$

with the definition of Eq. (8.27).

Let us note that this correction would disappear if the wall was flat at the point of impact ($a\to\infty$). However, if the periodic orbit becomes tangent to the disk ($\overset{0}{\varphi}_{N-1,N}\to\pi/2$) the contribution diverges because of diffraction.

F. Resummation

Going back to the Fredholm determinant in Eq. (8.7), we obtain

$$\det(\hat{I} - \hat{Q}) = \gamma \prod_{\mathcal{P}} \exp - \sum_{r=1}^{\infty} \frac{1}{r} \mathcal{K}(r\mathcal{P}) \tag{8.34}$$

with

$$\mathcal{K}(r\mathcal{P}) \equiv \mathcal{J}(r\mathcal{P}) \exp\left[\frac{ir}{2\kappa a} \sum_{j=1}^{P} \frac{1}{(\cos \varphi_{jj+1}^{0})^3} + \mathcal{O}\left(\frac{1}{\kappa^2}\right)\right] \tag{8.35}$$

The factor γ in Eq. (8.34) contains the possible contributions of paths where all the lengths go to zero at the critical point as for tr $\hat{Q}$. The factor γ is nonvanishing in the region of the complex plane κ, where the scattering resonances appear.

In order to obtain the scattering resonances, we consider the tail of the series when r is large. We observe that

$$\mathcal{K}(r\mathcal{P}) = (\Gamma_{\mathcal{P}})^r[1 + \mathcal{O}(|\Lambda_{\mathcal{P}}|^{-r})] \tag{8.36}$$

The summation rapidly converges for the factor associated with each unstable periodic orbit since $|\Lambda_{\mathcal{P}}| > 1$. In this case, we get

$$\det[\hat{I} - \hat{Q}(\kappa)] = A(\kappa) \prod_{\mathcal{P}} [1 - \Gamma_{\mathcal{P}}(\kappa)] \tag{8.37}$$

where $A(\kappa)$ is a function that is nonvanishing in a strip of the complex plane κ just below the real axis where the second factor admits its main zeros.

If the corrections in κ^{-1} were neglected we would recover the Selberg zeta function (6.26). At the leading approximation, the factor ζ_0^{-1} corresponds to $\Pi_{\mathcal{P}}(1 - \Gamma_{\mathcal{P}})$ while the product of the other factors with $m = 1, 2, 3, \ldots$ corresponds to the function $A(\kappa)$. It is known that the dominant scattering resonances with the longest lifetimes are the zeros of the factor ζ_0^{-1} at the leading approximation. As a consequence, the scattering resonances are given at the κ^{-1} approximations by the zeros of the inverse of the corrected Ruelle zeta function [20]

$$0 = \frac{1}{\zeta_0^c(\kappa)} \equiv \prod_{\mathcal{P}} (1 - \Gamma_{\mathcal{P}})$$
$$= \prod_{\mathcal{P}} \left\{ 1 - \frac{(-)^P}{|\Lambda_{\mathcal{P}}|^{1/2}} \exp\left[i\kappa \mathcal{L}_{\mathcal{P}} - i\frac{\pi}{2}\mu_{\mathcal{P}} + \frac{i}{\kappa}\mathcal{A}_{\mathcal{P}} + \mathcal{O}(\kappa^{-2})\right] \right\} \tag{8.38}$$

with

$$\mathcal{A}_{\mathcal{P}} \equiv \frac{1}{2a} \sum_{j=1}^{P} \frac{1}{(\cos \varphi_{jj+1}^0)^3} + \frac{3}{8} \sum_{j=1}^{P} \frac{1}{\ell_{jj+1}^0} + \lim_{r\to\infty} \frac{1}{r} \mathcal{C}_1(r\mathcal{P}) \tag{8.39}$$

where $\mathcal{C}_1(r\mathcal{P})$ is given by Eq. (8.28).

G. Application to the Disk Scatterers

The disk scatterers are examples of smooth billiards for which the walls are given by analytic curves of the plane.

1. Three-Disk Scatterer

The three disks of unit radius are assumed to form an equilateral triangle (see Fig. 11) and the centers of the disks are separated by a distance $R = 6$. The structure of the scattering resonances of this classically chaotic system has been elucidated by Gaspard and Rice [114], who used the transfer operator approach limited at the leading semiclassical approximation. Later, Cvitanović and Eckhardt [115] refined the calculation of the scattering resonances by the cycle expansion method still limited to the leading semiclassical approximation in κ^0.

At the κ^0 approximation, the scattering resonances are given as the zeros of the inverse Ruelle zeta function [Eq. (8.38)] with the κ^{-1}-coefficients set equal to zero: $\mathcal{A}_{\mathcal{P}} = 0$. Figure 29 depicts the scattering resonances located in the complex plane of the wavenumber κ at this approximation. Because of the threefold symmetry of the scatterer under the group C_{3v}, each resonance falls into one of the three symmetry representations: A_1 and A_2, which are simply degenerate, and E, which is doubly degenerate [114, 193]. In Fig. 29 we observe that there is a zone below the real axis that is empty of resonances. The formation of this gap can be explained thanks to the topological pressure and the inequality [Eq. (6.31)]. Indeed, the corresponding classical repeller has a partial Hausdorff dimension equal to $d_{\mathrm{H}} = 0.289538$, which is smaller than one

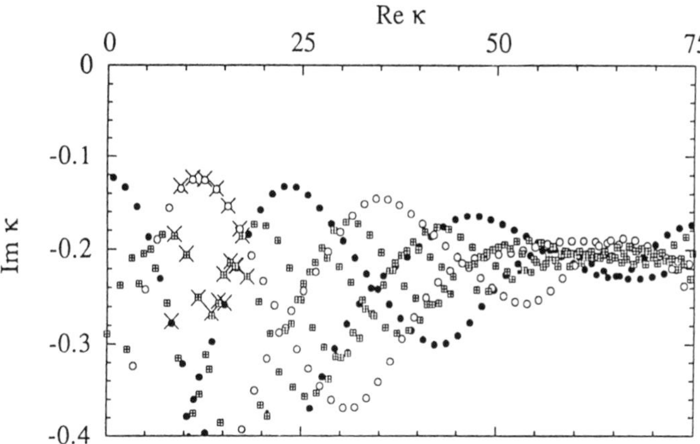

Figure 29. The quantum resonances of the three-disk scatterer ($R:a=6:1$) in the complex plane of the wavenumber κ, compared with the semiclassical approximation calculated with the cycle expansion in the fundamental domain up to period 6. The crosses are the exact poles. The semiclassical A_1 poles are denoted by filled circles, the A_2 poles by white circles, and the E double poles by crossed squares (adapted from [97]).

half. According to the results of Section III.D (see Fig. 17), the topological pressure at $\beta=\frac{1}{2}$ is then negative so that Eq. (6.31) constrains the resonances to have a nonvanishing half-width, which implies the formation of a gap. For two-dimensional billiards, this gap is given by [114]

$$\operatorname{Im}\kappa_n \leq \tilde{P}(\tfrac{1}{2}) = -\chi_{\text{gap}} \tag{8.40}$$

which is valid for $\operatorname{Re}\kappa_n \to \infty$. Now $\tilde{P}(\beta)=P(\beta)/v$ is the topological pressure per unit length, which is equal to the pressure per unit time divided by the particle velocity. In this regard, let us mention here that a velocity is associated to a quantum resonance according to $v_n = (\hbar/m)\operatorname{Re}\kappa_n$. In the three-disk scatterer, the pressure per unit length can be estimated in the limit of large separations R between disks of radius a as follows [93]:

$$\tilde{P}(\beta) \approx \frac{1}{R}\ln\left[\left(\frac{a}{2R}\right)^{\beta} + \left(\frac{\sqrt{3}a}{4R}\right)^{\beta}\right] \qquad \text{for} \qquad R/a \to \infty \tag{8.41}$$

Figure 30 shows the behavior of the quantum gap as a function of the disk separation R/a. Moreover, the quantum gap, which is the limit of the longest lived resonances, is compared with the classical escape rate (at velocity $v=1$) given by Eq. (3.52). We observe that the classical lifetime

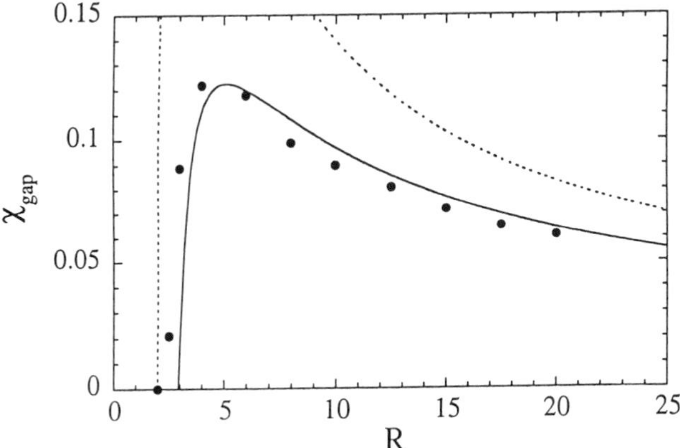

Figure 30. Quantum gap χ_{gap} [Eq. (8.40)] in the distribution of the scattering resonances of the three-disk scatterer versus the interdisk distance R for disks of radius $a = 1$. Comparison between the gap obtained from the exact quantum resonances (dots) and its semiclassical approximation [Eq. (8.40)], $\chi_{\text{gap}} = -\tilde{P}(\frac{1}{2})$, calculated from the zeta function $\zeta_{A_1}^{-1} \simeq 1 - t_0 - t_1$ truncated at the fundamental periodic orbit (solid line). The dashed line is the naive classical expectation, $\chi_{\text{gap}}^{(\text{cl})} = \tilde{\gamma}_{\text{cl}}/2 = -\tilde{P}(1)/2$. The difference between the quantum (solid line) and the quasiclassical (dashed line) is the effect of the Gaspard–Rice lengthening of the resonance lifetimes.

is significantly shorter than the quantum one in the presence of classical chaos. For a nonchaotic scatterer with a single periodic orbit, the classical and quantum lifetimes would coincide (at the leading semiclassical approximation) because the KS entropy would vanish in Eq. (6.31) and the inequality would be replaced by an equality. This lengthening of the resonance lifetimes in the presence of classical chaos may be attributed to the quantum mechanical interferences between the different periodic orbits of the repeller, as shown by Gaspard and Rice [114].

The presence of a gap also has a consequence on the distribution of the imaginary parts of the resonance wavenumbers

$$h(x) = \Pr\{-\text{Im}\, \kappa_n < x\} \tag{8.42}$$

where $x = \Gamma_n^0/v_n = -\text{Im}\, \kappa_n$ is known as the reduced half-width of the corresponding resonance. For the three-disk scatterer, Fig. 31 shows the cumulative distribution function [Eq. (8.42)]. This distribution should be compared with the distributions of the reduced half-widths of a scattering system with ν open channels obtained as chi-square probability dis-

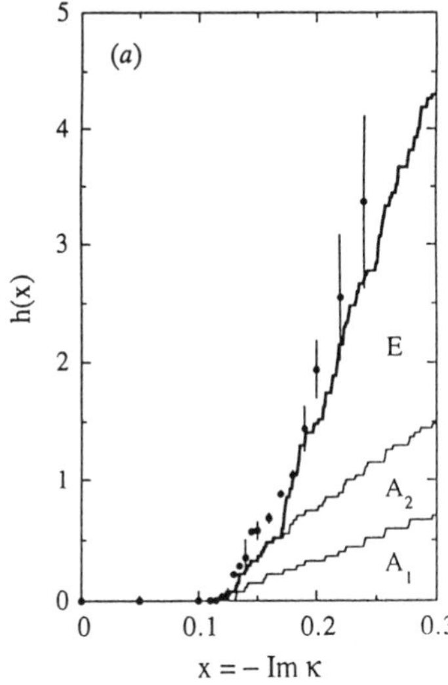

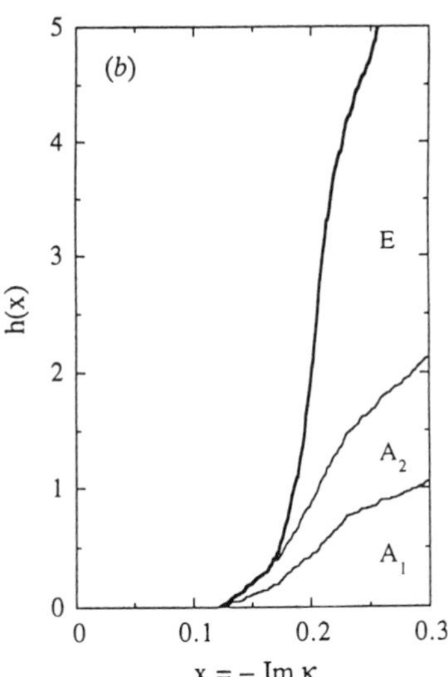

Figure 31. Cumulative distribution function $h(x)$ [Eq. (8.42)] of the imaginary parts $x = -\mathrm{Im}\,\kappa$ of the three-disk resonances for the geometry $R:a = 6:1$. (*a*) Comparison between the semiclassical approximation with cycle expansion truncated at maximum period 6 (solid lines) and the quantum results from [114] (dots with error bars), both calculated in the range $\mathrm{Re}\,\kappa = 8 - 35$. (*b*) Same semiclassical approximation as in (*a*) but calculated in the range $\mathrm{Re}\,\kappa = 0 - 200$. The symbols A_1, A_2, and E denote the contributions from each symmetry representations.

tributions of the parameter ν from the random matrix assumption [194]

$$\frac{dh}{dx} = \frac{x^{\nu/2-1}}{2^{\nu/2}\Gamma(\nu/2)} \exp\left(-\frac{x}{2}\right) \tag{8.43}$$

The Porter–Thomas distribution is the particular case of one open channel ($\nu = 1$). Figure 32 gives several chi-square distribution functions for comparison. We observe that none of those distributions strictly possesses a gap although a gap starts to appear when the number of open channels increases: $\nu \to \infty$. For the widely open three-disk scatterer with $R/a = 6$, we can partly explain the formation of a gap by the large number of open channels that is of the order of the maximum angular momentum involved in the scattering process, namely, $\nu \sim \kappa R$. To understand that the gap is independent of the wavenumber, the semiclassical theory is nevertheless required.

Recently, Alonso and Gaspard [20] applied the semiclassical quantization of the three-disk scatterer at the next-to-leading semiclassical

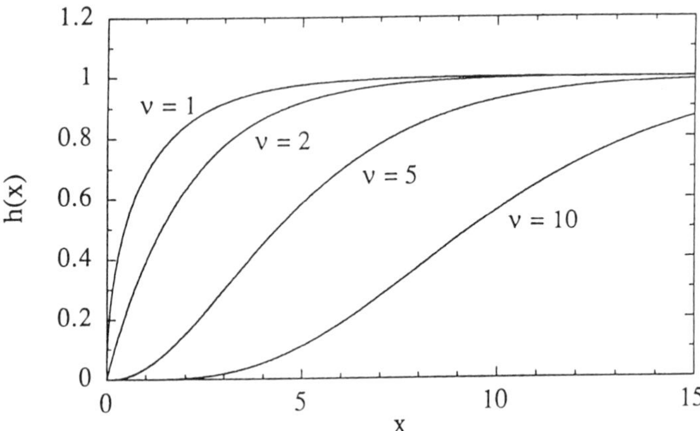

Figure 32. Cumulative functions $h(x)$ of the χ^2 distributions [Eq. (8.43)] for different numbers of open channels ν. These distributions are obtained for random-matrix models of the imaginary parts $x = -\text{Im}\,\kappa$ of the scattering resonances and should be compared with the corresponding distribution for the three-disk scatterer shown in Fig. 31.

approximation in κ^{-1} using the formula [Eqs. (8.38) and (8.39)]. We will now describe the content of this work. The coefficients were evaluated for every prime periodic orbit up to a maximum period of $P = 9$ bounces. The lengths $\mathscr{L}_{\mathscr{P}}$, the stretching factors $\{\Lambda_{\mathscr{P}}\}$, we well as the coefficients $\mathscr{A}_{\mathscr{P}}$ defined by Eq. (8.39) are plotted in Figs. 33 and 34. At the κ^{-1}-approximation, the main contribution is due to the degenerate

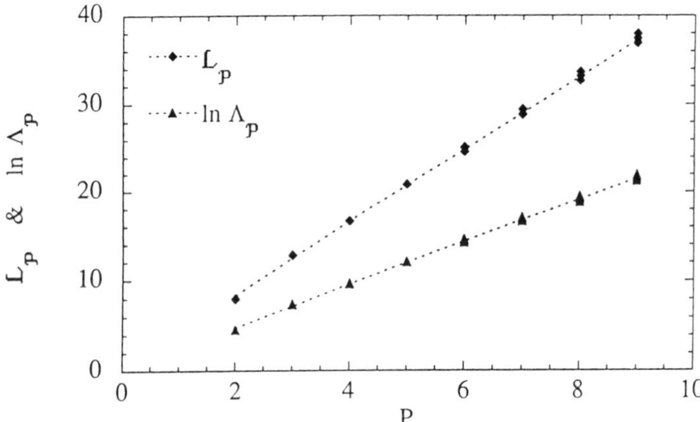

Figure 33. The lengths $\mathscr{L}_{\mathscr{P}}$ and the stability exponents $\ln \Lambda_{\mathscr{P}}$ versus the period P in number of bounces for the prime periodic orbits of the three-disk scatterer ($R:a = 6:1$).

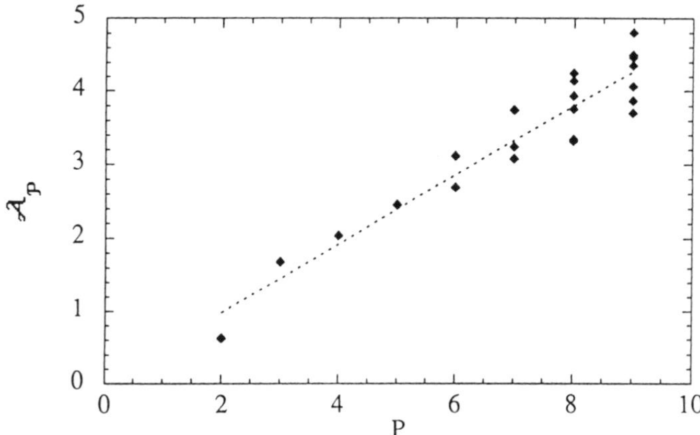

Figure 34. The coefficients $\mathcal{A}_{\mathcal{P}}$ [Eq. (8.39)] versus the period P in number of bounces for the prime periodic orbits of the three-disk scatterer ($R\!:\!a = 6\!:\!1$).

periodic orbits where one length goes to zero. This contribution has a very strong dependency on the incident angle of the trajectory at the point of impact. The contribution diverges in two circumstances: (1) when the periodic orbit becomes tangent to the disks; (2) when the border of the billiard becomes edgy. Here we recognize the effects of wave diffraction that are nonclassical effects. In this respect, the $\hbar$-expansion incorporates purely quantum mechanical effects in the semiclassical trace formula.

The semiclassical calculation is compared with exact quantum mechanical values obtained by Gaspard and Rice [114] and by Wirzba [195]. The numerical convergence of $\mathrm{Re}\,\kappa$ as P increases is depicted in Fig. 35 for the A_2-resonance $\kappa = 14.0070 - i0.13463$. We observe that the real parts are improved at the κ^{-1}-approximation and this improvement increases with $\mathrm{Re}\,\kappa$, as seen in Fig. 36, which shows the relative errors at the κ^{0}- and κ^{-1}-approximations for the A_1- and A_2-resonances. We observe that the error diminishes with $\mathrm{Re}\,\kappa$ faster for the corrected values. The accuracy improves by more than one order of magnitude at large wavenumbers. We see that the accuracy is considerably improved when the κ^{-1}-corrections are taken into account.

However, the imaginary parts have deteriorated at the κ^{-1}-approximation. A similar deterioration is observed in the two-disk scatterer where we have to go to the κ^{-2}-approximation to obtain an improvement on both the real and the imaginary parts of the resonances [19]. The reason for this behavior is that the coefficient of κ^{-1} in Eq. (8.38) is imaginary just like the leading term $i\kappa\mathcal{L}_{\mathcal{P}}$, while the coefficient of κ^{-2} is

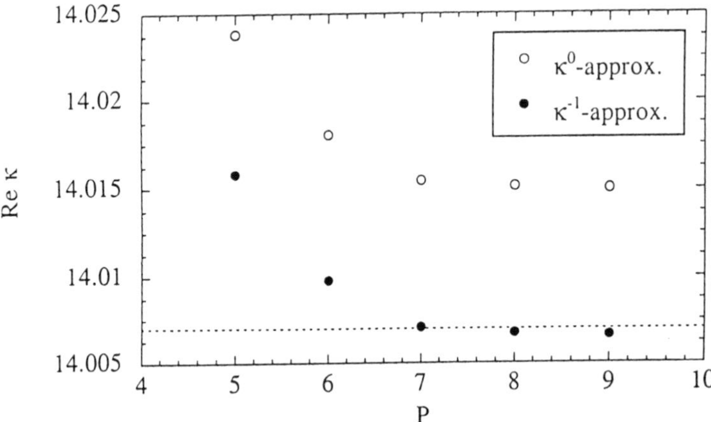

Figure 35. In the three-disk scatterer, Re κ versus the maximum period P of the prime periodic orbits included in the cycle expansion [Eq. (8.38)] used to calculate the A_1-resonance $\kappa = 14.0070 - i0.13463$ without (open circles) and with (filled circles) the κ^{-1}-correction. The dashed line is the location of the exact quantum mechanical value.

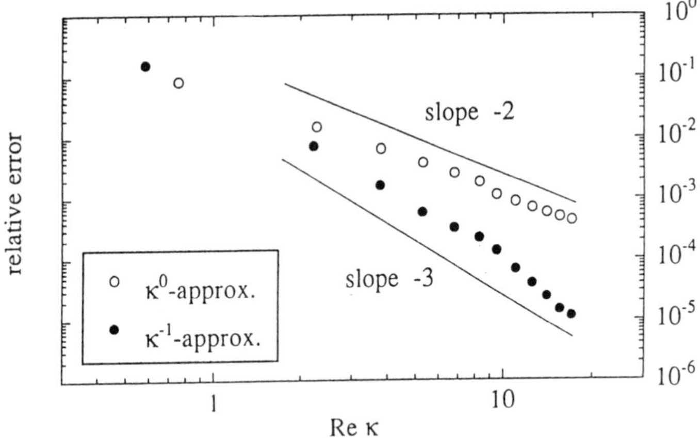

Figure 36. In the three-disk scatterer, relative error $|\mathrm{Re}(\kappa_{\mathrm{sc}} - \kappa_{\mathrm{exact}})/\mathrm{Re}(\kappa_{\mathrm{exact}})|$ on the real parts of the A_1 and A_2 resonances versus Re κ at the κ^0- and κ^{-1}-approximations, both calculated with cycle expansion truncated at maximum period $P = 9$.

real. As a consequence, the real parts are corrected at the κ^{-1}-approximation and the imaginary parts at the κ^{-2}-approximation. At the next-to-leading approximations, there may exist a few low-wavenumber resonances for which Eq. (8.40) is not satisfied. However, Eq. (8.40) still holds as an asymptotic property in the limit of large wavenumbers since the κ^{-n}-corrections become negligible in this semiclassical limit. For the two-disk scatterer, we have been able to obtain the corrections at the κ^{-2}-approximation where the accuracy can be improved up to two orders of magnitude, as summarized in Section VIII.G.2.

2. *Two-Disk Scatterer*

Going to the κ^{-2}-approximation is more complicated because the diagrams become numerous for both the regular periodic orbit and the degenerate ones [19]. However, in the case of the two-disk scatterer there are two main simplifications. (1) There is only one periodic orbit in the scatterer. (2) Many diagrams have vanishing contributions because $\varphi^0_{jj+1} = 0$ here. Thus we have been able to carry this numerical calculation up to the κ^{-2}-corrections for the two-disk scatterer. If the distance between the centers of the disks is $R = 6$ and the disks are of unit radius, we get

$$0 = 1 - \frac{1}{\Lambda^{1/2}} \exp[i\kappa\mathscr{L} + i\kappa^{-1}c_1 - \kappa^{-2}c_2 + \mathscr{O}(\kappa^{-3})] \tag{8.44}$$

where $\mathscr{L} = 8$ is the total length of the orbit, $\Lambda = 97.989795$ is its stability eigenvalue, and $c_1 = \frac{5}{8}$, $c_2 \approx -0.75$ are the two first correction coefficients. These coefficients have ben obtained independently by Turner [196]. Figure 37 depicts the resonances of the two-disk billiard in the

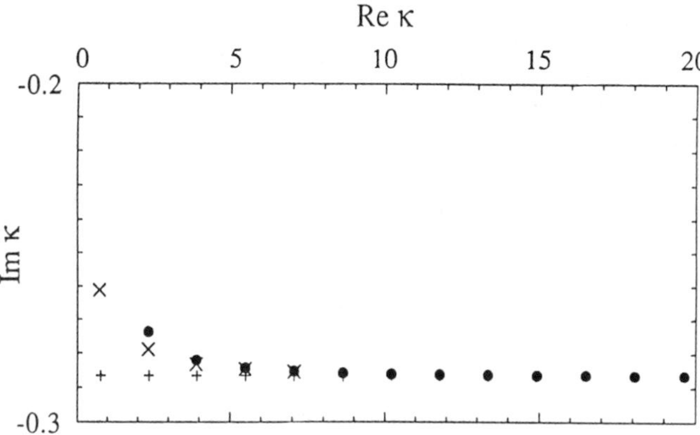

Figure 37. The A_1 quantum resonances of the two-disk scatterer ($R:a = 6:1$) in the complex plane of the wavenumber. The crosses are the exact poles from [195]. The pluses are the poles at the Gutzwiller κ^0-approximation, while the filled circles are the poles at the κ^{-2}-approximation [Eq. (8.44)].

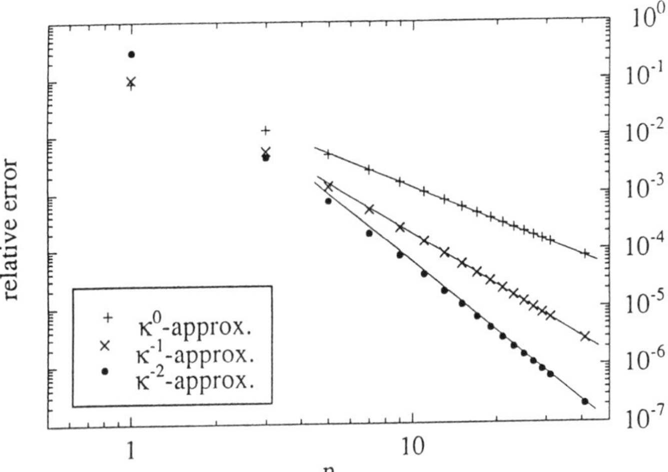

Figure 38. In the two-disk scatterer, relative error $|(\kappa_{sc} - \kappa_{exact})/\kappa_{exact}|$ on the A_1-resonances versus the quantum number $n \sim \mathrm{Re}\,\kappa$ labeling the resonances at the κ^0-, the κ^{-1}- and the κ^{-2}-approximations.

complex plane of κ. Figure 38 shows the relative error between the exact quantum mechanical values of the resonances obtained by Wirzba [195] and the values of the κ^0-, κ^{-1}-, and κ^{-2}-approximations. We observe the successive improvements in accuracy by more than two orders of magnitude at large wavenumbers.

We are now able to reproduce the structure of the resonances at low wavenumbers, which was overlooked in the standard Gutzwiller approximation [19]. In particular, the κ^{-2}-approximation can now explain the lengthening of the lifetimes at low energy. This lengthening is a new effect that arises only at the next-to-leading semiclassical approximations as well as in the absence of classical chaos, contrary to the preceding Gaspard–Rice lengthening mentioned in Section VIII.G.1.

This study on the disk scatterers shows that the semiclassical method can describe particular effects at its successive orders of approximation. These effects would remain unexplained in quantum mechanics, which would precisely locate the resonances but without providing an understanding of the resonance structures.

IX. MATRIX HAMILTONIANS

In many molecular systems, a set of degrees of freedom remain in quantum regimes with low quantum numbers while others reach semiclassical regimes. An example of such a situation occurs when vibrational motion takes place on two or more Born–Oppenheimer (BO) electronic surfaces that are coupled together [197]. Another example is a spin S in an inhomogeneous magnetic field, in which case there are $2S + 1$ coupled potential surfaces [198, 199]. Still another example is a Van der Waals molecule, such as benzene–helium or benzene–argon, where the rare gas adatom moves on different potential surfaces corresponding to different quantized vibrational states of the host molecule of benzene [200]. The separation of the orders of magnitude of the vibrational frequencies of the covalent bonds with respect to the Van der Waals bonds justifies the choice of an adiabatic scheme in a first approximation.

In the preceding sections, we have been concerned with the vibrational motion on a single BO surface. As soon as there are several coupled BO surfaces, new phenomena are possible, such as surface hopping and geometric phases at conical intersections.

One of our purposes in this section is to discuss how the presence of several BO surfaces modifies the results of the previous sections. In these sections, we saw that the semiclassical approximation essentially has four levels:

0. The quasiclassical level given by microcanonical ensembles of orbits according to the Weyl–Wigner method.
1. The leading semiclassical level given in terms of real periodic orbits and their linear stability.
2. The level of $\hbar$-corrections given as series in powers of $\hbar$, which are related to the nonlinear stabilities of the real periodic orbits.
3. The superadiabatic level with exponentially small corrections of the form $\exp(-C/\hbar)$ given in terms of complex periodic orbits.

We will see that, for matrix Hamiltonians, surface hopping is not allowed at the three first levels but can only be described at the fourth level in terms of complex periodic orbits [201]. Accordingly, nonadiabatic transitions between potential surfaces are not described by classical dynamics of real orbits. A recent work of Littlejohn and Weigert [198, 199] showed that the matrix Hamiltonian can be diagonalized to obtain a set of Hamiltonians that determine the motion on each individual potential surface. The transformed Hamiltonians are given as series of powers of the Planck constant. For each one of them, we can apply the semiclassical methods of the preceding sections. Therefore, we see that series in powers of the Planck constant appear at two points in the treatment of matrix Hamiltonians: (1) in the derivation of the Hamiltonians of each potential surface; (2) in the application of the semiclassical methods of the preceding sections to each one of them.

In Section IX.A, we will first present this reduction of the matrix Hamiltonian to a set of individual Hamiltonians. Subsequently, the problems of surface hopping and of geometric phases at conical intersections will be discussed and illustrated with examples.

A. Diagonalization of Matrix Hamiltonians

We suppose that the Hamiltonian is an $N \times N$ matrix of the form

$$\hat{\mathscr{H}} = \frac{\hat{\mathbf{p}}^2}{2} 1 + \mathrm{V}(\mathbf{q}) \tag{9.1}$$

where 1 is the identity matrix and V is a matrix of potential energy with the matrix elements $V_{\mu\nu}(\mathbf{q})$ $(\mu, \nu = 1, 2, \ldots, N)$. For simplicity, we do not treat the case where the off-diagonal elements also depend on the momentum $\hat{\mathbf{p}}$.

A spin in an inhomogeneous magnetic field is an example to which the following theory applies. The matrix potential is then given by

$$\mathrm{V}(\mathbf{q}) = -\mu_{\mathrm{B}} \mathbf{B}(\mathbf{q}) \cdot \mathrm{S} \tag{9.2}$$

where S is the spin angular momentum with multiplicity $N = 2S + 1$ and μ_{B} is the Bohr magneton. For the case of coupling between different BO surfaces, the potential matrix $\mathrm{V}(\mathbf{q})$ is a general matrix, each element of which depends on position $\mathbf{q}$.

Let us now proceed to the diagonalization of the Hamiltonian [Eq. (9.1)] essentially following a derivation by Littlejohn and Weigert [199]. In a first step, the potential matrix is diagonalized into its position-depending eigenvalues and eigenvectors

$$\sum_{\beta=1}^{N} V_{\alpha\beta}(\mathbf{q}) \tau_{\beta}^{(\mu)}(\mathbf{q}) = v^{(\mu)}(\mathbf{q}) \tau_{\alpha}^{(\mu)}(\mathbf{q}) \tag{9.3}$$

such that we obtain

$$V_{\alpha\beta}(\mathbf{q}) = \sum_{\mu=1}^{N} \tau_{\alpha}^{(\mu)}(\mathbf{q}) v^{(\mu)}(\mathbf{q}) \tau_{\beta}^{(\mu)*}(\mathbf{q}) \tag{9.4}$$

Since the potential matrix is Hermitian, the diagonalization is carried out by a unitary matrix $\mathrm{U}_0 = \mathrm{W}$ according to

$$\mathrm{W}^{\dagger}(\mathbf{q}) \mathrm{V}(\mathbf{q}) \mathrm{W}(\mathbf{q}) = \mathrm{v}(\mathbf{q}) \tag{9.5}$$

The unitary matrix is composed of the local eigenvectors of the potential matrix

$$W_{\mu\nu}(\mathbf{q}) = \tau_{\mu}^{(\nu)}(\mathbf{q}) \tag{9.6}$$

Accordingly, we have the local completeness and orthonormality relations

$$\sum_{\mu=1}^{N} |\tau^{(\mu)}\rangle\langle\tau^{(\mu)}| = 1 \leftrightarrow \sum_{\mu=1}^{N} \tau_{\alpha}^{(\mu)}\tau_{\beta}^{(\mu)*} = \delta_{\alpha\beta} \leftrightarrow \mathrm{WW}^{\dagger} = 1 \tag{9.7}$$

$$\langle\tau^{(\mu)}|\tau^{(\nu)}\rangle = \delta_{\mu\nu} \leftrightarrow \sum_{\alpha=1}^{N} \tau_{\alpha}^{(\mu)*}\tau_{\beta}^{(\nu)} = \delta_{\mu\nu} \leftrightarrow \mathrm{W}^{\dagger}\mathrm{W} = 1 \tag{9.8}$$

This unitary transformation is now applied to the full Hamiltonian. Because the kinetic energy operator does not commute with the position operator, we may expect that new terms in powers of $\hbar$ will arise. We have

$$\mathcal{H}^{(1)} = \mathrm{W}^{\dagger}\mathcal{H}\mathrm{W} = \mathrm{W}^{\dagger}\frac{\hat{\mathbf{p}}^{2}}{2}\mathrm{W} + \mathrm{v} \tag{9.9}$$

but the transformation depends on position so that

$$\mathrm{W}^{\dagger}\hat{\mathbf{p}}\mathrm{W} = \hat{\mathbf{p}} - i\hbar\mathrm{W}^{\dagger}\frac{\partial\mathrm{W}}{\partial\mathbf{q}} \tag{9.10}$$

Since W is unitary, the extra term in Eq. (9.10) is a Hermitian matrix that defines a matrix of potential vectors

$$\mathbf{A} = i\mathrm{W}^{\dagger}\frac{\partial\mathrm{W}}{\partial\mathbf{q}} \tag{9.11}$$

with the matrix elements

$$\mathbf{A}_{\mu\nu} = i\langle\tau^{(\mu)}|\nabla|\tau^{(\nu)}\rangle = -i\frac{\langle\tau^{(\mu)}|\nabla\mathrm{V}|\tau^{(\nu)}\rangle}{v^{\mu} - v^{\nu}} \tag{9.12}$$

where the last equality only holds for $\mu \neq \nu$. Accordingly, we obtain

$$\mathcal{H}^{(1)} = \tfrac{1}{2}[\hat{\mathbf{p}}1 - \hbar\mathbf{A}(\mathbf{q})]^{2} + \mathrm{v} \tag{9.13}$$

We observe that the diagonalization of the potential has introduced a gauge structure into the Hamiltonian. Changing the phase of the eigenvectors according to $|\tau^{(\mu)}\rangle \to \exp(ig)|\tau^{(\mu)}\rangle$ implies the following modification of the potential vector: $\mathbf{A} \to \mathbf{A} - \nabla g 1$.

Ordered in powers of $\hbar$, the new Hamiltonian [Eq. (9.13)] is composed of three terms as follows:

$$\begin{aligned}
\hat{\mathcal{H}}^{(1)} &= \hat{\mathcal{H}}_0^{(1)} + \hat{\mathcal{H}}_1^{(1)} + \hat{\mathcal{H}}_2^{(1)} \qquad \text{with} \\
\hat{\mathcal{H}}_0^{(1)} &= \frac{\hat{\mathbf{p}}^2}{2} 1 + \mathrm{v} \\
\hat{\mathcal{H}}_1^{(1)} &= -\frac{\hbar}{2}(\mathbf{A} \cdot \hat{\mathbf{p}} + \hat{\mathbf{p}} \cdot \mathbf{A}) \\
\hat{\mathcal{H}}_2^{(1)} &= \frac{\hbar^2}{2} \mathbf{A}^2
\end{aligned} \tag{9.14}$$

The leading Hamiltonian $\hat{\mathcal{H}}_0^{(1)}$ is already diagonal but the next-to-leading ones are not. Therefore, we must continue the transformation by successive unitary transformations

$$\hat{\mathcal{H}}^{(n+1)} = \hat{\mathrm{U}}_n^{\dagger} \hat{\mathcal{H}}^{(n)} \hat{\mathrm{U}}_n \tag{9.15}$$

with

$$\hat{\mathrm{U}}_1 = e^{i\hbar \hat{\mathrm{S}}_1}, \qquad \hat{\mathrm{U}}_2 = e^{i\hbar^2 \hat{\mathrm{S}}_2}, \ldots \tag{9.16}$$

such that $\hat{\mathrm{U}}_n$ eliminates the nondiagonal terms in $\hbar^n$. By this procedure, which is reminiscent of the Van Vleck contact transformation method [134], all the successive terms can be formally diagonalized. We obtain

$$\hat{\hat{\mathcal{H}}} = \cdots e^{-i\hbar^2 \hat{\mathrm{S}}_2} e^{-i\hbar \hat{\mathrm{S}}_1} \hat{\mathcal{H}}^{(1)} e^{i\hbar \hat{\mathrm{S}}_1} e^{i\hbar^2 \hat{\mathrm{S}}_2} \cdots \tag{9.17}$$

with

$$\begin{aligned}
\hat{\hat{\mathcal{H}}} &= \hat{\mathcal{H}}_0^{(1)} + \hbar\{\hat{\mathcal{H}}_1^{(1)} - i[\hat{\mathrm{S}}_1, \hat{\mathcal{H}}_0^{(1)}]\} + \hbar^2\{\hat{\mathcal{H}}_2^{(1)} - i[\hat{\mathrm{S}}_1, \hat{\mathcal{H}}_1^{(1)}] \\
&\quad - \tfrac{1}{2}\hat{\mathrm{S}}_1^2 \hat{\mathcal{H}}_0^{(1)} + \hat{\mathrm{S}}_1 \hat{\mathcal{H}}_0^{(1)} \hat{\mathrm{S}}_1 - \tfrac{1}{2}\hat{\mathcal{H}}_0^{(1)} \hat{\mathrm{S}}_1^2 - i[\hat{\mathrm{S}}_2, \hat{\mathcal{H}}_0^{(1)}]\} + \mathcal{O}(\hbar^3) \\
&= \hat{\hat{\mathcal{H}}}_0 + \hbar \hat{\hat{\mathcal{H}}}_1 + \hbar^2 \hat{\hat{\mathcal{H}}}_2 + \mathcal{O}(\hbar^3)
\end{aligned} \tag{9.18}$$

This calculation shows that

$$\hat{\mathrm{S}}_1 = \tfrac{1}{2}(\mathbf{R} \cdot \hat{\mathbf{p}} + \hat{\mathbf{p}} \cdot \mathbf{R}) \tag{9.19}$$

with the matrix elements

$$\mathbf{R}_{\mu\mu} = 0 \qquad \text{and} \qquad \mathbf{R}_{\mu\nu} = -i\frac{\mathbf{A}_{\mu\nu}}{v^{\mu} - v^{\nu}} \qquad \text{for} \qquad \mu \neq \nu \tag{9.20}$$

The matrix $\hat{\mathrm{S}}_2$ does not need to be calculated if we limit ourselves to the second order because the second-order terms are given by the diagonal elements of the corresponding matrices while $[\hat{\mathrm{S}}_2, \hat{\mathscr{H}}_0^{(1)}]$ has vanishing diagonal elements. Finally, the Hamiltonian diagonalized up to second order takes the following form

$$\hat{\tilde{\mathscr{H}}}_{\mu\nu} = \hat{h}^{(\mu)}\delta_{\mu\nu} \tag{9.21}$$

with

$$\hat{h}^{(\mu)} = \tfrac{1}{2}(\hat{\mathbf{p}} - \hbar\mathbf{A}_{\mu\mu})^2 + v^{(\mu)}(\mathbf{q}) + \frac{\hbar^2}{2}\sum_{\nu(\neq\mu)} \mathbf{A}_{\mu\nu}^{\dagger} \cdot \mathbf{A}_{\mu\nu} + \hbar^2 \sum_{\nu(\neq\mu)} \frac{\mathbf{A}_{\mu\nu}^{\dagger}\mathbf{A}_{\mu\nu}}{v^{(\mu)} - v^{(\nu)}} : \hat{\mathbf{p}}\hat{\mathbf{p}} + \mathscr{O}(\hbar^3) \tag{9.22}$$

where $v^{(\mu)}$ and $\mathbf{A}_{\mu\nu}$ depend on the positions $\mathbf{q}$ and where : denotes a contraction between the tensors on the left- and right-hand sides.

Because the Hamiltonian is now completely diagonal, the different components of the transformed wave function $\tilde{\Psi} = \hat{\mathrm{U}}^{\dagger}\Psi = (\tilde{\psi}_1, \ldots, \tilde{\psi}_N)$ are independent of each other in this new representation. Nevertheless, the effective coupling between the potential surfaces has been taken into account, first, because the eigenpotentials $v^{(\mu)}$ depend on the off-diagonal elements of the potential matrix V and, second, because these off-diagonal elements also enter the vector potentials $\mathbf{A}_{\mu\nu}$.

However, the Hamiltonian [Eqs. (9.21) and (9.22)] cannot explain nonadiabatic transitions between the potential surfaces, which are now decoupled. Consequently, the individual Hamiltonians $\hat{h}^{(\mu)}$ are all Hermitian and conserve the probability on each potential surface separately. This result leads us to conclude that the description of nonadiabatic transitions has been pushed into the nonanalytic terms in the Planck constant and that their amplitudes are exponentially small like $\exp(-C/\hbar)$ [201]. The examples of Sections IX.B and C will illustrate this point.

With respect to the dynamics on the separated surfaces, the individual Hamiltonians [Eq. (9.22)] can be treated by the semiclassical methods of the preceding sections, with the new feature that a potential vector appears as in electromagnetism. We also observe that the Hamiltonians

[Eq. (9.22)], which determine the classical periodic orbits, depend on the Planck constant. Besides this dependency, the $\hbar$-expansion of the preceding sections will introduce other terms in powers of $\hbar$ so that here the $\hbar$-corrections have the two origins (1) and (2) as mentioned earlier.

If we limit ourselves to the first order in $\hbar$, the classical equations of motion are modified by the presence of the vector potential $\mathbf{A}_{\mu\mu}(\mathbf{q})$. In particular, the time derivative of the positions is now related to momentum according to

$$\dot{\mathbf{q}} = \mathbf{p} - \hbar \mathbf{A}_{\mu\mu}(\mathbf{q}) + \mathcal{O}(\hbar^2) \tag{9.23}$$

As a consequence, the reduced action of a periodic orbit—which enters in the semiclassical trace formulas [Eqs. (6.7), (6.13), (6.63), or (7.15)] and the zeta functions [Eqs. (6.21) or (6.28)]—is now given by

$$S = \oint \mathbf{p} \cdot d\mathbf{q} = \oint [\dot{\mathbf{q}} + \hbar \mathbf{A}_{\mu\mu}(\mathbf{q}) + \mathcal{O}(\hbar^2)] \cdot d\mathbf{q} \tag{9.24}$$

Therefore, the vector potential introduces an extra phase into the trace formulas and the zeta function, which is of the same order in $\hbar$ as the Maslov index [see, e.g., Eq. (6.7)]. This extra phase, $\exp(i\gamma_C)$, is given in terms of the position-dependent eigenvectors $\tau^{(\mu)}(\mathbf{q})$ and potential matrix $\mathsf{V}(\mathbf{q})$ as follows [202–205]:

$$\begin{aligned}
\gamma_C &= \oint_{\partial C} \mathbf{A}_{\mu\mu}(\mathbf{q}) \cdot d\mathbf{q} \\
&= -\operatorname{Im} \oint_{\partial C} \langle \tau^{(\mu)} | d\tau^{(\mu)} \rangle \\
&= -\operatorname{Im} \int\!\!\int_C \langle d\tau^{(\mu)} | \wedge | d\tau^{(\mu)} \rangle \\
&= -\operatorname{Im} \int\!\!\int_C \sum_{\nu(\neq\mu)} \frac{\langle \tau^{(\mu)} | d\mathsf{V} | \tau^{(\nu)} \rangle \wedge \langle \tau^{(\nu)} | d\mathsf{V} | \tau^{(\mu)} \rangle}{(v^{(\mu)} - v^{(\nu)})^2}
\end{aligned} \tag{9.25}$$

where we use the definition [Eq. (9.12)] of the vector potential and $d\tau^{(\mu)} = d\mathbf{q} \cdot \nabla\tau^{(\mu)}$ between the first and the second lines, and Stokes' theorem is applied between the second and the third lines with $\wedge$ denoting the exterior product. From the third to the fourth line, we used the completeness relation [Eq. (9.7)] following an argument by Berry [205]. In our semiclassical context, the circuit C denotes a closed loop in position space traced by a classical periodic orbit. The phase [Eq. (9.25)] is well-known today as the geometric phase, after many fundamental contributions since the works of Mead and Truhlar [203] and, particularly, of Berry [205]. Let us remark that the geometric phase is a gauge invariant as shown by the last line of Eq. (9.25), but that the application of the above formalism rests on the occurrence of nonzero diagonal elements of **A**, together with the single-valuedness of the eigenvectors, $\tau^{(\mu)}(\mathbf{q})$, as $\mathbf{q}$ moves along the circuit C. As a consequence, we will see hereafter than the geometric phase appears only in problems where different potential surfaces are degenerate at some position $\mathbf{q}$, for example, at conical intersections, while the geometric phase does not arise in usual situations where the potential surfaces remain well separated.

In order to fix the ideas, we consider simple examples.

B. Two-Surface 1F Model and Surface Hopping

One of the simplest examples is given by the Hamiltonian

$$\hat{\mathcal{H}} = \frac{\hat{\mathbf{p}}^2}{2m} 1 + \mathrm{V} \tag{9.26}$$

with the potential [201]

$$\mathrm{V} = \begin{pmatrix} ax & b \\ b & -ax \end{pmatrix} \tag{9.27}$$

This real symmetric matrix can be diagonalized by an orthogonal transformation

$$\mathrm{W} = [\tau^{(+)}\tau^{(-)}] = \begin{bmatrix} \cos(\theta/2) & -\sin(\theta/2) \\ \sin(\theta/2) & \cos(\theta/2) \end{bmatrix} \tag{9.28}$$

with the angle $\theta = \mathrm{arctg}(b/ax)$, such that the eigenpotentials are $v^{(\pm)} = \pm \sqrt{a^2x^2 + b^2}$. Hence, the vector potential is given by

$$\mathbf{A} = \begin{pmatrix} 0 & -\frac{i}{2}\frac{d\theta}{dx} \\ \frac{i}{2}\frac{d\theta}{dx} & 0 \end{pmatrix} \qquad \text{with} \qquad \frac{d\theta}{dx} = -\frac{ab}{2(a^2x^2 + b^2)} \tag{9.29}$$

In this problem, we observe that the eigenvectors remain single-valued in the domain of variation of the angle $0 < \theta < \pi$ since it is never closing a whole circle. Therefore, the choice [Eq. (9.28)] of eigenvectors is appropriate for the potential [Eq. (9.27)]. As a consequence, the vector potential matrix has no diagonal components in Eq. (9.29) and the geometric phase is not present in this problem. The eigenvectors remain single-valued as long as the two potential surfaces do not intersect, which allows us to choose a gauge where the diagonal components of the potential vector vanish, whereupon the absence of the geometric phase in such nondegenerate systems is guaranteed.

The individual Hamiltonians are therefore given by

$$\hat{h}^{(\pm)} = -\frac{\hbar^2}{2m}\left[1 \pm \frac{\hbar^2 a^2 b^2}{4(a^2x^2 + b^2)^{5/2}}\right]\frac{d^2}{dx^2}$$

$$\pm\sqrt{a^2x^2 + b^2}\left[1 \pm \frac{\hbar^2 a^2 b^2}{8(a^2x^2 + b^2)^{5/2}}\right] + \mathcal{O}(\hbar^3) \qquad (9.30)$$

At the second order, the coupling between the surfaces leads to a renormalization of the mass into effective masses associated with each surface and depending on the position of the system, as well as a corresponding modification of the potential (see Fig. 39). These $\hbar$-corrections imply modifications in the energy eigenvalues and eigenfunctions of the corresponding potentials. In the present example, we have two different Hamiltonians. The upper Hamiltonian $\hat{h}^{(+)}$ is bounded so that it sustains quantum states that are stable, and its spectrum is discrete at this level of approximation. On the other hand, the lower Hamiltonian $\hat{h}^{(-)}$ is of the scattering type with nonvanishing accelerations at large distances (which is particular to this model).

The energy levels of the upper surface can be obtained with the tools of the preceding sections. At the leading order, we have

$$E_n = E_n^{(0)}(\hbar) + \hbar^2 E_n^{(2)}(\hbar) + \cdots \qquad (9.31)$$

with the following condition on the energy at lowest order

$$S^{(+)}(E_n^{(0)}) = 2\pi\hbar(n + \tfrac{1}{2}) \qquad (9.32)$$

where the reduced action is given by

$$S^{(+)}(E) = 2\int_{x_<^{(+)}}^{x_>^{(+)}} \sqrt{2m[E - v^{(+)}(x)]}\, dx$$

$$= 2\int_{x_<^{(+)}}^{x_>^{(+)}} \sqrt{2m(E - \sqrt{a^2x^2 + b^2})}\, dx \qquad (9.33)$$

$x_<^{(+)}$ and $x_>^{(+)}$ are the two turning points of the bounded potential $v^{(+)}(x)$. To be consistent, a calculation of the energy levels at the second order

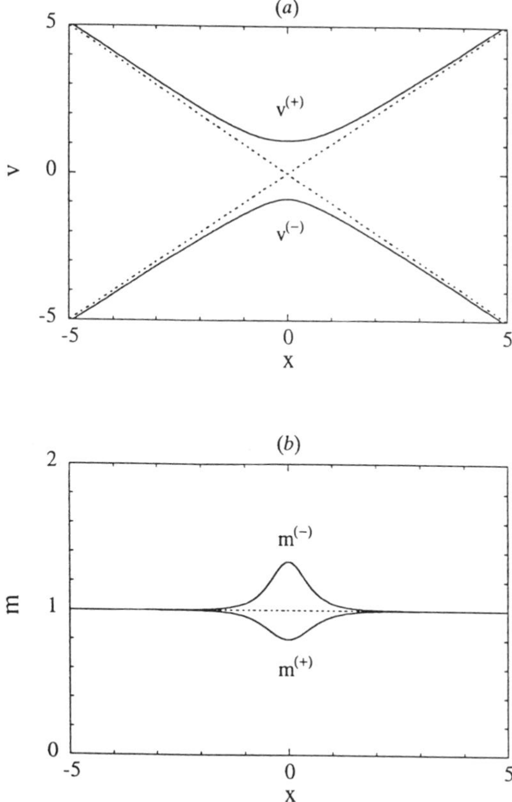

Figure 39. (*a*) Effective uncoupled potentials $v^{(\pm)}(x)$ and (*b*) effective masses $m^{(\pm)}(x)$ for the matrix potential [Eq. (9.27)] at the $\hbar^2$-approximation [Eq. (9.30)].

should include not only the corrections in $\hbar^2$ in the potential and the effective mass, but in addition the $\hbar^2$-corrections of the preceding sections.

Figure 40 shows the classical phase spaces of the upper and lower surfaces. At low energy, the motion is harmonic on $v^{(+)}$ but at high energies the potential seen by the system becomes $v^{(+)} \simeq a|x|$. In this limit $E \gg b$, the turning points are $x \simeq \pm E/a$, and the reduced action is obtained as

$$S^{(+)}(E) \simeq \frac{8\sqrt{2m}}{3a} E^{3/2} \tag{9.34}$$

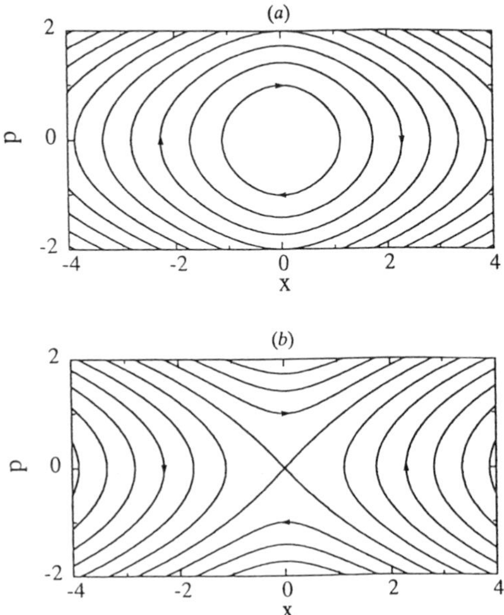

Figure 40. Phase portraits of classical motion in the (a) upper and (b) lower uncoupled potentials $v^{(\pm)}(x) = \pm\sqrt{a^2x^2 + b^2}$ at the $\hbar^0$-approximation.

so that the energy levels are

$$E_n \simeq \left[\frac{3\pi a\hbar}{4\sqrt{2m}}\left(n + \frac{1}{2}\right)\right]^{2/3} \qquad \text{for} \qquad E_n \gg b \tag{9.35}$$

We obtain discrete energy levels without imaginary parts because nonadiabatic transitions are neglected as we explained before. In reality, however, these states are metastable and their lifetime is exponentially small in $\hbar$, as we now explain. The lifetimes can be obtained by considering complex orbits. Indeed, we can jump from the upper surface to the lower one via complex paths. Figure 41 shows the geometry of the potential surfaces for a complex-valued x coordinate, where we see that the two surfaces join each other at $x = \pm ib/a$. Therefore we have a complex periodic orbit that oscillates between these degenerate points, and alternately visits the upper and the lower surface (see Fig. 42). In a semiclassical scheme, the surface hopping would occur each time the real periodic orbit oscillating in the bounded potential $v^{(+)}$ reaches the point $x = 0$, where the complex periodic orbit can make the connection to the lower surface $v^{(-)}$ so that the system escapes to large distances. By this

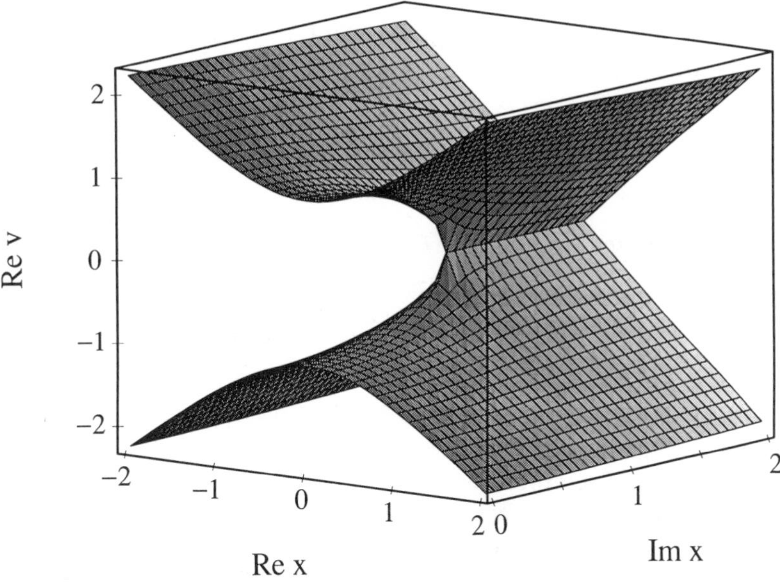

Figure 41. Real parts of the potentials $\operatorname{Re} v^{(\pm)}(x) = \operatorname{Re}(\pm\sqrt{a^2x^2 + b^2})$ at the $\hbar^0$-approximation represented in the complex plane of position $x = \operatorname{Re} x + i \operatorname{Im} x$.

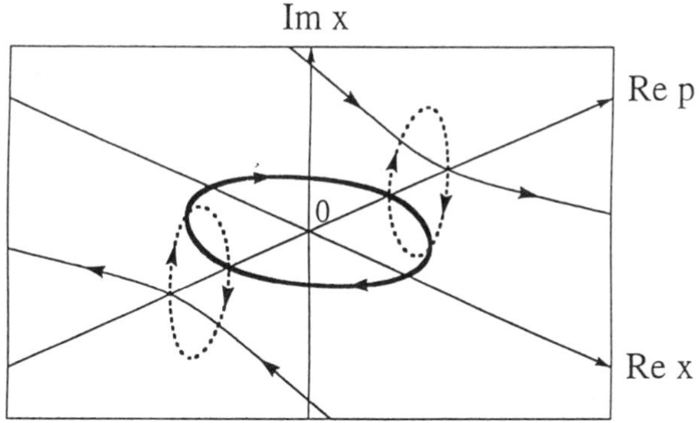

Figure 42. Real and complex periodic orbits involved in the process of surface hopping between the upper and lower branches of the complex surface of Fig. 41, depicted in the variables $\operatorname{Re} x$, $\operatorname{Re} p$, and $\operatorname{Im} x$. The closed real orbit around the origin describes the periodic oscillations in $v^{(+)}(x)$. The dashed orbits with $\operatorname{Im} x \neq 0$ are the complex periodic orbits responsible for surface hopping between $v^{(\pm)}(x)$. The two open real orbits describe the escape toward large separations on the surface $v^{(-)}(x)$.

nonadiabatic process, a lifetime is given to each energy level, which can be calculated from the imaginary part of the action of the complex periodic orbit, namely,

$$\operatorname{Im} S_c = \frac{2b}{a}\sqrt{2mE}\int_0^{\pi/2}\left(\sqrt{1+\frac{b}{E}\cos\xi} - \sqrt{1-\frac{b}{E}\cos\xi}\right)\cos\xi\, d\xi \tag{9.36}$$

where we have set $x = i(b/a)\sin\xi$. In the high-energy limit, we obtain

$$\operatorname{Im} S_c \simeq \frac{\pi b^2}{a}\sqrt{\frac{m}{2E}} \qquad E \gg b \tag{9.37}$$

According to Fig. 42 the reduced action [Eq. (9.36)] of the full complex periodic orbit is twice the reduced action to go from the upper to the lower potential. Therefore, $[1 - \exp(-|\operatorname{Im} S_c|/\hbar)]$ gives the probability of transition between the upper and lower surfaces, which is the square of the corresponding transition amplitude. Since the point $x = 0$, where the transition is possible, is crossed twice in a period of oscillation on $v^{(+)}$, the system has two opportunities to escape from $v^{(+)}$ in a period. Accordingly, the half-width of the quantum states is obtained as

$$\Gamma_n = \frac{\hbar}{\tau_n} \simeq \frac{2\hbar}{T^{(+)}(E_n)}\exp[-\hbar^{-1}|\operatorname{Im} S_c(E_n)|] \tag{9.38}$$

where the period is given by $T^{(+)}(E) = \partial_E S^{(+)}(E)$. Since this reasoning is based on a weak decay of the upper surface states, the above result holds in the energy range

$$\frac{\pi^2 m b^4}{2\hbar^2 a^2} \gg E \tag{9.39}$$

By using Eq. (9.37), we see that the lifetimes become shorter as the energy increases. Indeed, at higher energies, the coupling between the potential surfaces becomes ineffective and the upper surface rapidly loses its probability amplitude. In the limit $b = 0$, the two diabatic surfaces become independent and the system no longer has resonances. Figure 43 shows the distribution of the resonances in the complex energy surface. Time-dependent dynamics due to resonances of a similar type have been observed in femtosecond laser experiments by Zewail and co-workers [206] on the dissociation of NaI.

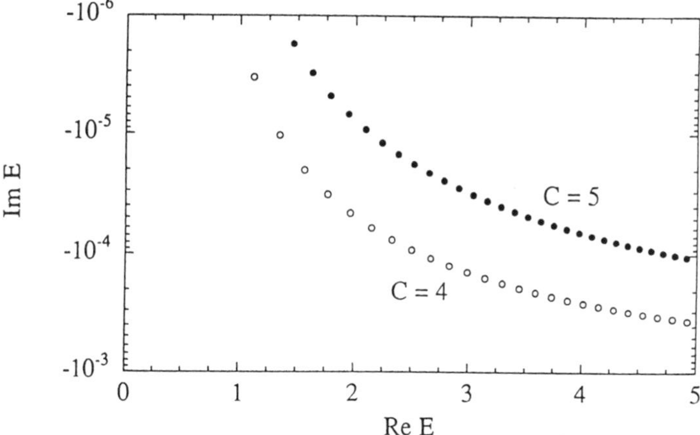

Figure 43. Complex energies of the scattering resonances of the matrix potential [Eq. (9.27)] obtained at the semiclassical approximation of Eqs. (9.32)–(9.33) for two different values of the parameter $C = b\sqrt{mb}/(\hbar a)$, namely, $C = 4$ and $C = 5$. The imaginary parts are due to tunneling and are given by Eqs. (9.36) and (9.38).

C. Conical Intersection and Geometric Phase

1. *Isotropic Conical Intersection*

According to a result of von Neumann and Wigner [207], the positions where the eigenvalues of a real symmetric matrix, such as V in Eq. (9.1), are doubly degenerate form a set of codimension two. Consequently, the potential surfaces given by the eigenvalues $v^{(\mu)}(\mathbf{q})$ of the potential matrix may intersect each other if the system has two degrees of freedom or more. A model of such a system is given by the Hamiltonian [Eq. (9.1)] with the potential matrix

$$\mathrm{V}(\mathbf{q}) = a\begin{pmatrix} x & y \\ y & -x \end{pmatrix} \tag{9.40}$$

whose eigenvalues are $v^{(\pm)} = \pm a\sqrt{x^2 + y^2}$. We see that the potential surfaces form a cone centered at $x = y = 0$, so that we speak of a conical intersection in this situation (see Fig. 44). Such a potential locally appears in triatomic molecules, such as Na_3 or O_3 [204, 208].

When going to polar coordinates ($x = r\cos\theta$, $y = r\sin\theta$), the potential matrix can be diagonalized by the same orthogonal transformation as Eq. (9.28), but here $\theta = \mathrm{arctg}(y/x)$. Now, the angle θ can vary over the whole circle: $0 < \theta \leq 2\pi$. After circling once around the origin, we remark that the matrix W does not come back to its initial value but that a minus sign

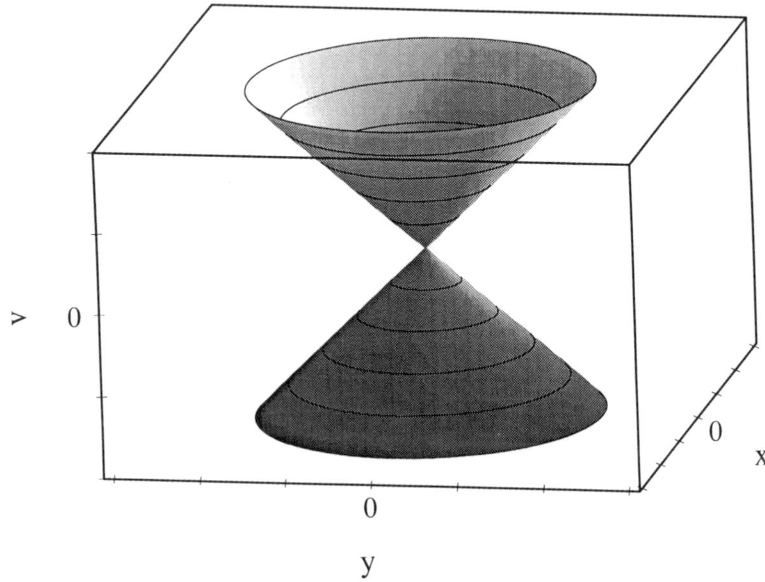

Figure 44. Conical intersection formed by the upper and lower uncoupled potentials $v^{(\pm)}(x, y) = \pm a\sqrt{x^2 + y^2}$ of the matrix potential [Eq. (9.40)] at the $\hbar^0$-approximation.

has now appeared

$$\mathrm{W} \rightarrow (-)^n \mathrm{W} \qquad \text{for} \qquad \theta \rightarrow \theta + 2\pi n \tag{9.41}$$

Therefore, the eigenvectors composing the matrix W are not single-valued in the two-dimensional potential [Eq. (9.40)]. To remedy this problem, we need to perform a gauge transformation, that is, to multiply the eigenvectors by a position-dependent phase: $\tau^{(\mu)'} = \exp(i\theta/2)\tau^{(\mu)}$. The new unitary transformation is now single-valued

$$\mathrm{W}' = e^{i\theta/2}\mathrm{W} \rightarrow \mathrm{W}' \qquad \text{for} \qquad \theta \rightarrow \theta + 2\pi n \tag{9.42}$$

As a consequence of the gauge transformation, the potential vector matrix acquires diagonal elements

$$\mathbf{A} = \begin{pmatrix} -\frac{1}{2}\nabla\theta & -\frac{i}{2}\nabla\theta \\ \frac{i}{2}\nabla\theta & -\frac{1}{2}\nabla\theta \end{pmatrix} \qquad \text{with} \qquad \nabla\theta = \frac{(-y, x)}{x^2 + y^2} \tag{9.43}$$

By applying the results of Section IX.A, the diagonalized Hamiltonians become

$$\hat{h}^{(\pm)} = \frac{1}{2}\left(\hat{p}_x - \frac{\hbar}{2}\frac{y}{x^2+y^2}\right)^2 + \frac{1}{2}\left(\hat{p}_y + \frac{\hbar}{2}\frac{x}{x^2+y^2}\right)^2 \pm a\sqrt{x^2+y^2} + \mathcal{O}(\hbar^2) \tag{9.44}$$

The classical equations of motion are

$$\begin{aligned} \ddot{x} &= +\dot{y}B_z - \frac{\partial v^{(\pm)}}{\partial x} + \mathcal{O}(\hbar^2) \\ \ddot{y} &= -\dot{x}B_z - \frac{\partial v^{(\pm)}}{\partial y} + \mathcal{O}(\hbar^2) \end{aligned} \tag{9.45}$$

where the pseudomagnetic field is given by

$$B_z = \hbar(\partial_x A_{y\pm\pm} - \partial_y A_{x\pm\pm}) = -\pi\hbar\delta(x)\delta(y) \tag{9.46}$$

We remark that the pseudomagnetic field is vanishing everywhere except at the origin where it forms a Dirac distribution. As a consequence, the classical dynamics does not see the effect of the conical intersection at the leading semiclassical approximation that neglects the $\hbar^2$ terms in Eq. (9.45). However, the quantization is dependent on the pseudomagnetic field, which gives a geometric phase to the quantum amplitudes of the orbits encircling the origin, like in the Aharonov–Bohm effect [202].

We observe that the isotropic Hamiltonian [Eq. (9.1)] with Eq. (9.40) commutes with a constant of motion, which is given by the z-component of the total angular momentum, composed of the orbital angular momentum plus a one-half pseudospin, each one of them projected along the z-axis

$$\hat{J}_z = \hat{L}_z 1 + \frac{\hbar}{2}\sigma_z \tag{9.47}$$

with

$$\hat{L}_z = x\hat{p}_y - y\hat{p}_x \qquad \text{and} \qquad \sigma_z = \begin{pmatrix} 0 & -i \\ i & 0 \end{pmatrix} \tag{9.48}$$

such that $[\hat{\mathcal{H}}, \hat{J}_z] = 0$.[2] This isotropic system is therefore separable. The presence of an extra one-half pseudospin shows that half-integer quantum numbers are entering the quantization [208].

We now turn to the semiclassical quantization at the leading approximation in order to determine the resonance states that are localized in the upper half-cone of the system. The classical equations of motion [Eqs. (9.45)–(9.46)] are integrable. If we neglect the terms in $\hbar^2$, the equations can be rewritten in polar coordinates as

$$
\begin{aligned}
&r^2\dot{\theta} = J_z \\
&\frac{\dot{r}^2}{2} + ar + \frac{J_z^2}{2r^2} = E
\end{aligned}
\tag{9.49}
$$

where the two constants of motion are the total angular momentum along the z-axis J_z and the energy E. We see that the classical orbits remain confined in the upper half-cone without approaching the intersection point $r = 0$ except if $J_z = 0$, but we will see below that this value is excluded by the quantization. As in the 1F system discussed above, we should expect that the quantum states supported by the classical orbits are actually metastable states with small lifetimes given by a complex periodic orbit.

Under these conditions, we can perform the semiclassical quantization by the WKB method. At the leading approximation, the Bohr–Sommerfeld conditions read

$$
\begin{aligned}
&S_r^{(+)} = \oint p_r\, dr = \oint \dot{r}\, dr = 2\pi\hbar(n + \tfrac{1}{2}) \\
&S_\theta^{(+)} = \oint p_\theta\, d\theta = \oint (r^2\dot{\theta} + \hbar A_{\theta++})\, d\theta = 2\pi\left(J_z - \frac{\hbar}{2}\right) = 2\pi\hbar m
\end{aligned}
\tag{9.50}
$$

where we used the facts that the radial component of the vector potential vanishes, $A_{r++} = 0$, while its angular component is given by $A_{\theta++} = -\frac{1}{2}$, and that the Maslov indexes are, respectively, $\mu_r = 1$ and $\mu_\theta = 0$. The radial quantum number is the integer $n \in \mathbb{N}$, while the quantum number of orbital angular momentum along the z-axis is the integer $m \in \mathbb{Z}$. As a

[2] There exists another representation where the matrix σ_z is the standard third Pauli matrix, which is diagonal. In this other representation, the matrix potential has the form

$$
\mathrm{V} = \alpha \begin{pmatrix} 0 & x - iy \\ x + iy & 0 \end{pmatrix}
$$

instead of Eq. (9.40). Both representations are connected by a unitary transformation.

consequence of the geometric phase, the total angular momentum is a half-integer multiple of the Planck constant $J_z = \hbar(m + \frac{1}{2})$. The quantization of the radial motion requires the evaluation of the integral

$$S_r^{(+)}(E) = 2\int_{r_<^{(+)}}^{r_>^{(+)}} \sqrt{2\left(E - ar - \frac{J_z^2}{2r^2}\right)}\, dr \tag{9.51}$$

The motion is harmonic at low energies, whereas it is controlled by the linear part of the potential at high energies. In these limits, we obtain

$$\begin{aligned} E_n &= \tfrac{3}{2}(aJ_z)^{2/3} + \hbar\sqrt{3}(a^2/J_z)^{1/3}(n + \tfrac{1}{2}) + \mathcal{O}(\hbar^2) \qquad \text{for} \quad n \simeq 0 \\ E_n &\simeq \left[\frac{3\pi\alpha\hbar}{2\sqrt{2}}(n + \tfrac{1}{2})\right]^{2/3} \qquad \text{for} \quad n \gg 1 \end{aligned} \tag{9.52}$$

Here, we remark that we could have worked in the gauge of Eq. (9.28), where the diagonal components of the potential vector matrix are vanishing, but we should have been careful to avoid missing the geometric phase. Indeed, in that case, the geometric phase is still present but is hidden in the eigenvectors $\tau^{(\pm)}$, which appear in the amplitude of the phase function: $\Psi \simeq \tau^{(\pm)} \exp(iS^{(\pm)}/\hbar - i\mu^{(\pm)}/2)$. Then, it is the eigenvector that produces the phase -1 when circling around the conical intersection rather than the reduced action $S^{(\pm)}$, which contains no potential vector. In the gauge adopted in this section, it is the reduced action that produces the geometric phase because it does contain the potential vector, while the eigenvectors $\tau^{(\pm)'}$ remain single-valued and do not produce an extra phase.

We now turn to an evaluation of the lifetimes of the metastable states supported by the upper half-cone. Figure 45 shows classical phase portraits of the radial motion on the upper and lower half-cones. We are looking for a complex path that will join a closed orbit on the upper surface to an open orbit but on the lower surface, both orbits being at the same energy E and angular momentum J_z. The complex orbit should make a contour around the branch point $r = 0$, where both surfaces join each other. There exists such a complex orbit with $\operatorname{Re} p_r = 0$ with a reduced action having the following imaginary part

$$\begin{aligned} \operatorname{Im} S_c &= 2\int_0^{r_<^{(+)}} \sqrt{2\left(\frac{J_z^2}{2r^2} + ar - E\right)}\, dr - 2\int_0^{r_<^{(-)}} \sqrt{2\left(\frac{J_z^2}{2r^2} - ar - E\right)}\, dr \\ &= 2\int_0^{r_<^{(-)}} \frac{8ar^2\, dr}{\sqrt{J_z^2 + 2ar^3 - 2Er^2} + \sqrt{J_z^2 - 2ar^3 - 2Er^2}} \end{aligned}$$

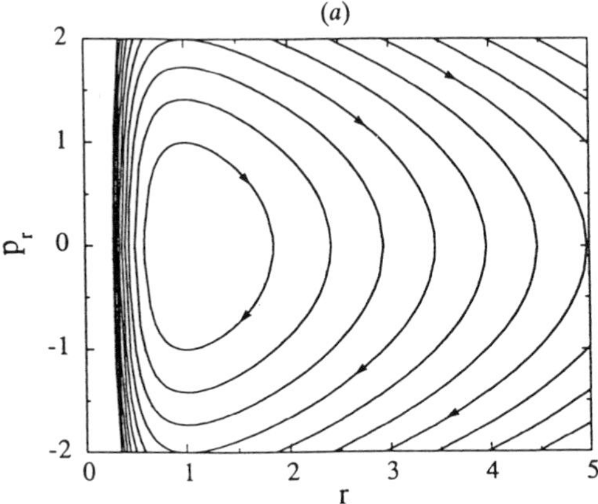

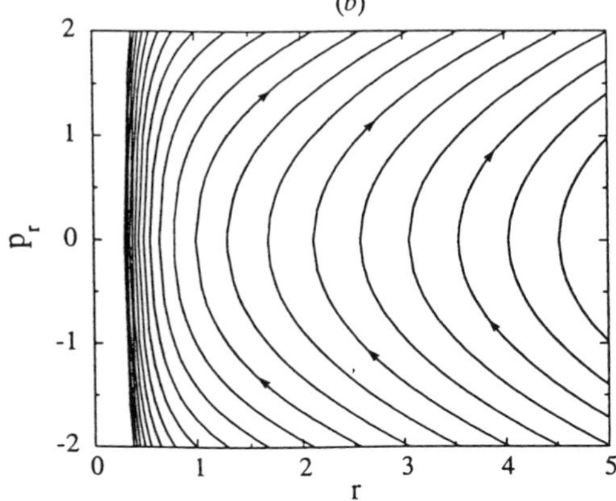

Figure 45. Phase portraits of classical motion in the (*a*) upper and (*b*) lower potentials of the conical intersection of Fig. 44 at the $\hbar^0$ approximations for $a = 1$, $J_z = 1$, and varying energy E in Eq. (9.49) (with $p_r = \dot{r}$).

$$+2\int_{r_<^{(-)}}^{r_<^{(+)}} \sqrt{J_z^2 + 2ar^3 - 2Er^2}\,\frac{dr}{r} \tag{9.53}$$

There is a cancelation between the two singularities at $r = 0$ so that the total integral is finite. At high energies, the integral can be estimated by

$$\mathrm{Im}\, S_c = \frac{\pi a J_z^2}{(2E)^{3/2}}\,[1 + \mathcal{O}(\sqrt{aJ_z/E^{3/2}})] \tag{9.54}$$

Since the turning point is approached only once in a period $T^{(+)}(E) =$

$\partial_E S^{(+)}(E)$ on the upper surface, the half-widths and the lifetimes are given by

$$\Gamma_n = \frac{\hbar}{\tau_n} \simeq \frac{\hbar}{T^{(+)}(E_n)} \exp[-\hbar^{-1}|\text{Im}\, S_c(E_n)|] \tag{9.55}$$

holding in the range $aJ_z^2 \gg \hbar E^{3/2}$.

The structures of the scattering resonances of the conical intersection are depicted in Fig. 46 in the complex plane of energy. We observe that the lifetimes increase with the total angular momentum J_z but decrease with energy. Indeed, at high angular momenta, there is a confinement effect in the upper half-cone due to the centrifugal force. On the other hand, as the energy increases the centrifugal barrier decreases and the conical intersection becomes more easily accessible so that the lifetimes shorten.

2. *Anisotropic Conical Intersection*

In general, the conical intersection will not be isotropic. A more general potential is given by

$$\hat{\mathcal{H}} = \frac{\hat{\mathbf{p}}^2}{2} 1 + \begin{pmatrix} ax & by \\ by & -ax \end{pmatrix} \tag{9.56}$$

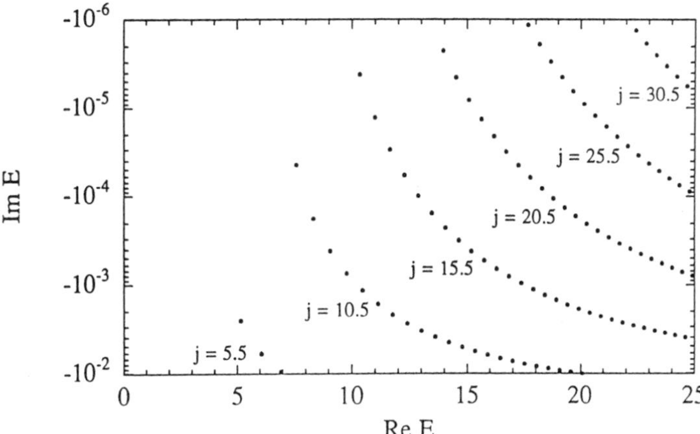

Figure 46. Complex energies of the scattering resonances of the isotropic conical intersection [Eq. (9.40)] obtained at the semiclassical approximation of Eq. (9.51) in units where $a = 1$ and $\hbar = 1$ and for different values of the quantum number $j = m + 1/2$ of angular momentum. The imaginary parts are due to tunneling and are given by Eqs. (9.53)–(9.55).

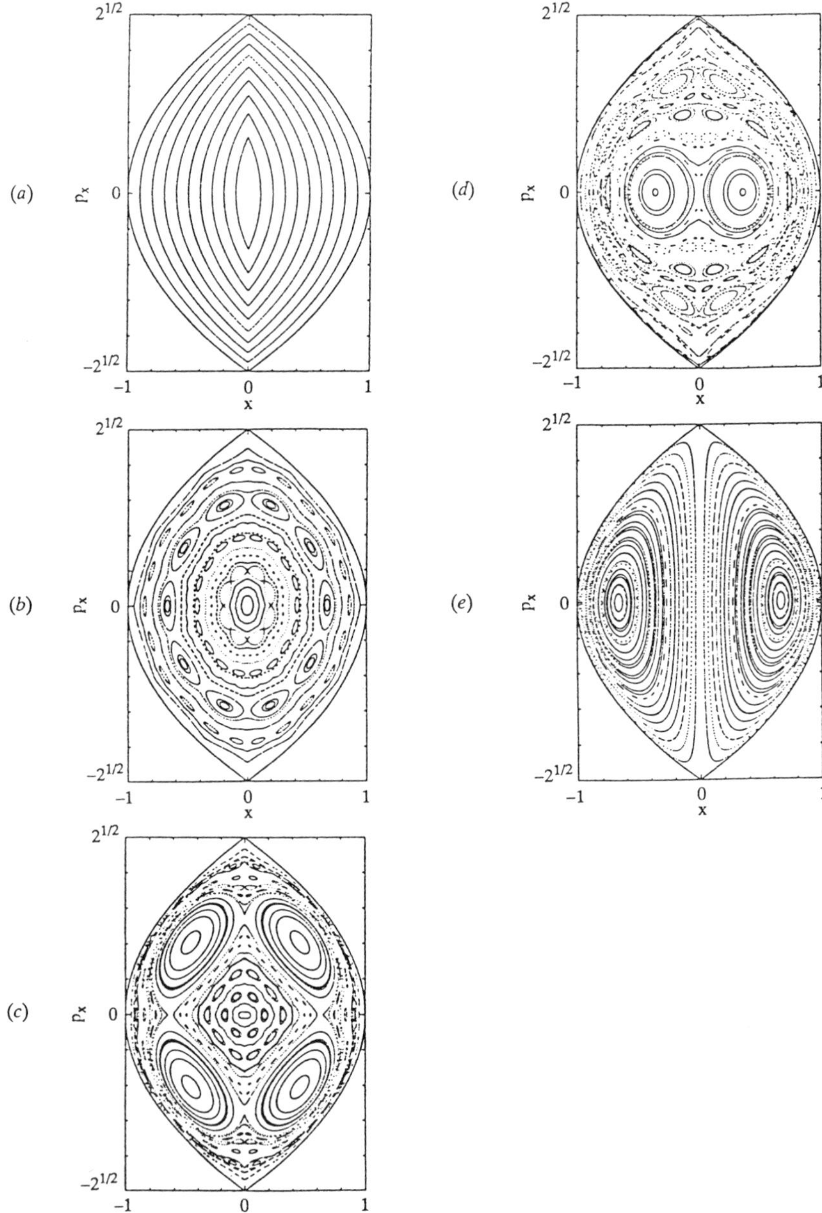

Figure 47. Poincaré sections $y = 0$ in the classical flow of the upper surface $v^{(+)} = \sqrt{a^2x^2 + b^2y^2}$ of the anisotropic conical intersection [Eq. (9.56)] for $a = 1$, energy $E = 1$, and different values of the other parameter that cannot be scaled out: (*a*) $b = 0$; (*b*) $b = 0.2$; (*c*) $b = 0.5$; (*d*) $b = 0.8$; (*e*) $b = 1$.

The classical motion on the upper half-cone is ruled by the potential $v^{(+)} = \sqrt{a^2x^2 + b^2y^2}$. Such a potential is known to appear locally in pyrazine [209], as well as in other molecules [62]. Figure 47 shows several Poincaré surfaces of section in the plane $y = 0$, where the linear momentum along the x-axis takes its values in the range $|p_x| \leq \sqrt{2(E - a|x|)}$. The system can be rescaled in such a way that energy is set equal to $E = 1$ and the only bifurcation parameter is the ratio a/b. In the limit $a/b = 1$, the system is integrable with a stable circular orbit. When a/b decreases, bifurcations occur, introducing a mild chaos in the system that remains dominated by quasiperiodic motions. At $a/b = 0.5$, the phase portrait shows the existence of a pair of local modes resulting from a period-doubling bifurcation. There are two stable orbits surrounded by quasi-periodic islands. At lower values of a/b, there is a new main quasi-periodic island that grows around the period-one orbit at the origin $x = p_x = 0$, and that invades the whole phase space as $a/b \to 0$. In the limit $a/b = 0$, the system is quasi-one-dimensional and is therefore also integrable. Between both the integrable limits $a/b = 1$ and $a/b = 0$, the system shows quasiperiodic motions with thin chaotic zones that are never covering the phase space. We may expect that the nonintegrability of the system when $0 < a/b < 1$ gives rise to an irregularity in the distribution of the scattering resonances of the conical intersection.

X. ATOMIC AND MOLECULAR HAMILTONIANS

In the preceding sections, we described the semiclassical methods of quantization. In the following sections, we would like to apply these methods to several atomic and molecular systems, especially of the scattering and reactive types. The purpose of this section is to discuss and compare the semiclassical properties of the atomic and molecular Hamiltonians.

A. Coulomb Hamiltonian

If we neglect spin interactions and others of the same order, the nonrelativistic Hamiltonian of a molecule or an atom is given by the Coulomb interaction between the electric charges of its constituent particles [197]

$$\mathcal{H} = \sum_a \frac{\hat{\mathbf{p}}_a^2}{2M_a} + \sum_i \frac{\hat{\mathbf{p}}_i^2}{2m} + \sum_{a<b} \frac{Z_a Z_b e^2}{r_{ab}} + \sum_{i<j} \frac{e^2}{r_{ij}} - \sum_{i,a} \frac{Z_a e^2}{r_{ia}} \tag{10.1}$$

where the indexes i, j label the electrons of mass m and the indexes a, b label the nuclei of masses M_a and atomic numbers Z_a. If the particles are exposed to an electromagnetic field, the Hamiltonian of the electromagnetic interaction and field must be added. Here, it is not our purpose to describe the details of this formalism [197], but we would like to summarize the treatment of this type of Hamiltonian in the case of atoms and molecules, and to describe some of the general properties we should expect.

B. Atomic Systems

Since atoms have a single nucleus, the solution of the atomic Hamiltonian describes the electronic motion. As the number of electrons increases, the methods of solution change. In certain circumstances, semiclassical methods turn out to be very effective. Let us summarize different cases.

It is well known that the one-electron Coulomb Hamiltonian is separable and can be exactly solved by wave function methods. On the other hand, the exact solution is also obtained by semiclassical methods, since the classical one-electron Coulomb Hamiltonian is integrable. As soon as we turn to two-electron systems, the solution becomes more complicated and numerical methods should be developed. Thanks to modern computers, the properties of the quantum Hamiltonian can be obtained with very high precision even at high electronic excitations. Nevertheless, semiclassical methods are important to provide an interpretation of the very complex spectra of highly excited states. One of the most remarkable new results is probably the numerical and experimental observation that the classical periodic orbits are emerging from the wave dynamics [34]. Such observations are based on the use of the Fourier transform to go from the energy domain to the time domain [18]. We will come back to this aspect later. For the moment, let us remark that the difficulty encountered in the solution of the quantum Hamiltonian by wave function methods appears to be related to the nonintegrability of the corresponding classical two-electron Coulomb Hamiltonian [23]. Numerical integration of the classical two-electron motion has shown that the phase space is a mixture of quasiperiodic and chaotic motions [100]. As soon as two electrons are present, it seems that the classical system will be of scattering type because of the mutual Coulomb repulsion between the electrons. Accordingly, the classical dynamics turns out to be much less stable than the quantum dynamics. In particular, ionization seems to be classically very common in energy domains where the quantum dynamics sustains bound states.

Semiclassical methods for atoms or ions with many electrons have been successfully developed since the pioneering works of Thomas and Fermi in which the electrons are supposed to form a low-temperature gas obeying the Pauli exclusion principle [30]. In this context of the many-body systems, let us mention that the indistinguishability of the fermions can be incorporated into the semiclassical trace formula [147, 182]. Nevertheless, we may expect that the quantum mechanical wave properties remain important at low temperatures.

On the other hand, the atom or the ion may enter a semiclassical regime at high excitations of one or several electrons. High excitations of one electron are known as Rydberg states, which have been the object of many recent studies [33, 34]. When a Rydberg atom is exposed to an external magnetic field, its classical motion becomes chaotic and its quantum spectrum irregular [210]. In the energy domain, the assignment of the spectral lines becomes impossible because of spectral irregularity. On the other hand, appropriate Fourier transforms to the energy domain show distinct peaks that can be assigned to classical periodic orbits [34]. Recently, similar studies have been devoted to doubly excited electronic states [36]. In such highly excited electronic states, it is natural to turn to semiclassical methods to understand the properties of the Coulomb Hamiltonian.

C. Molecular Systems

1. *Born–Oppenheimer Hamiltonian*

In molecules, not only the electrons but also the nuclei are in relative motion, thus adding vibrational and rotational degrees of freedom to the electronic ones. At low electronic excitations, the electronic degrees of freedom may be eliminated thanks to the Born–Oppenheimer (BO) method [56, 197]. This method exploits the difference in mass between the nuclei and the electrons. Accordingly, the nuclei are slowly moving in an effective potential produced by averaging over the fast motion of the electrons. Let us summarize the BO method. The Coulomb Hamiltonian [Eq. (10.1)] is separated as

$$\hat{\mathscr{H}}_{\text{molecular}} = \hat{T}_{\text{nuclei}} + \hat{V}_{\text{electrons, nuclei}} \tag{10.2}$$

where $\hat{T}_{\text{nuclei}}$ is the nuclear kinetic energy, while $\hat{V}_{\text{electrons, nuclei}}$ is the electronic Hamiltonian (kinetic and potential energies) plus the internuclear potential. Because of the fast electronic motion, the electronic Hamiltonian plays the role of an effective potential for the nuclear motion. In a discrete basis of electronic states, the electronic Hamiltonian $\hat{V}_{\text{electrons, nuclei}}$ may be represented by a matrix, similarly to the Hamiltonian [Eq. (9.1)], where the dynamical variables are now the nuclear positions $[\mathbf{q} = \{\sqrt{M_a}\mathbf{R}_a\}]$ and the nuclear momenta $[\hat{\mathbf{p}} = \{\hat{\mathbf{P}}_a/\sqrt{M_a}\}]$ [56]. The BO method is then similar to the diagonalization of the matrix

Hamiltonian, as explained in Section IX.A. The role of the small parameter $\hbar$ is played here by the ratio between the electronic to the nuclear masses.

The wavefunction is assumed to be of the form [56, 197]

$$\Psi(\{\mathbf{R}_a\}, \{\mathbf{r}_i\}) = \sum_{\mu} \psi_\mu(\{\mathbf{R}_a\})\tau^{(\mu)}(\{\mathbf{R}_a\}, \{\mathbf{r}_i\}) \tag{10.3}$$

where $\tau^{(\mu)}$ are the eigenvectors of the electronic Hamiltonian $\hat{V}_{\text{electrons, nuclei}}$. The Hamiltonian is therefore reduced to the form [Eq. (9.13)], where v is a diagonal matrix giving the effective potentials for the motion of the nuclei. Since the ratio between the electronic and the nuclear masses is in general very small, the vector potential matrix **A** may be neglected in a first approximation.

There will exist as many BO surfaces as there are discrete electronic states for fixed positions of the nuclei. The BO surfaces are thus assigned by electronic quantum numbers as long as they remain separated in energy. The BO potential is bonding if the electrons spend most of their time between the nuclei so that they can screen the Coulomb repulsion between nuclei, which is the case if the electronic wave function has no nodal surfaces between two nuclei. However, if there is such a nodal surface, there is no screening effect and the BO potential will be antibonding, leading to dissociation of the molecule. If a BO surface has a barrier between a well and a dissociation channel, tunneling may occur through the barrier so that the quantized states of the well may acquire a lifetime determined by the tunneling (probability) rate. This rate can be calculated semiclassically using complex periodic orbits.

At higher electronic excitations, the BO surfaces may come close to each other. In such situations, strong coupling may occur between several BO surfaces resulting in surface hopping and conical intersections with geometric phases, as we explained in Section IX. In particular, the BO surfaces cluster at the ionization thresholds where the electronic states are forming Rydberg series (as well as series of doubly excited electronic states). The potential surfaces may then overlap, which may lead to a breakdown of the classification scheme in terms of electronic quantum numbers. This breakdown (which also occurs at the level of the vibrational states within each electronic surface) is attributed to the nonintegrability of the electronic (respectively, vibrational) dynamics, which appears together with chaos in the corresponding classical dynamics. When the electrons move at large distances from the nuclei, the orbital periods of the electrons become comparable to the vibrational periods of the nuclei so that the BO scheme is no longer valid. In these highly excited electronic or vibrational regimes, semiclassical methods are important to find order in the spectral complexity thanks to the classical periodic orbits. Such studies are only at their beginning in the case of molecular Rydberg or doubly excited states [37], but they are already well developed for atomic systems, as we mentioned in Section X.B.

After this description of the electronic degrees of freedom, we turn to the vibrational and rotational motion sustained by a BO surface.

2. *Rotational–Vibrational Hamiltonian*

As soon as the electronic degrees of freedom are eliminated such that the vibrational motion is found to take place on a given BO potential surface, the BO Hamiltonian can be further simplified in terms of the translational, rotational, and vibrational degrees of freedom, to obtain the Watson form in the case of semirigid molecules [211]:

$$\hat{\mathcal{H}}_{\rm BO} = \frac{1}{2}\sum_{\alpha,\beta} \mu_{\alpha\beta}(\mathbf{q})(\hat{J}_\alpha - \hat{\pi}_\alpha)(\hat{J}_\beta - \hat{\pi}_\beta) + \frac{1}{2}\sum_k \hat{p}_k^2 - \frac{\hbar^2}{8}\sum_\alpha \mu_{\alpha\alpha}(\mathbf{q}) + v_{\rm BO}(\mathbf{q}) \tag{10.4}$$

where $\hat{p}_k = -i\hbar\,\partial/\partial q_k$ is the vibrational momentum conjugated to the vibrational positions q_k. The operators $\hat{J}_\alpha$ are the components of the angular momentum (with the indexes $\alpha, \beta = x, y, z$), and $\mu_{\alpha\beta}(\mathbf{q})$ are the elements of the inverse of the tensor of inertia of the molecule, which is dependent on the vibrational positions. Moreover,

$$\hat{\pi}_\alpha = -i\hbar \sum_{k,l} \zeta_{kl}^\alpha q_k \frac{\partial}{\partial q_l} \tag{10.5}$$

is usually called the vibrational angular momentum, where ζ_{kl}^α are the Coriolis interaction parameters. $v_{\rm BO}(\mathbf{q})$ denotes the BO potential and the last term is the Watson term due to rotation.

The rotational part of this Hamiltonian can also be studied semiclassically to reveal very interesting phenomena at the level of the molecular rotational bands [171, 212]. Here, semiclassical regimes can also be reached because the quantum numbers of angular momentum can climb up to numbers as high as $J = 50$ or more.

If we assume that the angular momentum is fixed, the vibrational motion can be studied on the corresponding potential surface. If the potential is bonding and possesses a minimum, we may introduce the normal modes of this stable equilibrium point and then reduce the vibrational part of the Hamiltonian to a sum of harmonic oscillators plus anharmonic corrections. A similar procedure exists near the saddle equilibrium point of an antibonding potential as we discussed in Section V.

D. Coulomb and BO Potentials: Comparison of Their Semiclassical Properties

There is a major difference between the Coulomb potential that determines the motion of the electrons around the nuclei and the kind of potentials that are obtained by the BO approximation. This difference has consequences on all properties of these systems. The Coulomb potential is long ranged and obeys well-known classical scaling laws [70]. If we scale all the positions and the time according to

$$\mathbf{r}_k \rightarrow \alpha \mathbf{r}_k \qquad \text{and} \qquad t \rightarrow \alpha^{3/2} t \tag{10.6}$$

with $k = (i, a)$ in the classical Hamiltonian corresponding to Eq. (10.1), the latter scales like $\mathcal{H} \rightarrow \mathcal{H}/\alpha$, while the actions scale like $S \rightarrow \alpha^{1/2} S$. This scaling law controls a lot of different properties. In particular, the periods of periodic orbits also scale like $T \rightarrow \alpha^{3/2} T$, which gives the famous Kepler law [70]. Furthermore, to any classical solution of the Hamiltonian flow at an energy E_1 there corresponds another solution at an energy E_2, which is scaled by a factor E_1/E_2, and which has a period $(E_1/E_2)^{3/2}$. A similar reasoning applied to the Bohr model of the hydrogen atom shows that the energy quantization $E_n = -1/2n^2$ is also a consequence of this scaling law, since the actions of the orbits are quantized according to $S_n \sim n$. Therefore, the scaling law is responsible for the accumulation of an infinity of energy levels at the ionization thresholds. Of course, the quantization breaks the scaling so that the property remains only in the semiclassical limit. The accumulation of levels forming Rydberg series is equivalently attributed to the infinite range of the Coulomb potential.

On the other hand, the BO potentials are short ranged and thus decrease faster at large distances. A typical Hamiltonian that is often considered, particularly for diatomic molecules, is the Morse Hamiltonian [Eq. (3.31)], where the decrease of the interaction is exponential at large separations [98]. We already mentioned that this potential is exactly solvable in classical and quantum mechanics: The quantum energy eigenvalues are given by Eqs. (4.32)–(4.33). Accordingly, the number of eigenstates sustained by the Morse potential is finite although it may be large. Using the rescaled energy $\varepsilon_n = E_n/D$, the spectrum is given by

$$\varepsilon_n = 1 - (\eta_n - 1)^2 \tag{10.7}$$

in terms of the variable $\eta_n = \hbar\omega_0(n + 1/2)/2D$ with $\eta_n \leq 1$. The spectrum

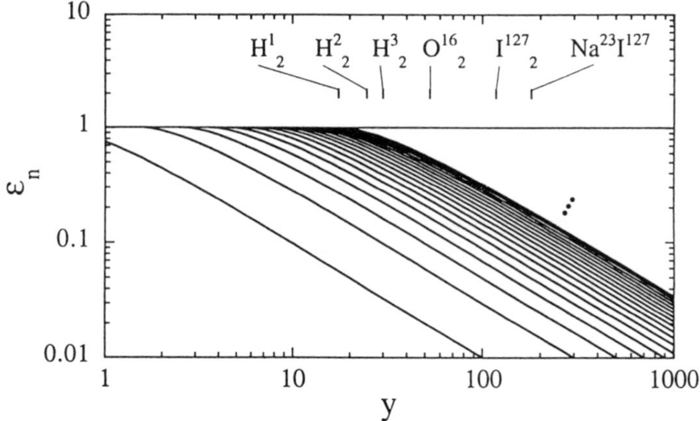

Figure 48. Quantum energy levels $\{\varepsilon_n = E_n/D\}$ [Eq. (10.7)] of the Morse oscillator versus the quantum variable $y = 2D/(\hbar\omega_0)$ for $n = 0$–17, together with the location of several diatomic molecules in this model. Note the accumulation of energy levels in the semiclassical limit $y \rightarrow \infty$. The purpose of this figure is to illustrate how the semiclassical limit is reached for the Morse oscillators. The limit does not exist within a single system but exists only within a class of systems.

is depicted in Fig. 48 versus the variable $y = 2D/(\hbar\omega_0)$, which in this case plays the role of the inverse of the Planck constant and gives the number of eigenvalues. We remark that the spectrum of the Morse oscillator may be homogenized in the sense of Eq. (4.4). Indeed, the energy levels correspond to an equal spacing between the values taken by the quantized variable $y = y_n = (n + 1/2)/(1 + \sqrt{1 - E/D})$.

If we assume a Morse potential for several diatomic molecules, we can localize the individual systems along the y axis in a plot of ε_n versus y, as in Fig. 48, where we observe that the number of eigenstates may become large in heavy molecules. The semiclassical limit is reached in the limit $y \rightarrow \infty$, where the number of eigenvalues is very large. The Morse potential is a theoretical idealization of a molecular potential, and more general potentials are used in practice to fit the experimental data. Nevertheless, the Morse potential already illustrates our point that there is no simple scaling law for molecular potentials contrary to the Coulomb potential.

XI. MOLECULAR VIBROGRAMS: SPECTROSCOPY IN THE TIME DOMAIN

In this and the following section, we report on the application of periodic-orbit theory to molecular systems. Transitions between electronic states, which can be bonding or nonbonding, may feature vibrational fine structure that can be associated with the nuclear motion on a BO potential energy surface. If the potential surface is known, and the number of degrees of freedom is limited, a detailed analysis in terms of the periodic orbits of the system may be undertaken. We give an example of such an analysis for the case of photodissociation of the HgI_2 molecule in Section XII.

Proceeding by the inverse route, one may extract information on the classical periodic orbits by analyzing molecular spectra. We will discuss this approach [214] in the present section, and give examples for several bound and unbound molecular systems.

A. The Principle of Vibrograms

As explained earlier, classical orbits are emerging from wave dynamics in the semiclassical limit of high excitations. The theory developed in the preceding sections provides us with the tools to establish explicitly the correspondence between the quantum mechanical quantities and the emerging classical ones. In this correspondence, a central role is played by the Gutzwiller trace formula [Eq. (6.13)] and, especially, by the semiclassical expression [Eq. (7.15)] for the photoabsorption cross-section. These semiclassical formulas differ for the different physical observables that are considered. In Eq. (7.15), we have the photoabsorption κ_ω. In a bounded system, the photoabsorption cross-section is discrete as

$$\kappa_\omega = \sum_{m,n} I_{mn}\delta(\omega - \omega_{mn}) \tag{11.1}$$

where I_{mn} are the transition intensities between the levels m and n separated by the Bohr frequency $\omega_{mn} = (E_n - E_m)/\hbar$. If the initial state of the molecule is at low temperature, the transitions occur essentially between the ground state $m = 0$ and some excited state. In this case, the photoabsorption cross-section essentially reproduces the energy spectrum of the molecule

$$\kappa_\omega = \hbar \sum_n I_{0n}\delta(E_\omega - E_n) \tag{11.2}$$

with $E_\omega = E_0 + \hbar\omega$.

The discrete energy levels of some molecules are known with precision so that we may consider the so-called stick spectrum

$$\sum_n \delta(E - E_n) \tag{11.3}$$

If the excited state is unbounded, the photoabsorption spectrum is continuous. Since diffuse bands usually occur at high energies [213], the molecule may often be supposed to lie in its ground state before absorption. In this case, the photoabsorption cross-section also reproduces the energy spectrum according to $E_\omega = E_0 + \hbar\omega$.

Let us point out that it is important to know the energy spectrum in order to apply semiclassical considerations. Otherwise, we could not use the semiclassical relations Eqs. (6.13) or (7.15). If thermal effects are important, the molecular spectrum will be composed of a large number of possible transitions as in Eq. (11.1). In this case, the photoabsorption spectrum is given by the autocorrelation function of the electric dipole operator averaged over a statistical ensemble represented by a density matrix [60]. The time evolution is therefore controlled by the Landau–von Neumann operator [187]. Accordingly, the photoabsorption spectrum as a function of the Bohr frequency gives information on the eigenvalues of the Landau–von Neumann operator rather than of the Hamiltonian operator. If the spectrum presents distinct discrete lines, both sets of eigenvalues are simply related to each other. However, since lines accumulate in the semiclassical limit, the Landau–von Neumann operator starts to be well approximated by the classical Liouville operator according to Section II.D [69]. The Liouville operator is known to have its own eigenvalues giving the classical frequencies in quasiperiodic motions as well as the classical relaxation and escape rates in unstable and chaotic phase-space regions. As a consequence, Bohr-frequency spectra tend to provide information on the quasiclassical eigenvalues in the semiclassical limit rather than on the quantum eigenvalues. In this chapter, which is devoted to semiclassical methods, we focus on the energy spectra that are amenable to the semiclassical analysis.

Let us assume that we know, for a molecule of interest, a spectral quantity $\sigma(E)$ of the kind described above. To obtain information on the classical periodic orbits from the energy spectrum, one may use the Fourier transformation to go from the energy domain to the time domain and define

$$\tilde{\sigma}(T) = \int_{-\infty}^{+\infty} \sigma(\varepsilon) \exp\left(-\frac{i}{\hbar}\varepsilon T\right) d\varepsilon \tag{11.4}$$

When referring to Eq. (6.13) or (7.15), we face the difficulty that the reduced actions of the different periodic orbits do not have a simple dependency on the energy, except in systems where the actions scale with energy, such as $S \sim |E|^\alpha$, with the exponent α being characteristic of the system. Such is the case for billiards where $\alpha = \frac{1}{2}$, for pure Coulomb systems where $\alpha = -\frac{1}{2}$, for quantum maps where $\alpha = 1$, and for the hydrogen atom in a magnetic field where a combined scaling exists between the energy and the magnetic field. However, the dependency is general in molecules so that a Fourier transform may introduce mixing between the behaviors at different energies and the periods of the orbits would not be available.

If the energy spectrum presents distinct diffuse bands [213], we could apply the Fourier transform [Eq. (11.4)] to each one of the diffuse bands. Here, the aforementioned difficulty would not arise if the classical actions of the periodic orbits remain nearly constant over the energy range of the diffuse band. However, a global Fourier transform, such as Eq. (11.4), would be useless if the energy spectrum extends over a large range, where classical actions and periods undergo significant variations.

As a solution to this problem, we propose a windowed Fourier transform, which we define by [214]

$$S(T, E; \Delta E) = \int_{-\infty}^{+\infty} \sigma(\varepsilon) \exp\left[-\frac{(\varepsilon - E)^2}{2\Delta E^2} - \frac{i}{\hbar} \varepsilon T\right] d\varepsilon \qquad (11.5)$$

where ΔE is the width of the window.

The function [Eq. (11.5)] leads to a natural graphical representation that we call a vibrogram. The vibrogram is constructed as follows. Since the function [Eq. (11.5)] is complex let us define another real function according to

$$f(T, E; \Delta E) = \frac{T}{\mathcal{N}(E, \Delta E)} |S(T, E; \Delta E)| \qquad (11.6)$$

where the normalizing factor $\mathcal{N}(E; \Delta E)$ is taken such that $\lim_{T\to 0} f(T)/T = 1$. We expect successive peaks $\{T_i(E)\}$ to appear in the function [Eq. (11.6)] when the time T varies for fixed energy E. The vibrogram is a representation of the locations in the (E, T) plane of the different peaks $\{T_i(E)\}$ as a function of the energy E.

These peaks may be interpreted as the periods of recurrences of the wavepacket dynamics. Under the following assumptions, the peaks may be interpreted moreover as the periods of the classical orbits emerging from the quantum dynamics:

1. The observable quantity has a periodic-orbit expansion of the Gutzwiller type:

$$\sigma(\varepsilon) \simeq \sigma_0(\varepsilon) + \sum_{p,r} A_{p,r}(\varepsilon) \cos\left[\frac{r}{\hbar} S_p(\varepsilon) - r\frac{\pi}{2}\mu_p\right] \qquad (11.7)$$

2. Within the energy window ΔE, the different quantities on the right-hand side of Eq. (11.7) are slowly varying with energy ε. In particular, the actions of the periodic orbits have an approximately linear dependency on energy ε while the amplitudes $A_{p,r}$ are approximately constant:

$$\begin{aligned} S_p(\varepsilon) &\simeq S_{p0}(E) + T_p(E)(\varepsilon - E) \\ A_{p,r}(\varepsilon) &\simeq A_{p,r}(E) \end{aligned} \tag{11.8}$$

where E is the center of the energy window of width ΔE.
3. There are enough energy levels inside the window ΔE.

Let us now discuss the behavior of the vibrogram if the above assumptions are not satisfied. If we suppose that the actions of the periodic orbits have a quadratic dependency of the form

$$S_p(\varepsilon) \simeq S_{p0}(E) + T_p(E)(\varepsilon - E) + \tfrac{1}{2}\, S_{p2}(E)(\varepsilon - E)^2 \tag{11.9}$$

in the Gutzwiller-type formula [Eq. (11.7)], a calculation shows that the function $S(T, E; \Delta E)$ is composed of a superposition of complex Gaussian functions of the following kind

$$\exp\left[-\frac{(T - T_p)^2}{2\hbar^2(\Delta E^2 - iS_{p2}/\hbar)} \right] \tag{11.10}$$

Consequently, the function [Eq. (11.5)] exhibits well-defined peaks centered on the classical periods at the condition that the second derivative of the action of each periodic orbit is small enough and satisfies the following inequality with respect to the time domain, using $S_{p2} = \partial_E T_p$:

$$\Delta T = \frac{\hbar}{\Delta E} \gg \sqrt{\hbar |\partial_E T_p|} \tag{11.11}$$

Otherwise, the structure of the peaks may be more complicated. A dependency on the energy of the amplitudes $A_{p,r}$ seems to have a less important influence on the structure of the peaks than the nonlinear dependency of the actions as in Eq. (11.9). Therefore, we expect that the vibrogram would correspond to the diagram of the periods versus energy of the classical orbits of the system introduced in Section III if the semiclassical assumptions (1)–(3) and conditions [Eq. (11.11)] are satisfied.

The transform [Eq. (11.5)] is also known as energy–time Husimi transform [214]. The window ΔE can be interpreted as a squeezing parameter. In one limit, we remain in the energy domain, while we recover the global Fourier transform [Eq. (11.4)] to the time domain in the other limit. Indeed, we have

$$\sigma(E) = \lim_{\Delta E \to 0} \frac{\exp(iET/\hbar)}{\Delta E \sqrt{2\pi}} S(T, E; \Delta E) \tag{11.12}$$

and

$$\tilde{\sigma}(T) = \lim_{\Delta E \to \infty} S(T, E; \Delta E) \tag{11.13}$$

Moreover, we have the alternative relation

$$S(T, E; \Delta E) = \frac{\Delta E}{\hbar \sqrt{2\pi}} \int_{-\infty}^{+\infty} \tilde{\sigma}(\tau) \exp\left[-\frac{\Delta E^2}{2\hbar^2} (\tau - T)^2 + \frac{i}{\hbar} E(\tau - T) \right] d\tau \tag{11.14}$$

which shows that the Husimi transform of the energy spectrum is equivalent to a windowed Fourier transform of the time spectrum with a time width $\Delta T = \hbar / \Delta E$. The form [Eq. (11.14)] suggests another alternative type of diagram known as the sonogram in acoustics, where the frequency (or here energy) content of a time signal in the window ΔT is represented as a function of time [214]. On such a sonogram, we can observe how the frequencies of the fast vibrational modes change in time due to the molecular distortions induced by the slow modes [214].

Let us conclude this introductory section with a remark on the semiclassical limit in molecular systems. Because of the BO approximation, traces of the limit $\hbar \to 0$ are lost as soon as we focus on a single BO surface, which implies that the electronic motion is treated quantum mechanically. If we nevertheless assume that this BO potential is given, we may formally consider the semiclassical limit for the vibrational motion on this given BO surface. However, the molecular potentials are of the Morse type and their energy levels do not accumulate in infinite series as for the Coulomb-type potentials. Therefore, we are obliged to use the windowed Fourier transform, at least in the case where the spectrum of a single molecular species is to be analyzed. However, it may be possible to study the semiclassical limit along a series of isotopomers with increasing nuclear masses as in Fig. 48. In this way, another kind of semiclassical scaling may be available.

B. Bounded Molecular Systems

For a discrete energy stick spectrum [Eq. (11.3)], we can also conveniently use a squared rather than a Gaussian window and define the function

$$f(T, E; \Delta E) = \left| \frac{T}{\mathcal{N}(E; \Delta E)} \sum_{E-\Delta E/2<E_n<E+\Delta E/2} \exp\left(-\frac{i}{\hbar} ET\right) \right| \quad (11.15)$$

where the sum extends over all the energy levels in the range of width ΔE around the energy E. Here the parameter $\mathcal{N}(E; \Delta E)$ is the number of energy levels in this range so that we have $\lim_{T\to 0} f(T)/T = 1$. This function can be used to draw the vibrogram associated with the stick spectrum.

1. *Morse-Type Model for* $I_2(\tilde{X}\,^1\Sigma)$

As a first example, we consider a Morse model of the ground state $\tilde{X}\,^1\Sigma$ of the diatomic molecule I_2 for which the dissociation energy is $D = 12{,}542\ \text{cm}^{-1}$, the ground-state frequency $\nu = 215.2\ \text{cm}^{-1}$, and the equilibrium distance $r = 2.67$ Å [215]. Figure 49 shows the vibrogram of I_2 superposed on the classical periods $T(E) = r\nu^{-1}\sqrt{D/|E|}$. The vibrogram is composed of the points where the function [Eq. (11.15)] exceeds certain thresholds, in order to display the values of T where this function has large peaks. We see that the vibrogram reproduces the classical

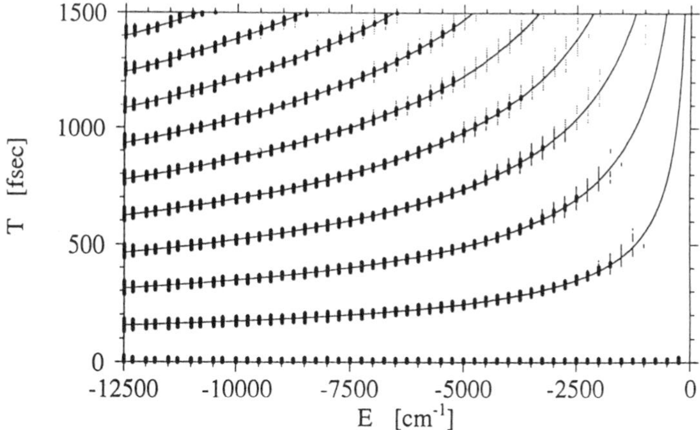

Figure 49. Vibrogram of a Morse model of $I_2(\tilde{X}\,^1\Sigma)$ calculated with the function [Eq. (11.15)] with $\Delta E = 1000\ \text{cm}^{-1}$. The solid lines are the classical periods. The big dots mark the points where the functions [Eq. (11.15)] is larger than a high value while the small dots do the same at a lower value of [Eq. (11.15)].

periods of the system. Because of the Heisenberg uncertainty relation between time and energy, only the structures of the vibrogram that have an area larger than $\hbar = 5308.7\ \mathrm{fs\ cm^{-1}}$ are meaningful. In this sense, the periodic orbits emerge from the wave dynamics but they do not underlie the wave dynamics.

2. $CS_2(\tilde{X}\,^1\Sigma_g^+)$: *Fermi Resonance*

In its ground electronic state, the molecule carbon disulfide (CS_2) is collinear and shares several other properties with carbon dioxide (CO_2). In particular, its vibrational spectrum is dominated by a Fermi resonance between the symmetric stretching and bending modes. This molecule has been the object of recent high-resolution spectroscopic studies, particularly those by Pique et al. [216–218] who obtained some 1000 vibrational energy levels in the energy range 0–20,000 $\mathrm{cm^{-1}}$. Below about 13,000 $\mathrm{cm^{-1}}$, the experimental spectrum of Pique et al. [218] essentially contains levels of excited symmetric stretching and bending modes. Above 13,000 $\mathrm{cm^{-1}}$, a transition is observed where the asymmetric stretching mode starts to interact strongly with the two other modes.

The spectroscopic analysis of the low-energy part of the spectrum (below 12,000 $\mathrm{cm^{-1}}$) led to the following effective Hamiltonian for the $\tilde{X}\,^1\Sigma_g^+$ ground electronic state of the CS_2 molecule [216, 217]

$$\hat{\mathcal{H}} = \hat{\mathcal{H}}_0 + \hat{\mathcal{H}}_{\mathrm{F}} \tag{11.16}$$

with

$$\begin{aligned}\hat{\mathcal{H}}_0 = {} & \omega_1(v_1 + 1/2) + \omega_2(v_2 + 1) + \omega_3(v_3 + 1/2) + x_{11}(v_1 + 1/2)^2 \\ & + x_{22}(v_2 + 1)^2 + x_{33}(v_3 + 1/2)^2 + x_{12}(v_1 + 1/2)(v_2 + 1) \\ & + x_{13}(v_1 + 1/2)(v_3 + 1/2) + x_{23}(v_2 + 1)(v_3 + 1/2) \\ & + y_{222}(v_2 + 1)^3 + g_{22}\ell_2^2\end{aligned} \tag{11.17}$$

and with the Fermi interaction

$$\hat{\mathcal{H}}_{\mathrm{F}} = \frac{k_{122}}{\sqrt{2}}\,(\hat{a}_1^\dagger \hat{a}_2 \hat{a}_2' + \hat{a}_1 \hat{a}_2^\dagger \hat{a}_2'^\dagger) \tag{11.18}$$

In Eqs. (11.17)–(11.18), v_1, v_2, v_3 stand for the vibrational quantum numbers of the corresponding symmetric stretching, bending, and asymmetric stretching modes. Since the CS_2 molecule is collinear in its ground state, the bending mode is doubly degenerate and $\ell_2 = v_2, v_2 - 2,$

$\ldots, -v_2$ is the corresponding quantum number of vibrational angular momentum. In terms of the creation–annihilation operators, we have $v_1 = \hat{a}_1^\dagger \hat{a}_1$, $v_3 = \hat{a}_3^\dagger \hat{a}_3$, $v_2 = \hat{a}_2^\dagger \hat{a}_2 + \hat{a}_2'^\dagger \hat{a}_2'$, and $\ell_2 = \hat{a}_2^\dagger \hat{a}_2 - \hat{a}_2'^\dagger \hat{a}_2'$, where $\hat{a}_2$ and $\hat{a}_2'$ are annihilation operators for the two bending modes. The Fermi interaction gives a coupling between energy levels of $\hat{\mathscr{H}}_0$ separated by $\Delta v_1 = \pm 1$ and $\Delta v_2 = \mp 2$. Table I provides the constants for this model as obtained by Pique et al. [216, 217].

In Fig. 50, we present the vibrogram obtained from the experimental spectrum of the $CS_2(\tilde{X}\,{}^1\Sigma_g^+)$ molecule. The experimental data have been kindly provided to us by Pique et al. and give the vibrational energies of a laser fluorescence spectrum of CS_2. Here, the vibrogram has been calculated by Fourier transform using a Gaussian window as defined in Eqs. (11.5)–(11.6) with $\Delta E = 4000\,\text{cm}^{-1}$. The maxima of the function [Eq. (11.6)] along the T axis are given by dots of different sizes depending on the height of the maxima.

In order to interpret the vibrogram, we studied the classical dynamics of the model [Eq. (11.16)–(11.18)]. In the classical limit, the vibrational quantum numbers become action variables according to

$$v_i + d_i/2 \leftrightarrow J_i \tag{11.19}$$

with $d_1 = d_3 = 1$, and $d_2 = 2$. Besides, we also have the vibrational angular momentum $\ell_2 \leftrightarrow L_2$ with $-J_2 \leq L_2 \leq J_2$. Associated with the action variables, we have the conjugated angle variables, θ_i and χ_2. The

TABLE I
Set of Parameters for the Low-Energy Model [Eqs. (11.16–11.18)] of CS_2

Parameter (cm^{-1})	Values by Pique et al. [216, 217]
ω_1	665.699
ω_2	399.835
ω_3	1561.347
x_{11}	−0.175
x_{12}	−1.243
x_{13}	−8.003
x_{22}	+0.770
x_{23}	−6.710
x_{33}	−6.333
y_{222}	−0.0124
k_{122}	40.908

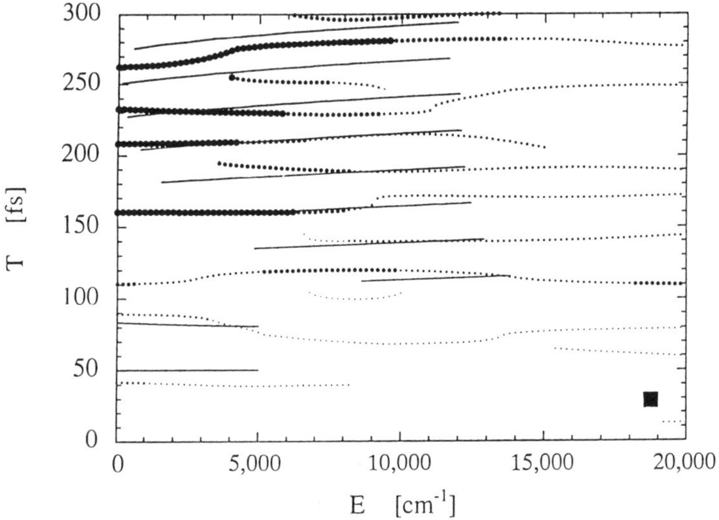

Figure 50. Vibrogram of $CS_2(\tilde{X}\,^1\Sigma_g^+)$ calculated with a Gaussian window according to Eqs. (11.5–11.6) with $\Delta E = 4000\,\mathrm{cm}^{-1}$. The experimental data by Pique et al. [216–218] contain about 900 levels in the range 0–20000 cm^{-1}. The size of the dots increases with the height of the peak in the vibrogram. In the lower part at small energies, the solid lines are the periods of the individual modes of the Hamiltonian [Eq. (11.20)] calculated according to $T_1 = 2\pi/\omega_1$ with $\omega_1 = \partial_{J_1}\mathscr{H}(J_1, J_2 = J_3 = 0)$, and $T_2 = 2\pi/\omega_2$ with a corresponding definition for ω_2. In the upper part, the solid lines are the periods of the periodic orbits of the Hamiltonian [Eq. (11.21)] calculated as explained in the text and reported from Fig. 52. The black square has an area equal to $\hbar = 5308.7\,\mathrm{cm}^{-1}$ fs.

classical Hamiltonian is therefore

$$\begin{aligned}\mathscr{H}_{\mathrm{cl}} &= \omega_1 J_1 + \omega_2 J_2 + \omega_3 J_3 + x_{11} J_1^2 + x_{22} J_2^2 + x_{33} J_3^2 + x_{12} J_1 J_2 \\ &\quad + x_{13} J_1 J_3 + x_{23} J_2 J_3 + y_{222} J_2^3 + g_{22} L_2^2 \\ &\quad + \frac{k_{122}}{\sqrt{2}} \sqrt{J_1 (J_2^2 - L_2^2)} \cos(\theta_1 - 2\theta_2) \end{aligned} \tag{11.20}$$

We have a Hamiltonian system with four degrees of freedom. The Hamiltonian is integrable. Indeed, the angle variables appear only in the combination $\theta_1 - 2\theta_2$ in the Fermi resonance term. Therefore, we may introduce the angular-momentum type variables appropriate to the Fermi resonance [219], namely, $I = (J_1 + J_2/2)/2$ and $\theta = \theta_1 + 2\theta_2$, together with $I_z = (J_1 - J_2/2)/2$ and $\psi = \theta_1 - 2\theta_2$. I and I_z correspond to operators of the angular momentum group SU(2) [219]. The Hamiltonian is now the function $\mathscr{H}_{\rm cl}(I, I_z, \psi, L_2, J_3)$. Accordingly, the actions I, L_2, and J_3 are constants of motion and there remains a single degree of freedom. A last quadrature shows that the system is integrable. This example illustrates the fact that a single Fermi resonance does not necessarily imply the presence of chaotic behavior. Extra perturbations are necessary to induce classical chaos [216, 217].

To simplify the discussion, we consider the two-degree-of-freedom subsystem where $L_2 = J_3 = 0$. This assumption is reasonable to discuss the low-energy part of the vibrogram in Fig. 50, since the quantum numbers v_3 and ℓ_2 are often vanishing in the experimental spectrum below 13,000 cm^{-1}. The classical Hamiltonian [Eq. (11.20)] becomes

$$\begin{aligned}\mathscr{H}_{\rm cl} =& (\omega_1 + 2\omega_2)I + (\omega_1 - 2\omega_2)I_z \\ &+ (x_{11} + 4x_{22} + 2x_{12})I^2 + (x_{11} + 4x_{22} - 2x_{12})I_z^2 + 2(x_{11} - 4x_{22})II_z \\ &+ 8y_{222}(I - I_z)^3 + k_{122}\sqrt{2(I + I_z)}(I - I_z)\cos(\psi) \qquad (11.21)\end{aligned}$$

A few words are now in order about the types of motion described by this Hamiltonian. A local mode of bending character is present around $\psi = 0$, $I_z = -I$ for all energies. On the other hand, a transition occurs at the energy $E = 3061.3$ cm^{-1}, where new local modes of symmetric stretching character emerge from $I = I_z$ at $\psi = \pi$. (Note that $I = I_z$ corresponds to pure symmetric-stretch motion.) This bifurcation from normal-to-local modes is due to the 2:1 Fermi resonance [219]. Figure 51 shows two types of phase portraits before and after the 2:1 bifurcation. The phase portraits are drawn in the (I_z, ψ) plane. The different trajectories correspond to different energies. Note that the angle θ is not represented in Fig. 51, although it is rapidly increasing as $\theta \simeq (\omega_1 + 2\omega_2)t$.

In a semiclassical approach, the energy levels would be obtained as a sum over periodic orbits as discussed in the preceding sections. In the integrable case, the trace formula we should consider is the Berry–Tabor formula [Eq. (6.63)] of Section VI.E. The periodic orbits of an integrable system are obtained from the condition that the ratio between the periods, T_ψ and T_θ, along the angles ψ and θ is a rational number

$$\frac{T_\theta}{T_\psi} = \frac{m}{n} \qquad (11.22)$$

This condition selects a discrete set of periodic orbits at each energy. The overall period is given here by $T = rmT_\psi = rnT_\theta$, where r is the number of repetitions of the fundamental period. In the harmonic limit of the model [Eq. (11.21)], for example, we have $T_\theta/T_\psi \simeq (2\omega_2 - \omega_1)/(2\omega_2 + \omega_1) \simeq$

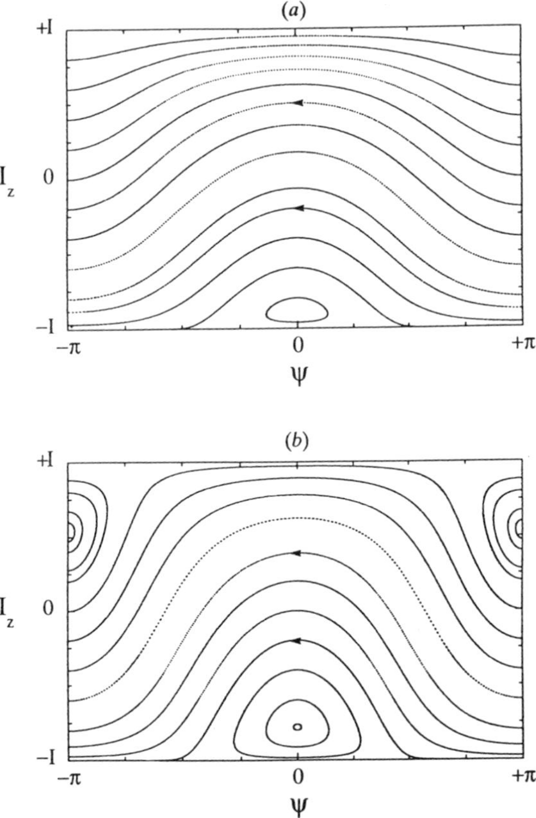

Figure 51. Phase portraits of the classical flow of the Hamiltonian [Eq. (11.21)] in the (ψ, I_z) plane for two different values of the action (*a*) before and (*b*) after the 2:1 bifurcation of the Fermi resonance of CS_2: (*a*) $I = 1$; (*b*) $I = 5$.

1/11, which shows that the smallest period should be expected at a relatively high value of the order of 250 fs with respect to the periods of the individual modes, namely, $T_1 = 2\pi/\omega_1 = 50$ fs and $T_2 = 2\pi/\omega_2 = 83$ fs. This phenomenon is referred to as the stroboscopic effect and is general in vibrograms of multifrequency systems, as discussed by Pique et al. [216, 217].

We have carried out a systematic search for the periodic orbits of lowest period beyond the harmonic limit. Figure 52 depicts the corresponding period-energy diagram of the Hamiltonian [Eq. (11.21)]. As energy increases with the anharmonic effects, a larger range of periods [Eq. (11.22)] is generated. This range is determined by two extreme

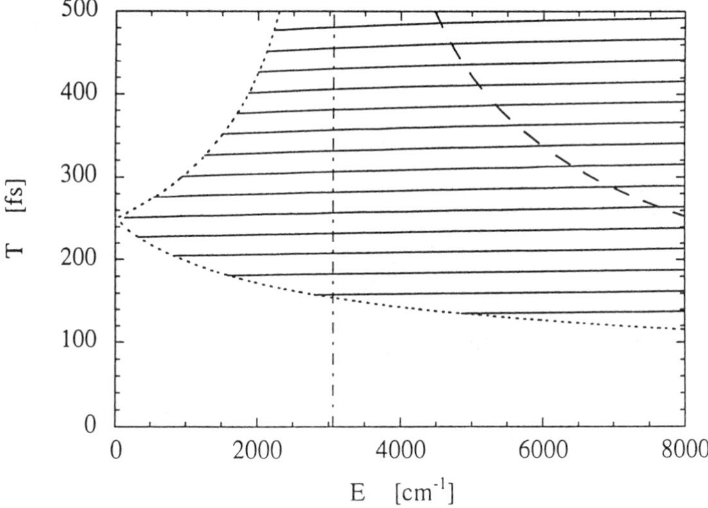

Figure 52. Period-energy diagram of the periodic orbits of the Hamiltonian [Eq. (11.21)] of CS_2. The dotted lines mark the borders of the main family of periodic orbits. The vertical dash–dotted line locates the critical energy $E = 3061.3\,\text{cm}^{-1}$ of the 2:1 bifurcation of the Fermi resonance where a new family of periodic orbits emerges, which have periods beyond the long-dashed line.

orbits in the phase portrait of Fig. 51. A lower limit is determined by the period of the small oscillations around the fixed point at $\psi = 0$ in Fig. 51(a) (represented by the lower dashed line in Fig. 52). On the other hand, the period of the orbit of pure symmetric stretching mode at $I = +I_z$ determines the upper limit (the upper dashed line in Fig. 52). The intrinsic periodic orbits satisfying the condition [Eq. (11.22)] are lying between these limits. In Fig. 52 we only depicted the periodic orbit with $m = 1$ and with repetition $r = 1$, since they are of smallest period near the lower limit. These periodic orbits are generated or absorbed by the two extreme orbits as the energy varies. Consequently, the solid lines representing their periods start abruptly on the limits. There are many more families of periodic orbits besides the $1/n$ given orbits, which are obtained by varying the integers m and r.

As energy increases, we meet the bifurcation at $E = 3061.3\,\text{cm}^{-1}$ (vertical dash–dotted line in Fig. 52) where a new fixed point appears at $\psi = \pi$, as seen in Fig. 51(b). At this bifurcation, the period of the upper extreme orbit ($I = +I_z$) has increased to infinity due to a critical slowing down, which is a typical phenomenon at bifurcations. Beyond the bifurcation, the fixed point ($\psi = \pi$) is surrounded by oscillations that correspond to periodic orbits of a new type (not represented in Fig. 52). These new periodic orbits exist beyond the long-dashed line in Fig. 52, which gives the period of infinitesimal oscillations around the fixed point at $\psi = \pi$.

The classical period-energy diagram can help us to interpret the vibrogram of Fig. 50. In the nearly harmonic regime at very low energies, we observe a stroboscopic effect with high maxima at periods significantly larger than the periods of the normal modes. Nevertheless, we also find small maxima at the periods of the normal modes taken individually: $T_1 = 50$ fs and $T_2 = 83$ fs. Still, at low energies, we find a series of lines of maxima in approximate correspondence with some periods $1/n$ of Fig. 52 (solid lines). In particular, the classical periods explain the observed recurrence at 160 fs. However, this correspondence disappears at higher periods. This deterioration may be due to various reasons: (1) the accumulation of higher periods m/n, (2) the presence of energy levels with $v_3 > 0$ in the experimental spectrum, (3) the finite number of energy levels that leads to a breakdown of the classical-quantum correspondence after long time periods. For those reasons, the observation of the 2:1 bifurcation remains elusive. Nevertheless, the main periodic orbits seem to be well accounted for by the vibrogram for this molecule.

Further examples of vibrograms are given in the following sections.

C. Unbounded Molecular Systems

Besides bound electronic states, molecules feature antibonding excited electronic states that give rise to dissociation and produce intrinsically continuous bands in ultraviolet (UV) absorption spectra [213]. In the case of diffuse bands, the vibrogram may also be useful to obtain the recurrence periods of the transient dynamics of dissociation. If the bands do not overlap, and since the bands already have a finite width, there is often no need to consider a Gaussian window in the Fourier transform so that we can apply the Fourier transform [Eq. (11.4)] to the individual bands.

1. CO_2

Figure 53 shows the UV photoabsorption spectrum of CO_2 obtained by Nakata et al. [220] in 1965. It is composed of different bands associated with different electronic states.

In the infrared (IR) and visible (vis) parts of the spectrum, which are not shown in Fig. 53, the motion takes place on the potential surface of the linear ground electronic state of the molecule ($\tilde{X}\,^1\Sigma_g^+$). This electronic state is bonding and thus gives rise to stable vibrational motion of

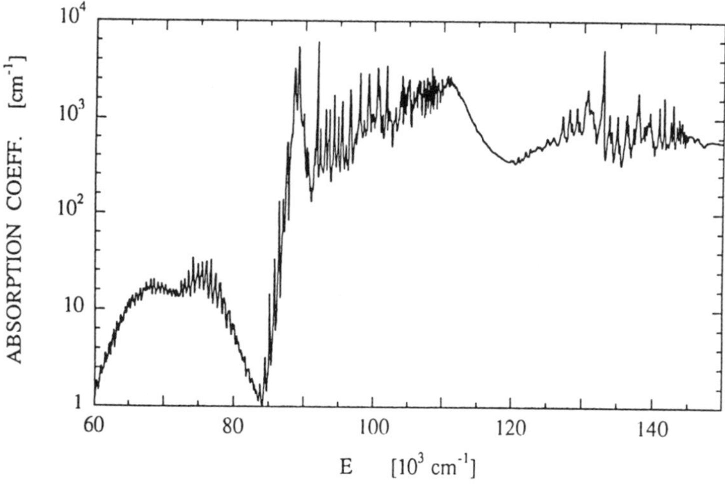

Figure 53. Ultraviolet photoabsorption spectrum of carbon dioxide (CO_2) obtained by Nakata, Watanabe, and Matsunaga in 1965 (adapted from [220]).

the molecule. The energy spectrum is discrete so that the spectral lines are sharp except for radiative (or Doppler and collisional) linewidth.

At higher energy, the motion on other electronic surfaces may be excited so that the UV spectrum reveals the different electronic states of the molecules in the form of successive bands as shown in Fig. 53. At low and medium energies, the electronic states appear in the form of distinct bands due to the series of vibrational states sustained by each electronic potential surface. The bands are either composed of discrete lines if the surface is bonding (which is not the case in Fig. 53) or are intrinsically continuous if the surface is antibonding or coupled to a dissociative continuum. The first two bands of Fig. 53 are known as $\tilde{A}\,{}^1\Delta_u$ and $\tilde{B}\,{}^1\Pi_g$ states, which have forbidden transitions with the ground surface [213]. The next band of high intensity is the $\tilde{C}\,{}^1\Sigma_u^+$ state, which features structures due to nontrivial vibrational motion in spite of the fact that the band is continuous. A potential surface proposed for the collinear motion of CO_2 on this surface is shown in Fig. 54 [221, 222]. The potential surface presents a saddle point between the two channels of dissociation

$$(\mathrm{O{-}C})^* + \mathrm{O}^* \leftarrow (\mathrm{O{-}C{-}O})^* \rightarrow \mathrm{O}^* + (\mathrm{C{-}O})^*$$

If the molecule has been excited to this potential surface, we have a process of direct dissociation that is very fast, of the order of 10 fs for CO_2. In spite of its rapidity, the vibrational motion is far from trivial, and

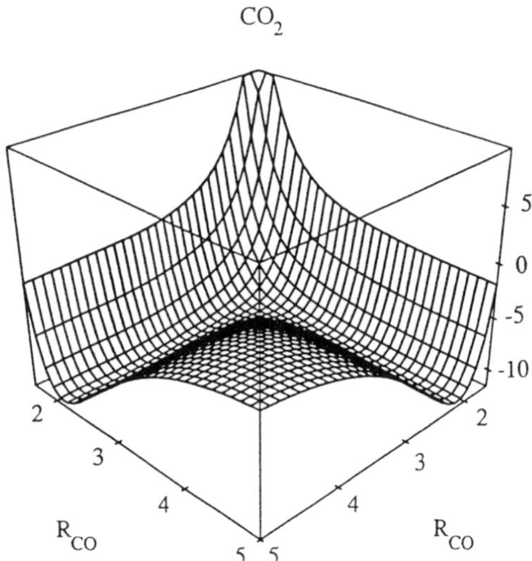

Figure 54. LEPS-type potential surface of the electronic state $CO_2(\tilde{C}\,^1\Sigma_u^+)$ versus the two distances r_{CO} in the collinear configuration with the parameter values adopted in [223].

may remain trapped and oscillate for a while in the region of the saddle equilibrium point. In a classical model of the dynamics, there exists a classical periodic orbit that corresponds to the symmetric stretch of CO_2 at an energy just above the saddle. This classical periodic orbit is trapped along the diagonal ridge of the potential energy surface, although it is unstable with respect to antisymmetric perturbations. Accordingly, it is possible to estimate the average time spent in the vicinity of the unstable periodic orbit using classical stability theory, as shown later. At higher energies just above the saddle, the classical motion becomes chaotic, which we also discuss later.

This example shows that the vibrational dynamics may contain metastable states of finite lifetime in the region of the saddle. These dynamical states can therefore be identified with the molecular transition complex so important in chemical kinetics. We will also show that the vibrational metastable states are at the origin of the peaks on top of the continuous bands. If the preceding reasoning is essentially classical we will show that it describes the phenomenon to a very good approximation, and thus gives a good example of the intuitive appeal of the semiclassical method.

At still higher energies, we find other bands forming Rydberg series leading to the ionization thresholds. The electronic states accumulate at the ionization thresholds of CO_2 for the formation of the corresponding ion

$$(CO_2)^* \rightarrow (CO_2^+)^* + e^-$$

In particular, the first threshold around 110,000 cm^{-1} corresponds to the ground state $\tilde{X}\,^2\Pi_g$ of the ion and the second visible threshold around 145,000 cm^{-1} to the state $\tilde{B}\,^2\Sigma_u^+$ of the ion [213]. The studies of Rydberg states and doubly excited electronic states in atomic species suggest the possibility of irregular spectra in these highly excited regimes of the molecules, which would correspond to classically chaotic behaviors.

The vibrogram of CO_2 was calculated by Eqs. (11.5)–(11.6) and is depicted in Fig. 55. In the lowest $\tilde{A}\,^1\Delta_u$ band, we find two recurrences at 60 and 150 fs, which also appear in an ordinary Fourier transform of the band. By contrast, the $\tilde{B}\,^1\Pi_g$ band is dominated by recurrences at 52 and 92 fs. In the intense $\tilde{C}\,^1\Sigma_u^+$ band there are more pronounced recurrences with two dominant ones at 50 fs and 60 fs. At higher energies, the vibrogram is more complicated with more recurrences. The origin of the large number of recurrences in the $\tilde{C}\,^1\Sigma_u^+$ band has been the object of recent works [62, 222, 223]. It has been shown by Schinke and Engel [223]

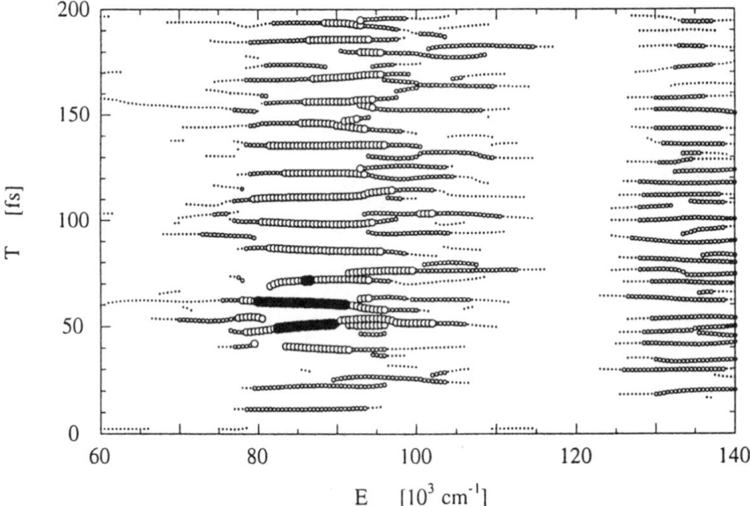

Figure 55. Vibrogram of CO_2 calculated from the experimental UV photoabsorption spectrum by Nakata et al. [220] shown in Fig. 53. The vibrogram is calculated with a Gaussian window according to Eqs. (11.5)–(11.6) with $\Delta E = 4000$ cm^{-1}. The size of the circles increases with the height of the peak in the vibrogram and visual contrast is increased by alternating black and white circles. See text for comments.

that the antibonding potential surface accommodates classically chaotic behavior. Accordingly, there exist several possible periodic orbits of short periods that can be at the origin of the two main periods. As we discussed above, the presence of recurrences does not contradict the antibonding character of the potential since there may exist trapped periodic orbits, particularly corresponding to symmetric stretch motion of the molecule, as shown by Heller and Pack [61, 224]. On the other hand, the possibility of several overlapping electronic potential surfaces is not excluded.

Note that the main recurrences of the three lowest bands are clustered around the values of 50 fs in spite of the fact that they are related to different bands. It seems as if the force constants were homogeneous for the different bands.

2. H_2S and D_2S

Another example is given for the molecules hydrogen sulfide (H_2S) and deuterium sulfide (D_2S). Figure 56 shows a diffusive band of the absorption spectrum of these isotopomers obtained experimentally by Lee et al. [225] (see also [226]). The standard Fourier transforms of these bands are given in Fig. 57. We observe recurrences at 25 fs for H_2S and at 37 fs for D_2S, which correspond to unstable vibrational motion in the region of the transition complex. Schinke et al. [62, 227] showed that there are two antibonding electronic surfaces that are coupled and form an unstable ridge with two conical intersections. According to our

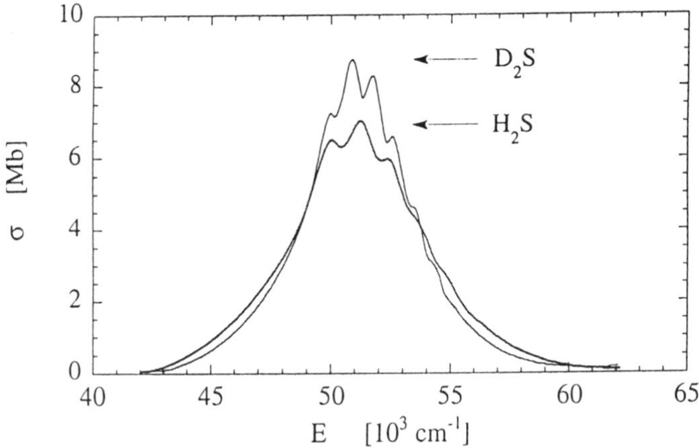

Figure 56. The first diffusive band in the photoabsorption spectra of H_2S and D_2S obtained experimentally by Lee et al. [225]. The band is due to the electronic states $^1B_1$ and $^1A_2$ as explained in [62]. Here, σ is the photoabsorption cross-section.

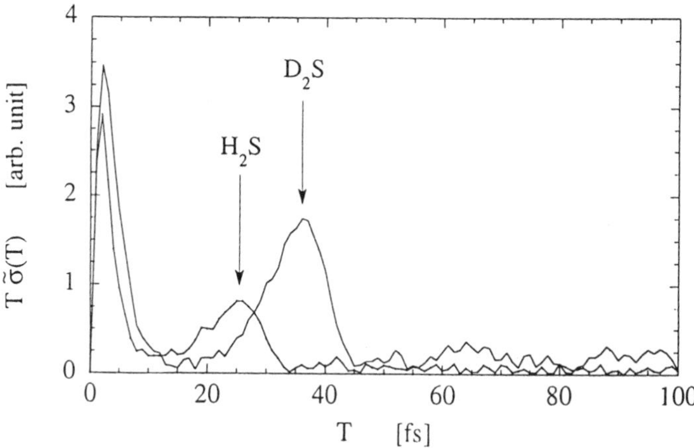

Figure 57. Fourier transforms $T \times \tilde{\sigma}(T)$ with $\tilde{\sigma}$ defined by Eq. (11.4) of the first diffusive bands of H_2S and D_2S of Fig. 56.

discussion in Section IX, we must consider, from the semiclassical point of view, the periodic orbits of each surface of the diagonalized potential separately. The lower surface is antibonding and may sustain unstable periodic orbits on its ridge. The diffusiveness of the band reflects the instability of the periodic orbit. It is interesting to note that the period is in the ratio $\sqrt{2}$ for one isotopomer with respect to the other, which corresponds to the doubling of the mass of the hydrogen nuclei.

3. *Remarks*

From the vibrograms presented in this section, we see that the periodic orbits emerging from the quantum dynamics are experimentally observable in absorption spectra.

In Section XII, we will discuss in detail the quantum dynamics of the molecular transition complex and of its metastable states. In particular, we will show that there are scattering resonances underlying the vibrational oscillatory structures of the diffuse bands. The scattering resonances can be calculated semiclassically in terms of the unstable periodic orbits that are trapped in the region of the molecular transition complex. In this context, the lifetime of the transition complex is determined by the Lyapunov exponents of the periodic orbits. This dynamical aspect was neglected until recently when femtosecond laser techniques became available to measure the decay times of the molecular transition complex.

XII. THE MOLECULAR TRANSITION STATE AND ITS RESONANCES

The molecular transition state is a fundamental concept in our understanding of chemical reactions. If the transition state was first considered in a statistical context in the classical work of Eyring [228], recent works have focused on its relationship to scattering theory and its foundation within quantum dynamics. In this regard, the transition state is described naturally by the set of metastable states associated with scattering resonances, that is, the poles of the S-matrix or equivalently of the Green function associated with the potential surface where the reaction dynamics takes place.

Different methods have been proposed and used to calculate scattering resonances, such as the complex rotation method [229], the coupled-channel approach [230], the R-matrix method [231], and the wavepacket propagation approach with absorbing boundaries [80]. Our purpose in this section is to apply the semiclassical method of periodic-orbit quantization as well as of equilibrium point quantization. These semiclassical methods are very well adapted for the calculation of the resonances of the molecular transition state in many circumstances. They turn out to be extremely efficient, and they provide relations with the classical and statistical approaches. In this sense, the semiclassical approach has a unifying role in the description of reaction dynamics, as was pointed out recently by Marcus [51].

This section is organized as follows. In Sections XII.A–F, we present a detailed analysis by Burghardt and Gaspard [21] of the transition complex in the photodissociation process of HgI_2, which has been the object of recent femtosecond laser experiments by Zewail and co-workers [232–234]. In a collinear model, we analyze the phase-space geometry of the classical repeller, as well as the bifurcations leading to classical chaos and the formation of a three-branch Smale horseshoe. Thanks to this result, we can list all the periodic orbits needed to describe the dynamics in the chaotic regime. Subsequently, we carry out equilibrium point and periodic orbit quantization to obtain the scattering resonances, which are compared with a quantum mechanical calculation by wavepacket propagation. The results are discussed in Section XII.F. In Section XII.G, we compare the classical repellers of symmetric ABA and asymmetric ABC triatomic molecules. The following Sections XII.H–K present extensions toward and inclusion of bending and rotation, as well as toward higher-dimensional systems in the framework of equilibrium point and periodic-orbit quantization.

A. The HgI_2 System

We present a detailed analysis of the transition complex in the photodissociation process

$$h\nu + \mathrm{HgI}_2(X\,{}^1\Sigma_g^+) \rightarrow [\mathrm{I}\cdots\mathrm{Hg}\cdots\mathrm{I}]^{\ddagger} \rightarrow \mathrm{HgI}(X\,{}^2\Sigma^+) + \mathrm{I}({}^2P_{3/2}) \quad (12.1)$$

This study has been motivated by the femtosecond laser experiments on mercuric iodide (HgI_2), which have recently been reported by Zewail and co-workers [232–234]. Our analysis is based on the potential energy surface (p.e.s.) proposed in [233, 234], that is, a damped Morse potential for two degrees of freedom (the two interatomic distances), assuming that the transition state is collinear. The dynamics of the HgI_2 system is modeled according to the following two-degree-of-freedom Hamiltonian

$$\hat{\mathcal{H}} = \hat{T} + \hat{V} = \frac{1}{2\mu_{\mathrm{HgI}}}(\hat{p}_1^2 + \hat{p}_2^2) - \frac{1}{m_{\mathrm{Hg}}}\hat{p}_1\hat{p}_2 + V(r_1, r_2) \quad (12.2)$$

with the potential

$$V(r_1, r_2) = D\left\{1 - \exp\left[-\beta\,\frac{e^{-\gamma r_1}(r_1 - r_e)/r_e + e^{-\gamma r_2}(r_2 - r_e)/r_e}{e^{-\gamma r_1} + e^{-\gamma r_2}}\right]\right\}^2 - D \quad (12.3)$$

where the parameters D, β, and r_e have further dependencies on r_1 and r_2 according to

$$\begin{aligned} D(r_1, r_2) &= D_0 + D_1 \exp[-(r_1 - r_2)^2/\sigma_D^2] \\ r_e(r_1, r_2) &= r_{e0} + r_{e1} \exp[-(r_1 - r_2)^2/\sigma_{r_e}^2] \\ \beta(r_1, r_2) &= \beta_0 + \beta_1 \exp[-(r_1 - r_2)^2/\sigma_\beta^2 - (r_1 + r_2 - 2r_e)^2/\sigma_{\beta_+}^2] \end{aligned} \quad (12.4)$$

with the parameters given in Table II. This potential is depicted in Fig. 58.

In terms of the semiclassical methods introduced above, the molecular transition complex is closely related to the classical repeller, that is, the ensemble of unstable periodic orbits in a hyperbolic system. If the repeller is composed of a single periodic orbit (the symmetric-stretch orbit), the lifetime of the metastable resonance states will be directly

TABLE II
Set of Parameters for the Dissociative Potential Energy Surface [Eqs. (12.3) and (12.4)] of HgI_2 [233, 234]

$D_0 = 2800\ \text{cm}^{-1}$	$D_1 = -1000\ \text{cm}^{-1}$	$\sigma_D = 1$ Å
$r_{e0} = 2.8$ Å	$r_{e1} = 0.2$ Å	$\sigma_{r_e} = 1$ Å
$\beta_0 = 7.1$	$\beta_1 = 5.5$ Å	$\sigma_\beta = 0.75$ Å, $\sigma_{\beta_+} = 2.5$ Å
$\gamma = 2.3$ Å^{-1}		

related to the Lyapunov exponent of this orbit, as pointed out in Section XI. This is the situation prevailing at low energies in the HgI_2 system, before a transition to a chaotic regime occurs. The chaotic, nonseparable regime at high energies is characterized by an infinite number of periodic orbits, and an individual periodic orbit is no longer related to a regular family of resonances. As we explained in Sections V and VI, the spectrum of resonance states then cannot be assigned in terms of "good" quantum numbers, such that perturbation expansions are not applicable.

In Section XII.D, we present the results of the periodic-orbit quantization in the low- and high-energy regimes. Note that in the regular regime, at low energies, a quantization condition can be derived by periodic-orbit quantization of the symmetric-stretch orbit or, alternatively, by the method of equilibrium point quantization, as described in Section XII.C.

The transition between the regular and chaotic regimes in the HgI_2 system is brought about by a series of bifurcations in classical phase

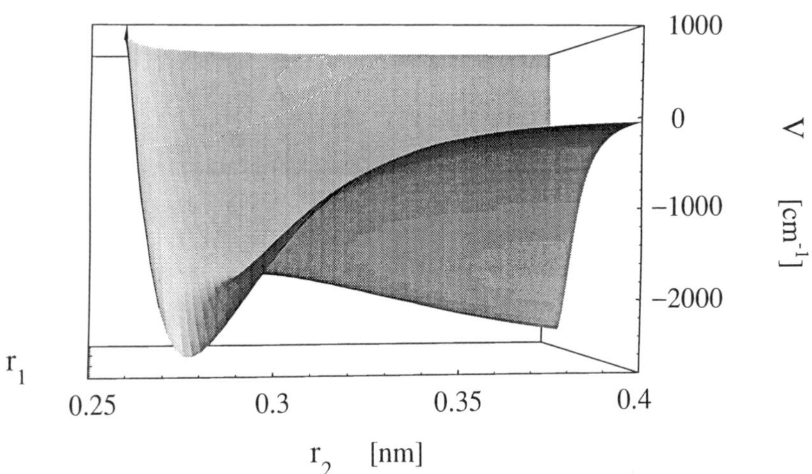

Figure 58. View in perspective of the potential surface [Eqs. (12.3)–(12.4)] of $[\text{I}\cdots\text{Hg}\cdots\text{I}]^\ddagger$.

space. Over a narrow energy band, the symmetric-stretch periodic orbit is stable, and is associated with an elliptic island in phase space. The area of the elliptic island is too small, though, to support a quantum state, such that its presence is mainly reflected in an enhanced lifetime of the nearby resonances in the low- and high-energy regimes.

The basic features of the classical dynamics in the HgI_2 system may be understood by constructing a Hamiltonian mapping, which makes use of an effective one-dimensional potential. By varying the parameters pertaining to this potential, one can model the features of the Hamiltonian flow in dependence on the overall energy. In particular, the chaotic dynamics at high energies corresponds to the phase-space structure of a Smale horseshoe, from which a symbolic dynamics can be derived. The classification of periodic orbits in terms of this symbolic dynamics is at the basis of the cycle expansion method [161], which can be applied to obtain the semiclassical resonances in the chaotic regime.

In the first of the following sections, we will introduce the Hamiltonian mapping, on the grounds of which we interpret the main features of the classical dynamics in the HgI_2 system. Section XII.E will be devoted to a quantum mechanical analysis of the HgI_2 system, which proceeds by a numerical integration of the Schrödinger equation for a wavepacket on the excited-state p. e. s. The autocorrelation function associated with the evolving wavepacket yields, by Fourier transformation, a spectrum from which the energies and widths of the resonance states may be deduced. The quantum mechanical results are compared with the semiclassical analysis.

B. Classical Dynamics and a Hamiltonian Mapping for the HgI_2 System

1. *From the Flow to the Mapping*

The classical dynamics of the HgI_2 system is modeled according to the Hamiltonian [Eqs. (12.2–12.4)]. The dynamics generated by this Hamiltonian is characterized by a variety of behavior, ranging from regular to chaotic. On increasing the total energy of the system, pronounced discontinuities occur in the classical dynamics, which are brought about by phase-space bifurcations due to the anharmonicity of the potential.

Three different energy regimes may be distinguished. In the first regime, the classical dynamics is regular, and there is one isolated, unstable periodic orbit, the symmetric-stretch trajectory. The second regime, which covers a very narrow energy bandwidth, is characterized by quasiperiodic motion in the transition state region. The third energy regime, which is delimited by the total dissociation threshold, corre-

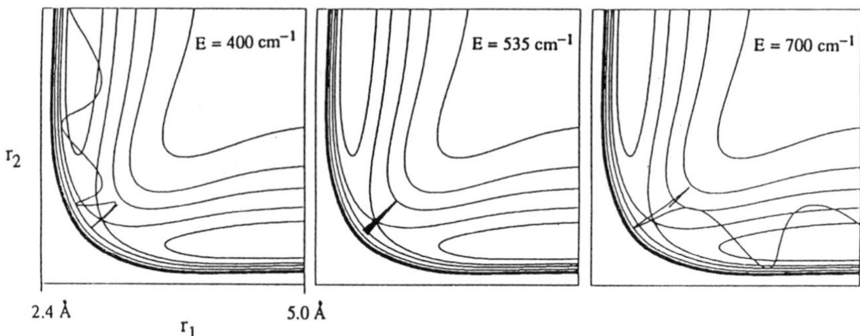

Figure 59. Trajectories with initial conditions slightly different from the symmetric stretch: $r_{10} = r_{20}$, $p_{10} = p_0 + \delta p_0$ (with $\delta p_0 \simeq 10^{-3} p_0$), and p_{20} being adjusted to meet the energy constraint. The energy is defined relative to the saddle point; the total dissociation threshold is at 1800 cm^{-1}. At 400 cm^{-1}, there is a single unstable periodic orbit of symmetric stretching. At 535 cm^{-1}, a stable elliptic island surrounds the symmetric stretching periodic orbit. At 700 cm^{-1}, the main elliptic island has been destroyed by a period doubling and the motion is mostly chaotic.

sponds to fully chaotic dynamics, with an infinite number of unstable and isolated periodic orbits. Figure 59 shows trajectories characteristic of the different energy regimes.

We have studied the linear stability of the symmetric-stretch periodic orbit according to the methods of Section III.A. Figure 60 shows how this periodic orbit changes from hyperbolic without reflection, to elliptic, and then to hyperbolic with reflection, as energy increases. This behavior is a

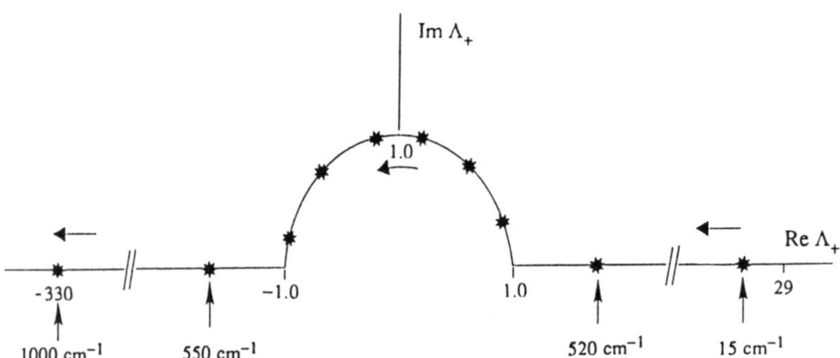

Figure 60. Stability properties of the symmetric-stretch orbit, varying from hyperbolic to elliptic to hyperbolic with reflection as a function of energy. The eigenvalue Λ_+ is plotted in the complex plane. For the first and third stability regimes, a few representative values are indicated; in the elliptic region, between 523 and 548 cm^{-1}, the points are equally spaced at 5 cm^{-1}.

result of the bifurcations that occur in the classical dynamics close to the symmetric-stretch motion.

The phase-space structures of the HgI_2 Hamiltonian can be studied by the phase portraits in a Poincaré surface of section transverse to the symmetric-stretch orbit (e.g., using the section plane $r_1 + r_2 = 2r_{eq}$, where $r_{eq} = 3$ Å is the saddle equilibrium point at $E_s = -1800\ \mathrm{cm}^{-1}$). It turns out that the Poincaré mapping induced by the Hamiltonian flow in this surface of section is qualitatively similar to a Hamiltonian mapping defined in Section III, with an effective potential of the form [21]

$$V(q) = \left(\alpha + \gamma \frac{q^2}{2}\right) \exp\left(-\frac{q^2}{2}\right) \quad \text{with} \quad \alpha, \gamma > 0 \tag{12.5}$$

and with $T = 1$ in Eq. (3.38). In this model, the position q corresponds to the position along the reaction path of the original system. The parameters α and γ are control parameters. Increasing the energy above the saddle equilibrium point in the flow corresponds to increasing the parameter γ while keeping α constant. Variation of these parameters induces bifurcations in the phase portraits. Depending on the choice of α and γ, the potential $V(q)$ will then represent a barrier, or a local well embedded in a barrier, such that locally bound dynamics may occur. (Note the similarity to vibrationally adiabatic models [235].) We observed in our simulations of the HgI_2 system that the sequence of bifurcations encountered as the energy increases is faithfully reproduced by the Hamiltonian mapping for $\alpha = 1$ as γ increases [21]. This observation can be interpreted by noting that an increase in the energy renders the motion more anharmonic, which goes along with an increase of the amplitude of the kicks in the mapping [Eq. (12.5)].

2. *Properties of the Mapping*

We start with the analysis of the fixed points and of their linear stability. Let us recall that the fixed points of the Poincaré mapping of a flow correspond to periodic orbits in the flow. The mapping [Eq. (12.5)] admits one or three fixed points depending on the parameter values.

By symmetry, the origin ($q = 0$, $p = 0$) is always a fixed point (labeled by the symbol 0), which corresponds to the symmetric-stretch orbit of the flow. This fixed point is the only one, and is hyperbolic (without reflection) as long as $\gamma < \alpha$. At $\gamma = \alpha$, it becomes elliptic, that is, a center with stability eigenvalues on the unit circle: $\Lambda_{\pm} = \exp(\pm i 2\pi\rho)$.

At this critical value, $\gamma = \alpha$, an antipitchfork bifurcation occurs (see also Section XII.G) and two new hyperbolic fixed points appear at

$q = \pm\sqrt{2 - 2\alpha/\gamma}$. These new fixed points (labeled by the symbols 1 and 2) correspond to two fundamental periodic orbits of the flow that always remain on the same side of the symmetric-stretch orbit during their period $[r_1(t) > r_2(t)$ for $p = 1$ and $r_1(t) < r_2(t)$ for $p = 2]$. This pair of periodic orbits, which were particularly studied by Pollak and Pechukas [236, 237] (and will in the following sometimes be referred to as the "off-diagonal" periodic orbits), borders the main elliptic island surrounding the symmetric-stretch orbit (see Fig. 59 at $E = 535\ \text{cm}^{-1}$). Since the fixed points 1 and 2 are hyperbolic, they have stable and unstable manifolds, which form a thin chaotic zone around the elliptic island due to their homoclinic intersections [238]. Smaller subsidiary elliptic islands may be present in this chaotic zone. Figure 61 illustrates the phase-space structure around the fixed points (at parameter values that refer to the discussion of the next paragraph).

As γ increases, the center fixed point 0 successively undergoes $m:n =$ 1:4, 1:3, 1:2 bifurcations, where its rotation number takes the values $\rho = m/n$. At the $m:n = 1:4$ bifurcation, a chain of four subsidiary elliptic islands separates from the main island. The area of the main island decreases in size on average till the period-doubling bifurcation $m:n =$ 1:2 at $\gamma = \alpha + 4$, where the fixed point becomes hyperbolic with reflection. The main elliptic island begins to split into two smaller islands at this period doubling, which initiates a cascade of further period doublings. Figure 61 gives examples of the phase-space structures that mark the transition between the quasiperiodic and chaotic regimes.

Since the three fixed points 0, 1, and 2 are hyperbolic above $\gamma = \alpha + 4$, they all have stable and unstable manifolds. Small subsidiary islands may subsist in phase space as long as these manifolds have not undergone their last homoclinic tangency at some critical value $\gamma = \gamma_c > \alpha + 4$. For $\alpha = 1$, this last homoclinic tangency occurs at $\gamma_c \simeq 9.5$.

Above this value $(\gamma = \gamma_c)$, the dynamics is purely chaotic and all the trapped trajectories are hyperbolic. The trapped trajectories are then in 1:1 correspondence with a symbolic dynamics (without constraint) based on the three symbols $\{0, 1, 2\}$, which label the three fundamental fixed points of the mapping (or fundamental periodic orbit of the flow). Figure 62 schematically shows that the dynamics is then similar to the horseshoe but with three branches as envisaged by Smale in 1967 [116]. The trapped orbits then form a fractal and chaotic repeller that controls the classical escape dynamics with respect to the molecular transition complex.

In Figure 63, we plotted an estimation of the topological entropy per iteration of the mapping [Eq. (12.5)] as a function of γ for $\alpha = 1$. The figure shows how the topological entropy, which gives us the rate of

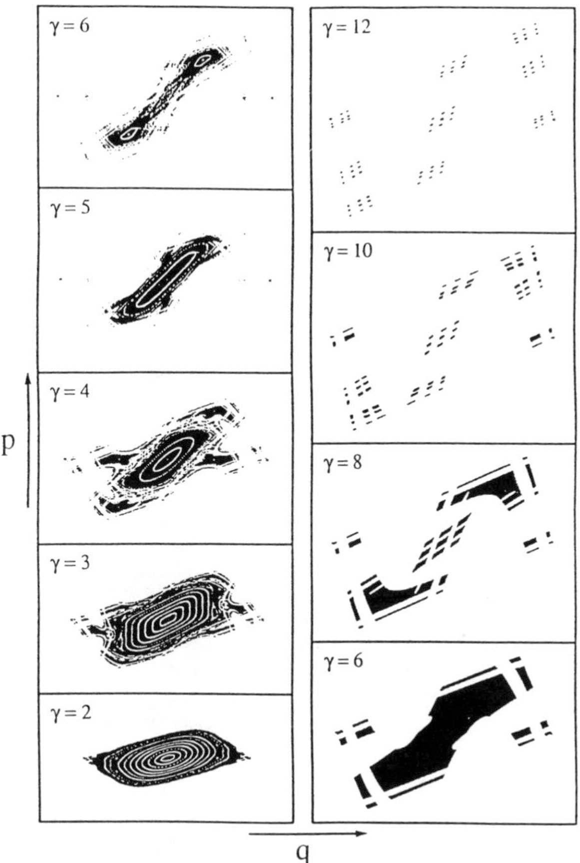

Figure 61. Initial conditions in the (q, p) plane of the orbits trapped at finite distance for the Hamiltonian mapping [Eq. (12.5)] for $\alpha = 1$ and increasing values of γ. In the left-hand column, the trapped orbits are represented by the initial conditions of the orbits that have survived 10 forward and 10 backward iterations (some trajectories are drawn in white inside the elliptic islands); in the right-hand column, trapped orbits are represented by those that have survived 2 forward and 2 backward iterations. Beyond $\gamma_c = 9.5$, the trapped orbits form a three-branch Smale horseshoe.

proliferation of periodic orbits, increases from the value zero taken below $\gamma = \alpha = 1$ up to its maximum value $\ln 3$ above γ_c.

3. *Back to the Flow*

A numerical study shows that the same scenario applies to the flow, which allows us to classify the set of periodic orbits in the chaotic regime.

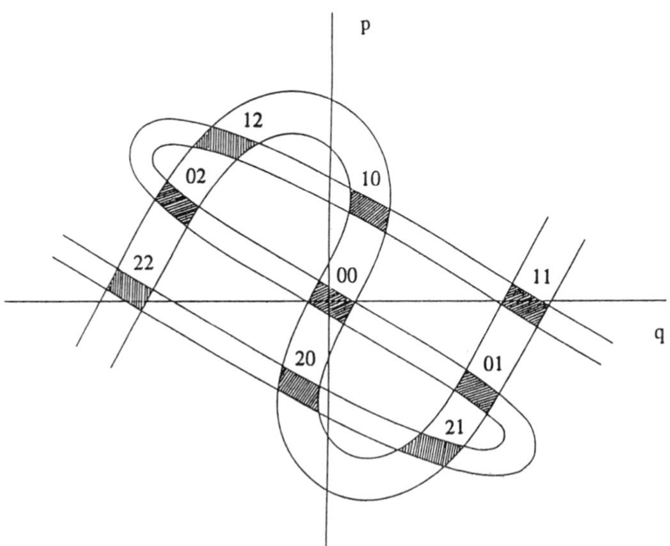

Figure 62. Schematic drawing of a three-branch Smale horseshoe as it appears in a Poincaré section transverse to the classical dynamics of the molecular transition state at high energies. The shaded intersections indicate regions where phase-space areas visited in a forward and a backward crossing in the Poincaré section. The partition that is thus generated may be labeled according to a symbolic dynamics, with a $2n$-symbol sequence for n forward/backward iterations. For a bi-infinite sequence, the intersections will correspond to single orbits of the transition state. These then represent the invariant set of the dynamics and are in 1:1 correspondence with the periodic orbits of the Hamiltonian flow.

In Fig. 64 we depict several of the periodic orbits and their corresponding symbolic sequences. The asymmetric-stretch orbit evidenced by Schinke and Engel [223] turns out to be assigned to sequence 12. There exist two further periodic orbits with a two-symbol sequence, namely, 01 and 02, which should be treated on the same footing as the asymmetric stretch orbit 12. According to the symbolic dynamics, the periodic orbits 0, 1, and 2 are the fundamental periodic orbits in the sense that they generate all the other periodic orbits, in terms of their topological combinations. By contrast, the asymmetric-stretch periodic orbit 12 is of higher period since it is topologically composed of 1 and 2. The orbits corresponding to the symbolic sequences 012 and 0102 have also been found in the CO_2 system [223]. We note that the periodic orbits 0, 02, 022, and 0222 have been identified in a two-degree-of-freedom model of the Hartley band of ozone by Johnson and Kinsey [239] who, moreover, made a tentative comparison with peaks in a Fourier transform of the spectrum of this band. Thanks to the three-branch horseshoe of Fig. 62 and the associated

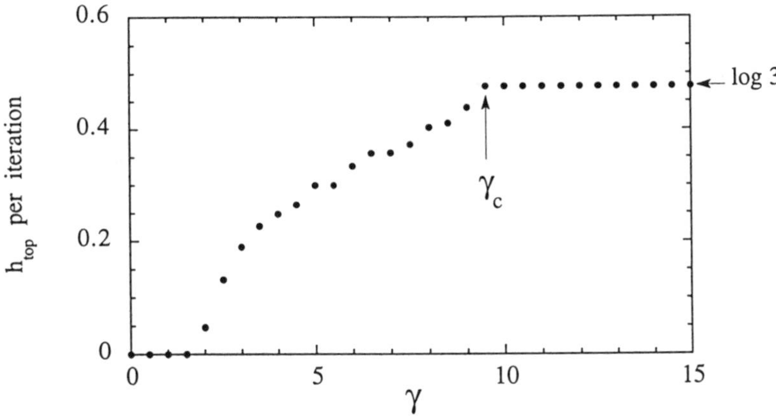

Figure 63. Topological entropy per iteration versus the parameter γ of the Hamiltonian mapping [Eq. (12.5)] with $\alpha = 1$. The topological entropy starts to be positive above $\gamma = 1$, which is the threshold of chaos. The topological entropy increases until the last homoclinic tangency occurs at $\gamma_c \simeq 9.5$. Beyond this last bifurcation, the invariant set is a full three-branch Smale horseshoe which is structurally stable with $h_{top} = \log_{10} 3$.

symbolic dynamics, we see that different and fragmentary results that are dispersed in the literature can be understood in a unified way. As we will show below, the classification in terms of symbolic sequences is an essential step before the application of the semiclassical method of quantization in the chaotic regime.

C. Equilibrium Point Quantization of the Transition State

We applied the method of Section V to the saddle equilibrium point of the potential surface of HgI_2. The cubic and quartic derivatives of the potential have been calculated in order to obtain the coefficients [Eqs. (5.5–5.7)] of the Dunham expansion. For the present two-degree-of-freedom system, the Dunham expansion truncated at order $\hbar^3$ is given by[3]

$$E_{nk} = E_s + \Delta E_2 + \hbar\omega(n + \tfrac{1}{2}) + x_{11}(n + \tfrac{1}{2})^2 + x_{22}(k + \tfrac{1}{2})^2$$
$$- i\hbar\lambda(k + \tfrac{1}{2}) + i\tilde{x}_{12}(n + \tfrac{1}{2})(k + \tfrac{1}{2}) \qquad (12.6)$$

The scattering resonances are labeled by two quantum numbers: $n, k = 0, 1, 2, \ldots$. The quantum number n corresponds to symmetric-stretching

[3] We note that Coriolis interactions are ignored in this collinear model.

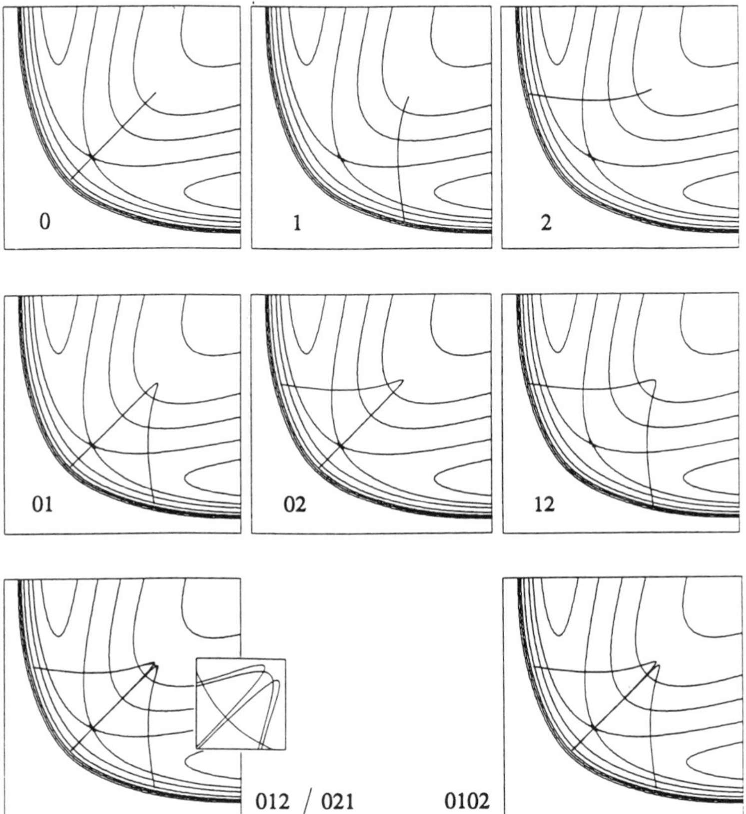

Figure 64. Periodic orbits of the HgI_2 system, at an energy of $1200\,cm^{-1}$. The symbolic sequences that are in correspondence with Fig. 62 are indicated. The first row gives the fundamental orbits, associated with one symbol $\omega_i = 0$, 1, or 2 (where the symbols ω_i refer to bi-infinite sequences $\{\omega_i\}_{i=-\infty}^{+\infty}$), and that correspond, respectively, to the symmetric stretch and the two off-diagonal orbits of the Hamiltonian flow. The second row gives all orbits that correspond to a two-symbol sequence, among them the asymmetric-stretch orbit 12. The last row gives examples of orbits corresponding to three- and four-symbol sequences. These orbits are very difficult to localize, due to their high instability, and, for the purpose of illustration, have been composed of the low-order orbits (while the other figures correspond to actual trajectories).

excitations of frequency ω, which are stable and, therefore, lead to an increase in energy. On the other hand, the quantum number k corresponds to asymmetric-stretching excitations of Lyapunov exponent λ, which are unstable and lead to shorter and shorter lifetimes. The numerical values of the coefficients are given in Table III for the Hamiltonian [Eqs. (12.2–12.4)].

TABLE III
Coefficients of the Dunham Expansion [Eq. (12.6)] for the Scattering Resonances of HgI_2

Scattering Resonances	Values (cm^{-1})
E_s	−1800
ΔE_2	−0.66393
$\hbar\omega$	91.84768
x_{11}	−1.22733
x_{22}	−1.16763
$\hbar\lambda$	49.06510
$\tilde{x}_{12}$	+1.70899

We observe a difference of more than one order of magnitude between the first- and second-order coefficients, which justifies the application of the Dunham expansion. The anharmonic zero-point energy ΔE_2 turns out to have a very small value. The coefficient x_{11} is negative so that the spacings decrease between the energies of the resonances, which is the behavior expected for a Morse-type potential. On the other hand, the coefficient $\tilde{x}_{12}$ is positive so that the imaginary parts of the complex energies increase with excitation of the quantum number n. Accordingly, the lifetimes also increase. Table IV contains the energies and lifetimes of the resonances of HgI_2 predicted by the Dunham expansion [Eq. (12.6)]. These resonances are also depicted in the complex energy plane as crosses in Fig. 65.

D. Periodic-Orbit Quantization of the Transition State

In Section VII, we derived a periodic-orbit formula [Eq. (7.15)] for the photoabsorption cross-section, which allows in principle a direct comparison with an experimental UV absorption spectrum. In our analysis for HgI_2 [21], we will focus, though, on the expression for the density of states [Eq. (6.13)], which provides no information about the intensity distribution in the spectrum, but which is equivalent to κ_ω as far as the information about the positions and widths of the resonances is concerned.

1. *The Low-Energy Regime*

To localize the resonances in the low- and high-energy regimes, we use the zeta function of Eq. (6.21). In the low-energy regime, where the classical dynamics is regular, the semiclassical quantization of the unstable symmetric-stretch orbit alone yields the resonances.

TABLE IV

Scattering Resonances of the Two-Degree-of-Freedom Hamiltonian [Eqs. (12.2–12.4)] of HgI_2[a]

n	Equilibrium Point		Periodic Orbit		Quantum Mechanical	
	E (cm^{-1})	τ (fs)	E (cm^{-1})	τ (fs)	E (cm^{-1})	τ (fs)
0	44.7	110.1	45.7	110.3	47.8	111.1
1	134.1	114.2	135.3	115.6	136.9	117.6
2	221.0	118.5	222.2	120.8	225.2	117.3
3	305.5	123.2	306.6	132.3	308.6	129.5
4	387.5	128.3	388.3	149.4	391.5	144.9
5	467.1	133.8	466.1	202.0	472.2	160.8
6	544.2	139.9			551.2	163.8
7	618.9	146.5	639.6	473.9	624.7	164.1
8	691.1	153.7	709.2	340.7	708.2	176.8
9	760.8	161.7	779.7	271.6	777.9	186.2
10	828.1	170.6	851.0	220.8	846.9	213.0
11			921.0	183.0	915.2	219.7
12			989.0	158.7	981.1	203.3
13			1055.7	150.7	1045.8	175.1
14			1119.5	165.5	1109.9	168.2
15			1178.9	189.6	1168.0	150.2
16			1234.7	208.1	1233.9	162.3
17			1287.3	215.1	1284.7	213.2
18					1329.0	219.7
			1337.2	213.5	1338.8	239.2
19			1384.9	208.5	1378.3	191.4
20			1430.6	207.9	1434.2	161.1

[a] The first pair of columns contains the resonances calculated by equilibrium point quantization with the Dunham expansion [Eq. (12.6)]. The second pair of columns contains the resonances obtained by periodic-orbit quantization. In the low-energy regime, the resonances were calculated as the zeros of the periodic-orbit quantization condition [Eq. (12.8)] by a Newton–Raphson method. In the high-energy regime, the resonances are obtained as the zeros of the periodic-orbit quantization condition [Eq. (12.14)]. The third pair of columns contains the quantum mechanical resonances calculated by fitting spectra obtained by wavepacket propagation (see Section XII.E) by a rational fraction $f_m(E) = P_m(E)/Q_m(E)$, with $P_m(E)$ and $Q_m(E)$ being polynomials of order $m = 220$ [21]. The poles of $f_m(E)$ (i.e., the zeros of $[f_m(E)]^{-1}$) were identified by a Newton–Raphson method, combined with an algorithm for rational function extrapolation [245], which we extended to complex data. We remark that, for the resonance $n = 18$, the fitting by a rational function of high order $m = 220$ is giving a spurious value. The energies E and lifetimes τ of the resonances are given for each pair of columns.

For a single periodic orbit in a two-degree-of-freedom system ($f = 2$), Eq. (6.21) reduces to

$$Z(E) = \prod_{k=0}^{\infty} \left\{ 1 - \frac{\exp[(i/\hbar)S(E) - i(\pi/2)\mu]}{|\Lambda(E)|^{1/2}\Lambda(E)^k} \right\} \tag{12.7}$$

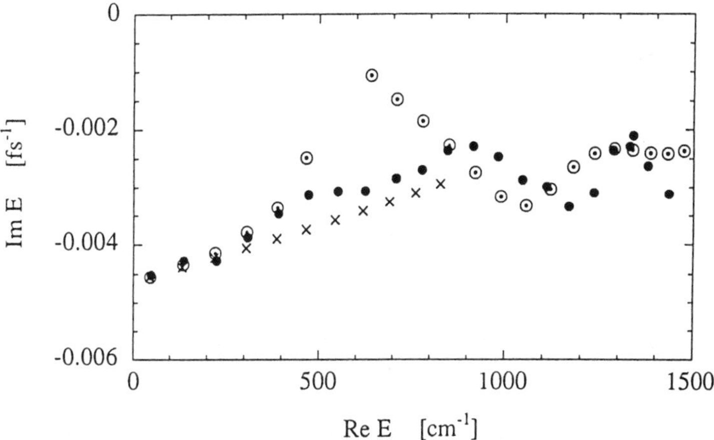

Figure 65. The scattering resonances of the molecular transition state of HgI_2 (collinear model) of Table IV. The quantum mechanical values (filled circles) are derived from wavepacket propagation. The equilibrium-point quantization values (crosses) are calculated from the Dunham expansion [Eq. (12.6)] with the coefficients of Table III. The periodic-orbit quantization values (dotted circles) are calculated from Eq. (12.8) at low energies but from the zeta function [Eq. (12.14)] at high energies. The imaginary part of the complex energies is given in units of inverse lifetime.

The zeros of $Z(E)$ are given by the zeros of its factors, which results in the condition

$$\frac{i}{\hbar} S(E) - i \frac{\pi}{2} \mu - u(E)\left(\frac{1}{2} + k\right) = 2\pi i n \tag{12.8}$$

where $u(E) = \ln \Lambda(E)$. The product in Eq. (12.7) associated with the quantum number k reflects that we use a harmonic approximation with respect to the potential transverse to the periodic orbit. The second quantum number n appearing on the right-hand side of Eq. (12.8) refers to the quantization in the direction parallel to the periodic orbit. The Morse index was found to be $\mu = 2$. (Note that the Morse index is even in the hyperbolic case, and odd in the hyperbolic-with-reflection case.)

In the harmonic case, where the action is given by $S = T_0 E$ and the eigenvalues of the stability matrix are energy independent, $u = \lambda T_0$, we obtain the following energy expression:

$$E_{nk} = E_s + \hbar\omega\left(n + \frac{\mu}{4}\right) - i\hbar\lambda\left(k + \frac{1}{2}\right) \tag{12.9}$$

where $\omega = 2\pi/T_0$. These complex energies define a progression of equally spaced resonances in the complex energy plane (in the lower one half of the second Riemann sheet), which correspond to scattering resonances. It is evident that the quantum number n defines the "stable" mode (corresponding to the symmetric-stretch vibration, in the harmonic approximation), while the quantum number k corresponds to the "unstable" mode (dissociation along the "reaction coordinate") as in Section XII.C. The latter quantum number k is associated with the lifetime, which is the inverse of the Lyapunov exponent λ, $\tau = \lambda^{-1}$. The Lyapunov exponent gives a measure of the exponential escape of trajectories from the vicinity of the periodic orbit, and it is intuitively clear that the lifetime of a quantum state corresponding to this repeller with a single periodic orbit should be inversely proportional to the classical escape rate (see also Heller [190]). This discussion establishes the relationship to Section XII.C. Equation (12.9) is equivalent to the truncation of the Dunham expansion [Eq. (12.6)] at the harmonic approximation, that is, at first order in $\hbar$, since the Maslov index here equals $\mu = 2$.

In the anharmonic case, one may express the energy dependence of the action and stability eigenvalues in terms of a Taylor expansion about the saddle point (at $E = 0$):

$$\begin{aligned} S(E) &= T_0 E + S_2 E^2 + S_3 E^3 + \cdots \\ u(E) &= \lambda T_0 + u_1 E + u_2 E^2 + \cdots \end{aligned} \tag{12.10}$$

In principle, these expansions can be used to obtain the eigenenergies beyond the harmonic approximation [Eq. (12.9)]. However, we would not recover the complete Dunham expansion [Eq. (12.6)] because the motion transverse to the periodic orbit is treated harmonically in the zeta function [Eq. (12.7)], which is restricted to the leading approximation of the $\hbar$-expansion [Eqs. (6.41) and (6.42)]. Note that the coefficient x_{22} describing anharmonicities in the motion transverse to the periodic orbit cannot be derived from Eq. (12.8). On the other hand, the anharmonicities in the direction parallel to the periodic orbit are taken into account by the energy dependence of the action $S(E)$ and the stability parameter $u(E)$.[4]

In order to calculate the scattering resonances, we have numerically determined the coefficients of the expansion [Eq. (12.10)], by using a polynomial fit to the classical action obtained by integration (see Fig. 66) [21]. In the analysis based on Eq. (12.8), we have considered only the leading resonances (longest lifetimes), which are located close to the real axis (corresponding to $k = 0$). The resonances that have thus been derived for the low-energy regime (below the formation of stable islands and the onset of chaos) are listed in Table IV and reported in Fig. 65.

2. *The Intermediate Regime*

Between 523 and 548 cm^{-1} above the saddle point, the symmetric-stretch orbit is stable and the phase space contains a main elliptic island. The periodic-orbit quantization becomes difficult in this regime because the stability eigenvalues are close to the value 1, which may cause the presence of small denominators in the trace formula [Eq. (6.13)].

Some insight is provided by the following qualitative argument. According to the Bohr–Sommerfeld quantization, an elliptic island of area A in the Poincaré surface of section can support a number of quantized states equal to $A/2\pi\hbar$. If $A \gg 2\pi\hbar$, the quantized states are long-lived but metastable because of dynamical tunneling out of the elliptic island, as explained in Section VI.D.2. On the other hand, if the area A is much smaller than $2\pi\hbar$, as it turns out to be the case here, the island cannot be the support of any long-lived metastable state. If a metastable state subsists its lifetime may be expected to be short since it is determined by the dynamics in an area of the order of $2\pi\hbar$ around the elliptic island where the dynamics is unstable (see Section VI.F). Comparing different triatomic molecules, it seems that the main elliptic island is of small area in light–heavy–light molecules as is already the case with HgI_2, where the mass ratio is not extreme. On the other hand, the island has a large area and may sustain many metastable states in heavy–light–heavy complexes, such as ClHCl [240], IHI [241], or the muonic (Mu)

[4] More precisely, $S(E)$ depends on ω, x_{11}, y_{111}, $z_{1111}, \ldots$, while $u(E)$ depends on these same coefficients together with λ, x_{12}, y_{112}, $z_{1112}, \ldots$. However, the coefficients x_{22}, y_{122}, $y_{222}, \ldots$, concern the nonlinear stability of the periodic orbit and, therefore, they do not enter into Eq. (12.10) and are absent from the leading approximation of the Gutzwiller trace formula. Nevertheless, these further coefficients contribute to the $\hbar$-corrections of the Gutzwiller trace formula so that we can recover the results of equilibrium point quantization in this way, which shows consistency between the equilibrium point and the periodic-orbit quantization in the low-energy regime.

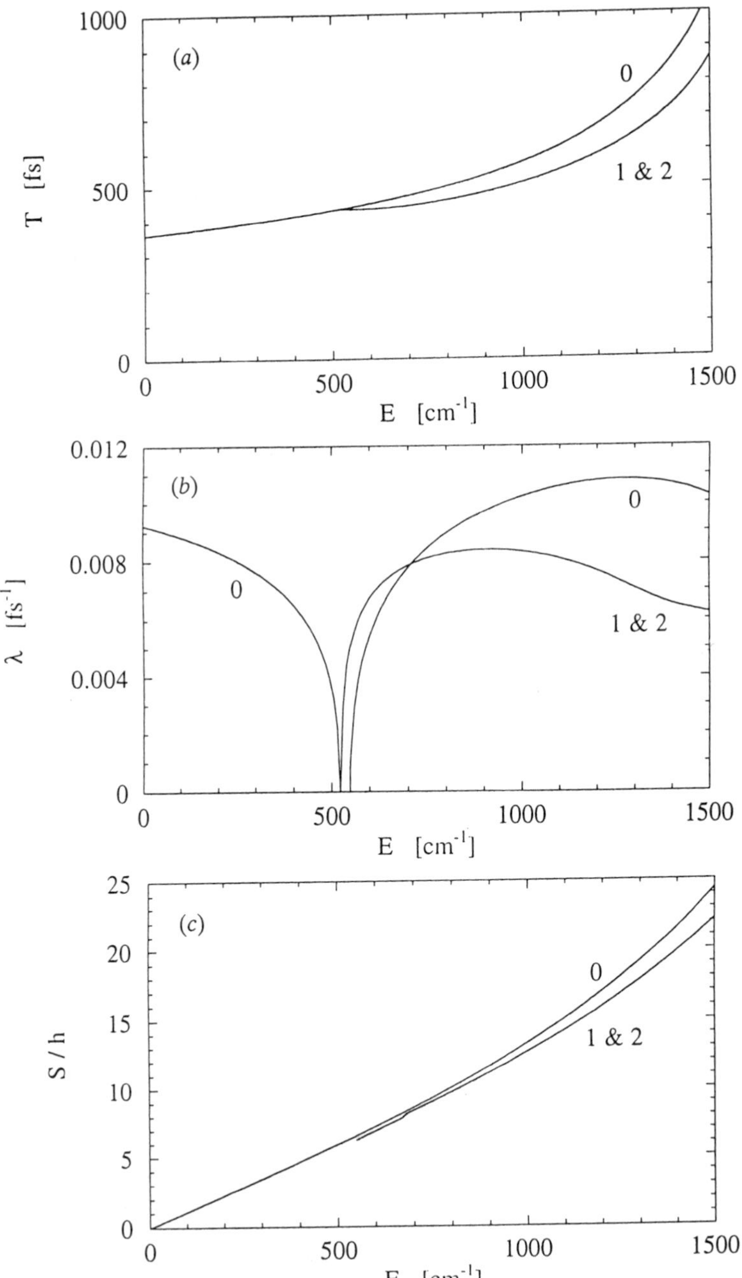

Figure 66. (*a*) Periods, (*b*) Lyapunov exponents, and (*c*) reduced actions for the three fundamental orbits 0, 1, and 2 of the HgI_2 Hamiltonian system.

molecule HMuH [242], which is the six-body system $[(p^+, e^-), (\mu^+, e^-), (p^+, e^-)]$ (see also [243]).

3. *The High-Energy Regime*

In the high-energy regime, the phase-space dynamics is described by a symbolic dynamics based on the three symbols $\{0, 1, 2\}$ labeling the three periodic orbits we called fundamental. As these periodic orbits generate the whole repeller by topological combination, the number of periodic orbits increases exponentially with a period, so that the so-called topological entropy per crossing of the Poincaré section is equal to $\ln 3$.

Under these circumstances, the cycle expansion described in Section VI.C.4 may be applied to calculate the resonance states. As in Section XII.D.1, we only consider the dominant resonances with longest lifetimes that are given by the factor $k = m_1 = 0$ of the Selberg zeta function [Eq. (6.21)]. Accordingly, we consider the inverse of the first Ruelle zeta function

$$\begin{aligned}\zeta_0(E)^{-1} &= \prod_p [1 - t_p(E)] \\ &= 1 - t_0 - t_1 - t_2 - (t_{01} - t_0t_1) - (t_{02} - t_0t_2) - (t_{12} - t_1t_2) - \cdots\end{aligned} \tag{12.11}$$

where

$$t_p(E) = \frac{\exp[(i/\hbar)S_p(E) - i(\pi/2)\mu_p]}{|\Lambda_p(E)|^{1/2}} \tag{12.12}$$

At high energies, the periodic orbits with periods higher than 1 are very well approximated as combinations of the fundamental periodic orbits. According to the discussion of this shadowing mechanism in Section VI. C.4, we have

$$t_{01} \simeq t_0t_1 \qquad t_{02} \simeq t_0t_2 \qquad t_{12} \simeq t_1t_2, \ldots \tag{12.13}$$

so that our approximate quantization condition reduces to

$$\begin{aligned}0 &= 1 - t_0(E) - t_1(E) - t_2(E) \\ &= 1 - \frac{\exp[(i/\hbar)S_0(E) - i(\pi/2)\mu_0]}{|\Lambda_0(E)|^{1/2}} - 2\frac{\exp[(i/\hbar)S_1(E) - i(\pi/2)\mu_1]}{|\Lambda_1(E)|^{1/2}}\end{aligned} \tag{12.14}$$

where we took into account the $r_1 \leftrightarrow r_2$ symmetry, which implies $t_1 = t_2$.

As in Section XII.D.1, the quantities $S_p(E)$ and $\Lambda_p(E)$ were fitted to polynomials (see Fig. 66). The Morse indexes are $\mu_0 = 3$ and $\mu_1 = 2$. Note that the index for the symmetric-stretch orbit is odd, as this orbit is hyperbolic with reflection above the quasiperiodic zone (cf. Section XII.B).

The semiclassical resonances are given in Table IV as well as in Fig. 65. Note that an analysis based only on the symmetric-stretch orbit would yield lifetimes that are shorter than those obtained by taking into account the "off-diagonal" orbit. Part of this effect is due to the fact that the off-diagonal orbits have a smaller Lyapunov exponent than the symmetric-stretch orbit. Another part has its origin in the interference between the quantum amplitudes of the different fundamental periodic orbits of the repeller. As we explained in Section VI.C.3 with Eq. (6.31), this lengthening of the lifetimes predicted by Gaspard and Rice [114] occurs in the presence of classical chaos and has similarities with the phenomenon of Anderson localization of a wave moving in a random potential [244].

At this point of the analysis, we would like to return to the problem of nonseparability. The validity of the Dunham expansion [Eq. (12.6)] shows that the motion is effectively separable just above the saddle equilibrium point. This result is supported by the periodic-orbit quantization in the low-energy regime, where a unique unstable periodic orbit constitutes the (nonfractal) repeller. In Section XII.D.1, we showed that the quantization in terms of the action and the linear stability parameter of the unstable periodic orbit is equivalent to an effective separability between the motions parallel and transverse to the periodic orbits. Thus, a repeller with a single orbit is intrinsically separable.

However, as soon as the repeller is chaotic and generated by several fundamental periodic orbits, the system in general becomes nonseparable. Indeed, the progression of the resonances given by the zeros of a function such as Eq. (12.14) will in general be very complicated with a mixing of different families of resonances. Let us point out that there is nevertheless a limiting situation where an effective (partial) separability may be recovered in the presence of a chaotic repeller. This is the case of a degeneracy in the sense that all periods become nearly equal to each other so that $S_p(E) = S_{p0} + TE$, while $\Lambda_p(E) = \Lambda_{p0}$ for all periodic orbits p, such that the zeros of the zeta function are regularly spaced as in $E_n = E_c + 2\pi\hbar n/T - i\Gamma/2$. We will return to this discussion later.

E. Wavepacket Propagation

The results obtained from the semiclassical analysis may be verified by a quantum mechanical approach using wavepacket propagation, by a numerical integration of the Schrödinger equation. This approach provides a precise description of the experimental situation where the ground-state wavepacket is "transported" to the excited-state potential energy surface by a laser pulse. We use an approximative treatment in so far as we neglect the effect of the pump pulse (which will slightly broaden the ground-state wavepacket [233, 234]), and simply monitor the time evolution of the wavepacket on the excited-state surface.

In Sections II and VII we outlined how the experimentally observable absorption spectrum κ_ω can be related to the energy Green's function. Thus, the spectrum may be obtained by a Fourier transform of the time-correlation function for the excited-state wave function $|\Phi(t)\rangle = \hat{\mathbf{d}}|\phi_0\rangle$ [cf. Eq. (7.11)]. Note that time-reversal symmetry implies the property $\langle\Phi(0)|\Phi(t)\rangle = \langle\Phi(0)|\Phi(-t)\rangle^*$ (this is similar to the statement of time-reversal symmetry for the S-matrix). Thus, if we integrate only over positive times, with the observation of the physical system starting at $t = 0$, the spectrum will be given by

$$\kappa_\omega = \frac{\omega}{3\hbar c\varepsilon_0} \operatorname{Re} \int_0^{+\infty} e^{i\omega t} e^{iE_0 t/\hbar} \langle\Phi(0)|\Phi(t)\rangle \, dt \tag{12.15}$$

where Re denotes the real part of the integral.

The numerical integration of the Schrödinger equation is based on the split-time propagation scheme introduced by Feit et al. [246], combined with the Fourier technique by Kosloff [80, 247] to calculate the effect of the kinetic-energy operator in momentum space. Figure 67(a) shows the (real part of the) time correlation function for a wavepacket that is initially centered in the vicinity of the saddle point [21]. A rapid decrease of $\langle\Phi(0)|\Phi(t)\rangle$ is observed within the first 100 fs, followed by four recurrences, spaced at an interval of about 430 fs. This interval agrees well with the period of the classical fundamental orbits, that is, the symmetric-stretch and "off-diagonal" orbits at an energy just above the quasiperiodic zone. (The classical periods at an energy of 550 cm^{-1} are 441 fs for the symmetric-stretch orbit, and 435 fs for the off-diagonal orbit; these periods tend to diverge at higher energies, as may be inferred from Fig. 66). At this energy, the spectral intensity for the chosen initial conditions is close to the maximum.[5] After the fourth recurrence, the amplitude of the autocorrelation function has approximately dropped to zero.

[5] The average energy $\langle\Phi|\mathscr{H}|\Phi\rangle$ of the wavepacket is below the total dissociation threshold, which does not correspond to the experimental situation described in [232–234], where the wavepacket has a large excess energy on excitation from the ground state, and which may be an explanation why the resonance structure had not been observed [248].

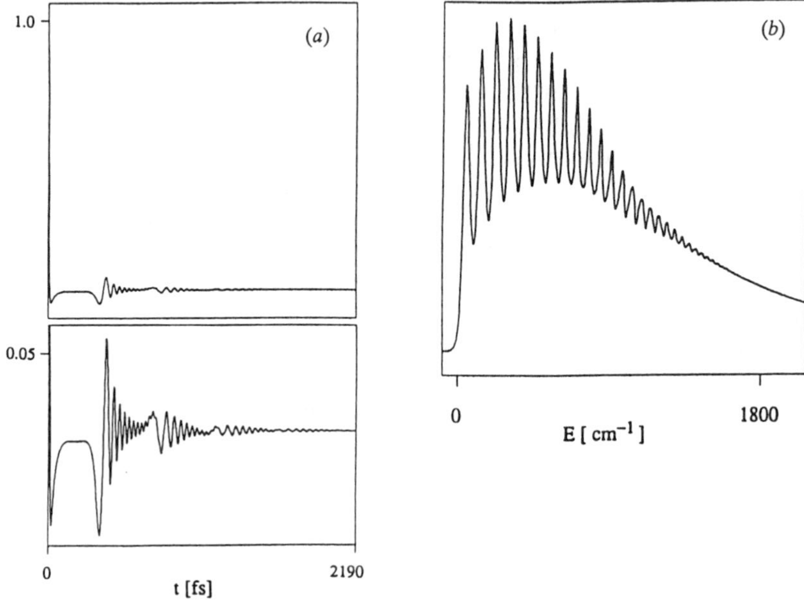

Figure 67. (*a*) Autocorrelation function $\langle\Phi(0)|\Phi(t)\rangle$, corresponding to the time evolution of the transition state of the HgI_2 molecule. The recurrences are of low intensity, as may be inferred from the fact that most of the wavepacket has left the transition state region at $t = 350\,fs$ (i.e., at a time that is smaller than the period of the classical fundamental orbits). The lower part of the figure shows the details of the recurrences on a larger scale. (*b*) The real part of the spectrum obtained by Fourier transformation of the autocorrelation function (*a*).

Figure 67(*b*) shows the real part of the Fourier transform of the autocorrelation function. The overall envelope of the spectrum is determined by the rapid initial decay of $\langle\Phi(0)|\Phi(t)\rangle$, while the fine structure, which is typical of a vibrational progression, is due to the recurrences [61, 190, 224]. The spectrum appears very regular, and in its overall appearance resembles the spectrum of a Morse oscillator, with lines broadened due to the finite lifetime. The lifetimes associated with the individual lines have been determined approximately as described in the caption of Table IV; the frequencies and lifetimes associated with the individual lines are listed in this table. According to the semiclassical formula [Eq. (7.15)], the spectrum is the sum of different contributions. The background contribution is quasiclassical due to the microcanonical average of the observable corresponding to $|\Phi(0)\rangle\langle\Phi(0)|$. When superposed on this background, the contribution from the periodic orbits produces the oscillatory structures of vibrational origin, which are related to the underlying scattering resonances of the molecular transition complex. In the HgI_2 system, the different periods characteristic of the classically chaotic regime are not apparent in the spectrum, in contrast to other molecules like ozone (O_3) or CO_2, where distinct periodic orbits seem to emerge from the vibrogram of the photoabsorption spectra [223, 239]. The apparent regularity of the HgI_2 spectrum seems to be due to the fact that the classical properties of the symmetric-stretch and off-diagonal orbits are, on the whole, very similar, such that their contribution to the recurrences of the quantum-mechanical autocorrelation function are not well separated; thus, the fundamental orbits do not give rise to interleaved progressions of resonances in the energy domain, but rather generate a sequence of resonances that still resembles the progression due to a single periodic orbit.

F. Discussion of the Transition State Dynamics of HgI_2

One of the main results obtained in the above sections is that the lifetimes of the molecular transition complex of HgI_2 are found to have values between 100 and 200 fs, which are comparable with the experimental results by Zewail et al. [232–234].

According to Table IV and Fig. 65, the semiclassical and quantum mechanical results are in close agreement at low energies. At high energies, the agreement is reasonable. The quantum spectrum does feature a tendency towards enhanced lifetimes, as predicted semiclassically. Remember that, according to Section XII.C.3, this tendency is due to the fact that the off-diagonal orbits contribute to the resonances and given rise to an interference effect, as discussed above.

Between 500 and 800 cm^{-1}, a marked discrepancy exists between the quantum mechanical and the periodic-orbit results in the vicinity of the classically quasiperiodic zone, where the periodic-orbit quantization predicts a very marked line-narrowing effect. In fact, the quantum dynamics has a smoothing effect not only on the narrow quasiperiodic zone but also on its neighborhood, which is classically of weak instability. As a result, the quantum dynamics is more unstable than expected from the periodic-orbit quantization. As explained in Section VI.F, the periodic-orbit quantization, which is applicable to isolated, unstable orbits, will encounter the problem of small denominators close to a bifurcation, and will break down when the periodic orbit becomes stable, such that the amplitude factor diverges. An effect of this kind can be felt even as bifurcation is approached, which results in an increase of the predicted lifetimes. The periodic-orbit quantization can be modified in the vicinity of the classical bifurcations by deriving an amplitude factor that goes beyond the *linear* stability of the orbit, as discussed in Section VI.F. We intend to extend our analysis along these lines in a future publication [21]. On the other hand, the equilibrium point quantization seems to be more successful in this intermediate region for the following reason. The Dunham expansion [Eq. (12.6)] has a domain of validity that is restricted to the vicinity of the saddle-point energy. Therefore, Eq. (12.6) cannot reproduce all the effects of the bifurcations that occur at a higher energy falling outside the immediate vicinity of the saddle point. Accordingly, the equilibrium point quantization describes the aforementioned quantum smoothing of the classical dynamics with more success than periodic-orbit quantization.

However, if we attempted a "naive" continuation of equilibrium point quantization (i.e., disregarding the occurrence of bifurcations) into the nonseparable regime at high energies, its accuracy would deteriorate. In this classically chaotic regime, it is the periodic-orbit quantization that provides an accurate description of the energies of the resonances while the corresponding lifetimes show fluctuations that are also described but with a lower accuracy.

In conclusion, we observed that the semiclassical method adequately explains the main features of the transition complex of this model of HgI_2, particularly with regard to the nonseparable classically chaotic regime, where new phenomena arise such as the Gaspard–Rice lengthening [114]. In the following sections, we will see how the semiclassical method extends to more general situations.

G. Comparison of the Classical Repellers of ABA and ABC Molecules

In this section, we would like to generalize the Hamiltonian mapping [Eq. (12.5)] to the case of an asymmetric triatomic molecule ABC. We are still concerned with the two-degree-of-freedom dynamics of a triatomic molecule with a fixed angle of bending. The motion takes place on a dissociative potential with a saddle equilibrium point. We may expect that the scenario is very similar to the case of a symmetric ABA molecule. Indeed, the works by Pollak et al. [236, 237] show the existence of three disymmetric fundamental orbits in the transition complexes of FH_2 and HCl_2.

On these grounds, we generalize the potential [Eq. (12.5)] by inclusion of a new term breaking the $r_1 \leftrightarrow r_2$ or $q \leftrightarrow -q$ symmetry characteristic of ABA molecules

$$V(q) = \left(\alpha + \beta q + \gamma \frac{q^2}{2}\right) \exp\left(-\frac{q^2}{2}\right) \qquad \text{with} \qquad \beta \neq 0 \quad (12.16)$$

which now applies to the transition complex of asymmetric ABC molecules. The Hamiltonian is then given by Eq. (3.36) with $T=1$. Here, γ is taken, as before, as the bifurcation parameter.

For nonvanishing values of β, the bifurcations of this new mapping are very similar to those of the symmetric one, with some differences. In Fig. 68 we see that the hyperbolic fixed point at $q=0$ (corresponding to the symmetric-stretch periodic orbit) becomes a center through an antipitchfork bifurcation at $\gamma=\alpha>0$, when $\beta=0$. At the antipitchfork bifurcation two new fixed points emerge that correspond to the two off-diagonal periodic orbits.

The symmetry of this bifurcation is broken for $\beta\neq 0$, as shown in Fig. 69. The hyperbolic fixed point remains hyperbolic but is displaced on the side. A saddle-center bifurcation occurs on the other side. Above the bifurcation, the center is lying between two hyperbolic fixed points as in the case $\beta=0$. This scenario explains the numerical observations of Pollak et al. [236, 237]. At higher values of γ, that is, at higher energies in the corresponding flow, there is also a last homoclinic tangency leading to the formation of a three-branch Smale horseshoe with a triadic symbolic

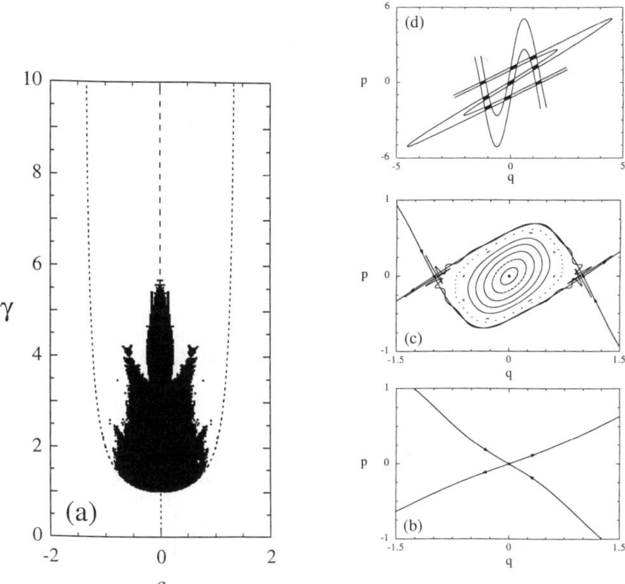

Figure 68. (*a*) Bifurcation diagram in the (q,γ) plane of the trapped orbits of initial conditions $p=0$ for the Hamiltonian mapping [Eq. (12.16)] of an ABA molecule with $\beta=0$ and $\alpha=1$. (*b*) Phase portrait in the (q,p) plane at $\gamma=0.8$ before the antipitchfork bifurcation. (*c*) Same as (*b*) but at $\gamma=2$ slightly after the antipitchfork bifurcation. (*d*) Same as (*b*) and (*c*) but at $\gamma=12$ when the three-branch Smale horseshoe is formed.

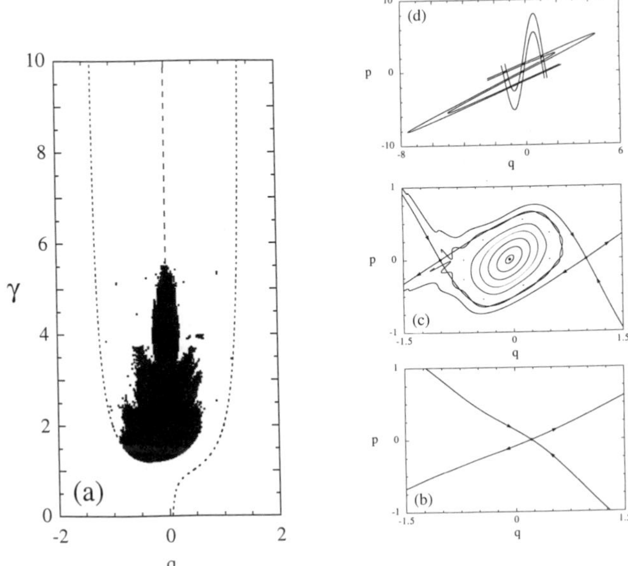

Figure 69. (*a*) Same as Fig. 68 but for an ABC molecule with $\beta = 0.05$ and $\alpha = 1$ in Eq. (12.16). (*b*) Phase portrait in the (q, p) plane at $\gamma = 0.8$ before the saddle-center bifurcation. (*c*) Same as (*b*) but at $\gamma = 2$ slightly after the saddle-center bifurcation. (*d*) Same as (*b*) and (*c*) but at $\alpha = 1$, $\beta = 3$, and $\gamma = 15$ when an asymmetric three-branch Smale horseshoe is formed.

dynamics of topological entropy ln 3. For large values of β, which would correspond to a very asymmetric molecule, this last homoclinic tangency may never occur.

This analysis shows that the mapping [Eq. (12.16)] qualitatively describes the two-degree-of-freedom classical dynamics along the reaction coordinate for dissociating triatomic molecules whether they are symmetric or not. If the potential is bonding, the parameter γ should be taken negative. In this case, the same mapping would describe classical dissociation taking place on a bonding potential above the energy of dissociation. However, the bifurcation scenario is then qualitatively different as we may expect for a bonding potential.

H. Inclusion of Bending and Extension to Larger Molecules: Equilibrium Point Quantization

As the preceding analysis of the classical dynamics has shown, the classical repeller is composed of a single periodic orbit at energies just above the saddle in the two-degree-of-freedom model. This single periodic orbit can be described by the Hamiltonian developed in Taylor series around the equilibrium point. In this separable regime, the periodic-orbit analysis is equivalent to an equilibrium point analysis. Therefore, we can obtain the scattering resonances of the transition complex using the equilibrium point quantization of Section V. While the search for periodic orbits may be cumbersome in high-dimensional systems, the equilibrium point quantization is, by contrast, straightforward.

Near a saddle equilibrium point with s stable directions of frequencies $\{\omega_i\}$ and u unstable directions of Lyapunov exponents $\{\lambda_j\}$, the application of successive Van Vleck contact transformations reduces the Hamiltonian to the form

$$\begin{aligned}\hat{\mathcal{H}} = E_0 &+ \sum_i \omega_i \frac{\hat{p}_i^2 + q_i^2}{2} + \sum_j \lambda_j \frac{\hat{p}_j^2 - q_j^2}{2} \\ &+ \sum_{i_1<i_2} \tilde{x}_{i_1 i_2} \frac{\hat{p}_{i_1}^2 + q_{i_1}^2}{2} \frac{\hat{p}_{i_2}^2 + q_{i_2}^2}{2} \\ &+ \sum_{j_1<j_2} \tilde{x}_{j_1 j_2} \frac{\hat{p}_{j_1}^2 - q_{j_1}^2}{2} \frac{\hat{p}_{j_2}^2 - q_{j_2}^2}{2} \\ &+ \sum_{ij} \tilde{x}_{ij} \frac{\hat{p}_i^2 + q_i^2}{2} \frac{\hat{p}_j^2 - q_j^2}{2} + \mathcal{O}(6)\end{aligned} \tag{12.17}$$

The scattering resonances are obtained by the quantization rules

$$\begin{aligned}\frac{\hat{p}_i^2 + q_i^2}{2} &\to \hbar\left(n_i + \frac{1}{2}\right) \\ \frac{\hat{p}_j^2 - q_j^2}{2} &\to -i\hbar\left(k_j + \frac{1}{2}\right)\end{aligned} \tag{12.18}$$

In Section V we showed that this method is equivalent to the equilibrium point quantization. Accordingly, the scattering resonances are given by

$$\begin{aligned}E_{n_1 \ldots n_s k_1 \ldots k_u} = E_0 &+ \sum_i \hbar\omega_i\left(n_i + \frac{1}{2}\right) - i \sum_j \hbar\lambda_j\left(k_j + \frac{1}{2}\right) \\ &+ \sum_{i_1<i_2} \hbar^2 \tilde{x}_{i_1 i_2}\left(n_{i_1} + \frac{1}{2}\right)\left(n_{i_2} + \frac{1}{2}\right) \\ &- \sum_{j_1<j_2} \hbar^2 \tilde{x}_{j_1 j_2}\left(k_{j_1} + \frac{1}{2}\right)\left(k_{j_2} + \frac{1}{2}\right)\end{aligned}$$

$$-i\sum_{ij}\hbar^2\tilde{x}_{ij}\left(n_i+\frac{1}{2}\right)\left(k_j+\frac{1}{2}\right)+\mathcal{O}(\hbar^3) \qquad (12.19)$$

We obtain the resonances including the anharmonic corrections to the lifetimes. Note that the coefficients $\tilde{x}$ are identical to the coefficients x of Eqs. (5.4–5.6) modulo factors ± 1 or $\pm i$.

For a nonlinear triatomic molecule, we may consider different types of equilibrium points depending on the stability of the symmetric stretching, asymmetric stretching, and bending modes, as summarized in Fig. 70. Let us label the three normal modes of symmetric stretching, bending, and asymmetric stretching, respectively, by 1, 2, and 3. In the case of a stable minimum of the potential, the energy eigenvalues are given by

$$E = E_0 + \hbar\omega_1(n_1 + 1/2) + \hbar\omega_2(n_2 + 1/2) + \hbar\omega_3(n_3 + 1/2) + \mathcal{O}(\hbar^2) \qquad (12.20)$$

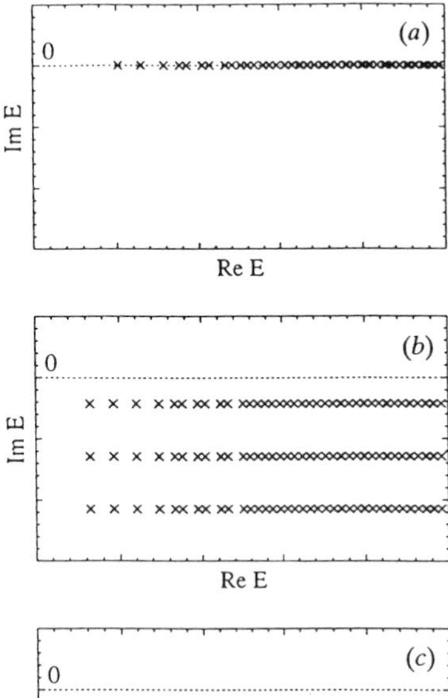

Figure 70. Energy eigenvalues for 3F harmonic potentials with different kinds of stabilities: (*a*) eigenenergies [Eq. (12.20)] for a minimum-type potential with frequencies $(\omega_1, \omega_2, \omega_3)$; (*b*) eigenenergies [Eq. (12.21)] for a saddle-type potential with frequencies $(\omega_1, \omega_2, -i\lambda_3)$; (*c*) eigenenergies [Eq. (12.22)] for a saddle-type potential with frequencies $(\omega_1, -i\lambda_2, -i\lambda_3)$.

in the case of a saddle with stable bending, by[6]

$$E = E_0 + \hbar\omega_1(n_1 + 1/2) + \hbar\omega_2(n_2 + 1/2) - i\hbar\lambda_3(k_3 + 1/2) + \mathcal{O}(\hbar^2) \tag{12.21}$$

and, in the case of a saddle with unstable bending, by

$$E = E_0 + \hbar\omega_1(n_1 + 1/2) - i\hbar\lambda_2(k_2 + 1/2) - i\hbar\lambda_3(k_3 + 1/2) + \mathcal{O}(\hbar^2) \tag{12.22}$$

The case of three unstable modes is not considered.

In the harmonic approximation of Eqs. (12.20–12.22), the scattering resonances are clustered on lines in the complex energy surface. We see that each new unstable mode decreases the imaginary part of the resonances as well as the lifetimes. If the maximum imaginary part is given by $\text{Im}\, E = -\hbar\lambda_3/2$ for the case of a saddle with a stable bending, then it will decrease to $\text{Im}\, E = -\hbar\lambda_2/2 - \hbar\lambda_3/2$ if the bending becomes unstable.

The bending motion can also be taken into account in the periodic-orbit quantization with similar effects on the resonances as described by Fig. 70. The zeta function Eq. (6.24) should be used instead of Eq. (6.21) if the bending motion is stable. In this case, a phase factor is added to each amplitude of the periodic orbits and the scattering resonances are shifted. Moreover, each previous scattering resonance is replaced by a family of resonances corresponding to the excitation of bending. If, on the other hand, the bending motion is unstable, the lifetimes will be modified and the families of scattering resonances due to bending excitations would progress downward in the complex energy surface. In both cases, there is an effect of the quantum mechanical zero-point energy that influences the energy levels even in the absence of classical bending [101]. For the dissociative potential surface of HgI_2, the bending is supposed to be stable with a period of about 1 ps. The consequence would be that the density of resonances may double or triple in Fig. 65 leading to strong overlapping of the resonances. This overlapping may be another reason why distinct peaks do not appear in the experimental photoabsorption spectrum of HgI_2.

[6] Such a model has very recently been used to interpret the Hartley absorption band of ozone [254] with the result that the lifetime of the transition state is $\lambda_3^{-1} \simeq 3.2$ fs.

I. Inclusion of Rotational Motion: Equilibrium Point Quantization

We may always apply successive Van Vleck contact transformations to the full molecular Hamiltonian including the rotational degrees of freedom in the separable regime above a saddle equilibrium point. In this way, Sepulchre and Gaspard obtain the rotational structure of each vibrational state described hereabove. In the present context of the molecular transition state, the vibrational states are understood as the metastable states given by the poles of the Green function at complex energies. For a saddle-type potential of a linear molecule where the symmetric stretching and bending modes are stable while the asymmetric stretching mode is unstable, Sepulchre and Gaspard [249] obtained

$$\begin{aligned} E = {} & E_0 + \hbar\omega_1(n_1 + 1/2) + \hbar\omega_2(n_2 + 1) + \text{vibr. anh.} \\ & + BJ(J+1) - \alpha_1 J(J+1)(n_1 + 1/2) - \alpha_2 J(J+1)(n_2 + 1) \\ & + \text{higher orders} \\ & - i\hbar\lambda_3(k_3 + 1/2) - i\tilde{\alpha}_3 J(J+1)(k_3 + 1/2) \\ & + \text{vibr. anh.} + \text{higher orders} \end{aligned} \tag{12.23}$$

In this equation, the vibrational quantum numbers take the values n_1, n_2 $k_3 = 0, 1, 2, 3, \ldots$. The quantum number of total angular momentum is denoted by J, and B is the molecular rotational constant given by the inverse of the moment of inertia of the molecular complex at the saddle equilibrium point. The bending-type vibrational angular momentum projected along the molecular axis is supposed to be vanishing $\ell_2 = 0$, otherwise $J(J+1)$ should be replaced by $[J(J+1) - \ell_2^2]$. The constants α_1, α_2, $\tilde{\alpha}_3$ describe the vibrational–rotational interaction of the resonances and are given in terms of the frequencies ω_1, ω_2, λ_3; of the moments of inertia; and of the cubic derivatives f_{ijk} of the potential with respect to the normal mode positions; each quantity being evaluated at the saddle equilibrium point [250]. When the asymmetric stretching motion becomes unstable, one of the constants, α_i, becomes imaginary, producing a vibrational–rotational interaction on the lifetimes of the scattering resonances.

Whether the vibrational–rotational interaction has a stabilizing or destabilizing effect on the molecular complex depends on the sign of $\tilde{\alpha}_3$. Figure 71 shows the rotational structure of the resonances in two different cases. We may expect that the rotational structure of the resonances would be of the same order of magnitude as in discrete energy spectra of stable molecules. For triatomic hydrides such as HAH, where the

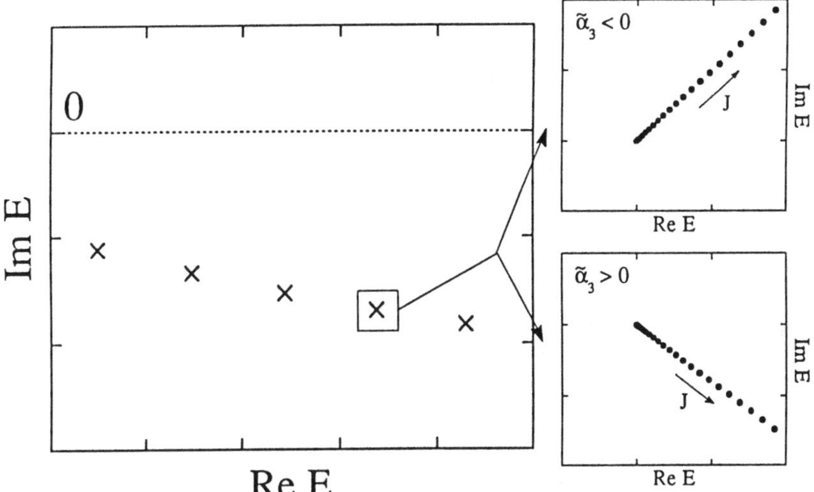

Figure 71. Rotational structure of a resonance of the molecular transition state. The vibrational resonance is resolved into a series of rovibrational resonances according to Eq. (12.23). The cases of rotational stabilization ($\tilde{\alpha}_3 < 0$) and of rotational destabilization ($\tilde{\alpha}_3 > 0$) are shown as inserts.

rotational constants are large and of the order of $B \sim 10\,\text{cm}^{-1}$ and $\alpha_i \sim 0.5\,\text{cm}^{-1}$, we may expect rotational families $J = 0$–20 of resonances to extend over an energy range of $\Delta\,\text{Re}\,E \sim 4000\,\text{cm}^{-1}$ and of $\Delta\,\text{Im}\,E \sim 100\,\text{cm}^{-1}$, which corresponds to changes of lifetimes of the order of $\Delta\tau \sim 4$ fs around $\tau \sim 10$ fs for $0 < J < 20$. On the other hand, for a heavy molecule such as HgI_2, the rotational constants are of the order of $B \sim 0.01\,\text{cm}^{-1}$ and $\alpha_i \sim 10^{-5}\,\text{cm}^{-1}$, which would give rotational families $J = 0$–20 extending over an energy range of $\Delta\,\text{Re}\,E \sim 4\,\text{cm}^{-1}$ and lifetimes with extremely small variations of $\Delta\tau \sim 0.008$ fs around $\tau \simeq 100$ fs [249].

J. Resonances in Nonseparable and Statistical Regimes

The preceding considerations are based on the existence of a single saddle equilibrium point in the potential. Note that it is already surprising to find as complex behavior as classically chaotic motions at high energies in such a system. We have seen that, in a nonseparable regime, the equilibrium point quantization is no longer valid and we must use the periodic-orbit quantization because the eigenenergies start to interfere with each other. In such regimes, we observed in the context of the three-disk scatterer that the resonances may form complicated structures, which obey statistical laws, such as Eq. (8.42). Therefore, the widths of the resonances are expected to fluctuate in nonseparable regimes.

In general, the potential may be complicated with several saddle equilibrium points playing the role of bottlenecks and delimiting a region of classical trajectories where chaotic and quasiperiodic motions are mixed. In two-degree-of-freedom systems, the formation of quasiperiodic islands may lead to long-lived metastable states determined by dynamical tunneling. In higher-dimensional systems, Arnold diffusion complicates the classical picture, but the quantum resonances should still be expected to be long-lived in potentials that would produce globally stable quasiperiodic motions when restricted to two degrees of freedom. This reasoning is based on the fact that the diffusion time of Arnold diffusion is very long compared to the quantum mechanical lifetimes [108]. In such systems, the scattering resonances should be distributed as predicted by random matrix theories [142, 194].

We suppose that the molecular complex can decay into different channels $c = 1, 2, \ldots, \nu$ at the energy of the resonance $E_r = \varepsilon_r - i\Gamma_r/2$ of velocity v_r. According to the Breit–Wigner theory, each channel contributes to Γ_r by a partial half-width Γ_{rc} so that the total half-width is the sum of the partial half-widths over all the open channels: $\Gamma_r = \Sigma_c \Gamma_{rc}$ [251]. The average number of open channels at an energy ε above the energy threshold of dissociation, $\varepsilon^{\ddagger}$, is given in the harmonic approximation by

$$\nu(\varepsilon) \simeq C(\varepsilon - \varepsilon^{\ddagger})^{f-1} \tag{12.24}$$

which may be interpreted as the integrated density of states transverse to the reaction coordinate if the system is assumed to be at a molecular transition state with f degrees of freedom. The parameter C is a positive constant while $f - 1$ is the number of degrees of freedom that are transverse to the reaction coordinate. According to the RRKM statistical theory, all the partial half-widths are assumed to take approximately the same value $\Gamma_{cr} \simeq 1/[2\pi\rho_{\text{av}}(\varepsilon)]$, where $\rho_{\text{av}}(\varepsilon)$ is the average level density of the reactant species, which is supposed to be a quasibounded system. Therefore, the average value of the total half-widths at the energy ε is given by

$$\bar{\Gamma}(\varepsilon) \simeq \frac{\nu(\varepsilon)}{2\pi\rho_{\text{av}}(\varepsilon)} \simeq A\left(\frac{\varepsilon - \varepsilon^{\ddagger}}{\varepsilon - \varepsilon_0}\right)^{f-1} \tag{12.25}$$

where A is another positive constant and $\varepsilon_0 < \varepsilon^{\ddagger}$ is the ground-state energy. This equation is at the basis of statistical reaction rate theories, which are of application in large molecules where the number of degrees of freedom is so large that a detailed analysis, as seen in this section, is impossible in practice [65]. The average reaction rate would then be given by the basic RRKM formula $\bar{k}(\varepsilon) = \bar{\Gamma}(\varepsilon)/\hbar = \nu(\varepsilon)/[h\rho_{\mathrm{av}}(\varepsilon)]$.

Nevertheless, the detailed semiclassical analysis can help us to clarify certain points and connections. In particular, the three-disk scatterer studied in section VIII has scattering resonances with half-widths growing like $\Gamma_r \sim \sqrt{\varepsilon}$ with energy ε. This result can be recovered by following the preceding reasoning. Indeed, in the disk billiards, the wavepacket escapes from the scatterer through the holes between the disks. Therefore, the number of open channels is given by the width w of these holes divided by the de Broglie wavelength, $\lambda \sim 1/\sqrt{\varepsilon}$ so that we obtain $\nu(\varepsilon) \sim w\sqrt{\varepsilon}$. On the other hand, the average level density of the bounded system obtained by closing the exit channels is given by Eq. (6.3) and is therefore constant. Hence, by interpreting Γ_r as a total half-width in the Breit–Wigner theory we recover the behavior of Γ_r for the disk scatterers by using Eq. (12.25). In this regard, the disk scatterers provide examples were the RRKM theory can be compared with an exact treatment of the reaction rate. In contrast, the saddle-type potential surfaces studied in Sections XII.A–I have properties differing from those predicted by the RRKM theory (although energy-independent rates may be formally obtained in Eq. (12.25) by setting $\varepsilon_0 = \varepsilon^{\ddagger}$). These differences may be attributed to the fact that the resonances sustained by saddle-type potential surfaces describe ultrafast reactions rather than the slower reactions usually treated with RRKM theory.

Thanks to random matrix theories, information is also available concerning the fluctuations of the half-widths around their average value [Eq. (12.25)]. Following random matrix theories, the reduced total half-widths $\Gamma_r^0 = \Gamma_r/v_r$ are expected to have a χ^2-distribution with a number of parameters equal to the number of open channels ν, namely, Eqs. (8.42) and (8.43). This gross behavior has been recently improved after comparison with resonance data on formaldehyde [252]. In Section VIII we showed that these fluctuations may obey a gap law in some semiclassical regimes. The semiclassical and random matrix approaches therefore complete our knowledge of reaction rates in a consistent way. (See [253] for further information on the statistical approach.)

K. General Discussion

In Section XII, we have seen applications of the equilibrium point and periodic-orbit quantization methods to the resonances of the molecular transition state. In the preceding sections, we already considered different possible systems where resonances can be obtained by semiclassical methods, and which may be important in the context of the molecular transition state. One of them is the matrix Hamiltonian of Section IX, describing a conical intersection where the geometrical phase introduces half-integer quantum numbers. In this case, the resonances are long-lived, as determined by tunneling. In contrast, on a single potential surface with a saddle point, the dissociation is classically allowed and the resonance lifetimes are much shorter, taking values in the range of 1–200 fs from light molecules such as, O_3, H_2O, or CO_2, to heavier molecules, such as HgI_2. We presented a detailed analysis in the case of HgI_2, which features a transition between separable and nonseparable regimes [21]. We described in detail the classical repeller, its bifurcations towards chaos, and its ternary symbolic dynamics. These results are very general and we may reasonably suppose that they extend to other dissociating triatomic molecules. We also considered how to include the effects of rotation in separable regimes [249]. For nonseparable systems, we showed the relevance of our work on the three-disk scatterers, which is a simplified model of unimolecular dissociation dynamics. In the nonseparable regimes, statistical features appear in the structure formed by the resonances that are partially described by semiclassical quantization of classically chaotic systems and by random matrix theories. For larger molecules, we briefly envisaged how statistical considerations may become essential to establish the connection between the resonances of the molecular transition complex and the unimolecular reaction rates. In the following section, we will turn to highly excited states in atomic and solid-state systems.

XIII. SEMICLASSICAL ELECTRONIC REGIMES IN ATOMIC AND SOLID-STATE SYSTEMS

A. Highly Excited Electronic States

Close below their ionization thresholds, atoms and molecules are found in highly excited electronic states where one or several electrons are in semiclassical motion at large distances from the nuclei. The UV photo-absorption spectrum of CO_2 illustrates the generality of this phenomenon (see Fig. 53).

Excited states with a single electron at a large distance are called Rydberg states. Such atomic and molecular states have been the object of many studies [33–35, 37, 210]. When the electron moves far from the nuclei and the other electrons, its motion in the absence of external fields is governed by an effective field due to the ionic core. This effective field is a perturbed Coulomb force field, with the perturbations due to medium-range interactions determined by the multipole moments of the ionic core. Moreover, fast crossing of the electron through the ionic core may occur, which involves important interactions that also have to be taken into account. In this picture, the semiclassical motion of the electron is composed of a succession of segments of Coulomb-type orbits, separated by scattering events with the ionic core [37]. The energy levels of the Rydberg states form perturbed hydrogen-like series

$$E_n = E_i - \frac{1}{2(n - \mu_n)^2} \tag{13.1}$$

where E_i is the energy of the ionization threshold, n is the principal quantum number, and μ_n is the so-called quantum defect [255], which depends on other quantum numbers, such as angular momentum. The calculation of the quantum defect μ_n in terms of the perturbing interactions is the purpose of quantum defect theories, which are discussed at length in the literature [255]. In the absence of an external magnetic field, in which case the unperturbed quantum Coulomb problem is exactly solvable, the periodic-orbit approach does not provide a great advantage over conventional wave function methods.

In contrast, the periodic-orbit method has proved to be both conceptually and technically important, over the last years, for Rydberg states in strong external electric and magnetic fields [34, 35, 210]. In particular, the dynamics of the hydrogen atom in a magnetic field is classically chaotic. Phase space portraits and the periodic orbits of such systems have been studied by different groups [18, 34, 210]. Methods have been proposed to Fourier transform the energy spectra into periodic-orbit spectra, especially by Wintgen [18]. Since then, such methods have been applied to experimental spectra obtained in several laboratories, which revealed not only how the classical periodic orbits emerge from the quantum wave dynamics but also how nonlinear properties typical of classical dynamics, such as bifurcations and Poincaré resonances, have their roots in long and complex series of quantum energy eigenvalues [34, 35].

Highly excited states of atoms or molecules, where two electrons are excited (the so-called doubly excited states), have also been the object of recent studies [23, 36, 99–101]. Such doubly excited states are particularly important in two-electron atomic systems, which are isoelectronic with helium: H^-, He, Li^+, Be^{2+}, B^{3+}, C^{4+}, ... The Hamiltonian of these three-body systems is given by

$$\mathscr{H} = \frac{\hat{\mathbf{p}}_1^2 + \hat{\mathbf{p}}_2^2}{2} - \frac{Z}{r_1} - \frac{Z}{r_2} + \frac{1}{r_{12}} \tag{13.2}$$

which has been studied by different groups [99–101, 256–259]. The properties of several families of classical orbits are known for this Hamiltonian (see Fig. 72). The classical motion is known to be chaotic and ionizing in some cases while it is quasiperiodic in others. The linear stability of these periodic orbits revealed important features (see Table V) [259]. This information on the classical dynamics was used to carry out periodic-orbit quantization, which showed that accurate values of the eigenenergies can be obtained for helium and the hydrogen negative ion. In some respect, the three-body Coulomb systems have similarities with the triatomic molecular systems. Nevertheless, important differences have been discovered, which have their origin in the essential singularity of the Coulomb potential at three-body collisions. As a consequence, the Wannier symmetric stretching motion is infinitely unstable at zero angular momentum and the most important periodic orbit corresponds to the antisymmetric stretching motion [259], contrary to the situation in triatomic molecules as analyzed in the preceding section. Here, we will focus on the case of the H^- ion that was studied by Gaspard and Rice [101]. Let us also mention that exotic three-body Coulomb systems have been or may be envisaged from the semiclassical point of view, such as

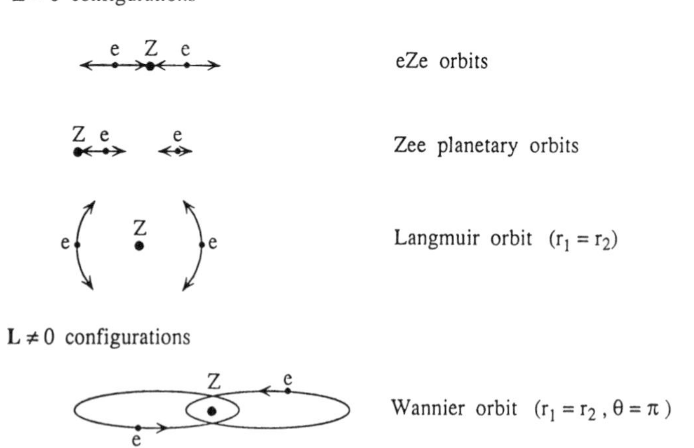

Figure 72. Different families of classical orbits of the two-electron atomic species for vanishing or nonvanishing orbital angular momentum L.

TABLE V
Spectrum of the Nontrivial Lyapunov Exponents of the Classical Orbits for the Different Atomic Species Isoelectronic with He at Zero Angular Momentum, $\mathbf{L}=0$ [100, 259][a]

Species	eZe Orbits	Fundamental Zee Orbit	Langmuir Orbit
H^-	(+, 0)	∄	(0, +)
He	(+, 0)	(0, 0)	(0, 0)
Li^+	(+, 0)	(0, 0)	(0, +)
Be^{2+}	(+, 0)	(0, 0)	(0, +)
B^{3+}	(+, 0)	(0, 0)	(0, +)
C^{4+}	(+, 0)	(0, 0)	(0, +)
N^{5+}	(+, 0)	(0, 0)	(0, +)
O^{6+}	(+, 0)	(0, 0)	(+, +)
F^{7+}	(+, 0)	(0, 0)	(+, +)
Ne^{8+}	(+, 0)	(0, 0)	(+, +)
Na^{9+}	(+, 0)	(0, 0)	(+, +)
Mg^{10+}	(+, 0)	(0, 0)	(+, +)
Al^{11+}	(+, 0)	(0, 0)	(+, +)
Si^{12+}	(+, 0)	(0, 0)	(+, +)
P^{13+}	(+, 0)	(+, 0)	(+, +)
S^{14+}	(+, 0)	(+, 0)	(+, +)
Cl^{15+}	(+, 0)	(+, 0)	(+, +)
Ar^{16+}	(+, 0)	(+, 0)	(+, +)
⋮	(+, 0)	(+, 0)	(+, +)

[a] For the eZe and Zee collinear orbits, $(\operatorname{sgn}\lambda_\alpha, \operatorname{sgn}\lambda_\theta)$ is given, where λ_α is the Lyapunov exponent for perturbations in the hyperangle between r_1 and r_2, $\tan\alpha = r_1/r_2$, such that the collinear configuration is maintained, while λ_θ is the Lyapunov exponent for the bending-type perturbation in the bending angle θ between the vectors $\boldsymbol{r}_1$ and $\boldsymbol{r}_2$. For the Langmuir orbit, we give $(\operatorname{sgn}\lambda_1, \operatorname{sgn}\lambda_\alpha)$ where λ_1 is the Lyapunov exponent for perturbations perpendicular to the orbit and to the hyperangle α, with λ_α the same as before. The collinear eZe orbits are conjectured to form a fully chaotic repeller along the whole isoelectronic series.

the positronium negative ion $(e^-e^+e^-)$; the antiprotonic species, $(p^+\bar{p}^-e^-)$ or $(p^+\bar{p}^-p^+)$; and the muonic species $(Z\mu^-e^-)$, $(p^+\mu^-p^+)$, or $(t^+\mu^-d^+)$ [23, 260, 261].

B. Hydrogen Negative Ion

The hydrogen negative ion (H^-) is isoelectronic to helium, but differs from helium in a number of important respects. For example, the H^- ion has many autoionizing states and only one bound state, which lies very close to the ionization threshold, whereas the helium atom has many bound states. The stability of the hydrogen negative ion was first established, theoretically, by Bethe and Hylleraas in 1929–1930 [262, 263]; earlier studies of H^- using the Bohr–Sommerfeld quantization conditions failed to predict the stability of H^-. The modern semiclassical quantization methods have overcome the difficulties met in the earlier attempts, in particular thanks to the obtention of the correct Maslov indexes, and thanks to our understanding of the way the different periodic orbits interfere in the zeta functions to produce the quantum spectrum [23, 99, 101].

First, we will describe the classical dynamics of H^- and then we will turn to its semiclassical quantization.

1. *Classical Dynamics of H^-*

The classical motion of the ion may be considered in the different configurations of Fig. 72.

1. Consider the collinear motion in which both electrons are on opposite sides of the nucleus (the eZe configuration). This eZe collinear motion is classically unbounded and fully chaotic for H^- as well as for He. In the course of this motion, the electrons undergo collisions with the nucleus. If such a collision occurs between one electron and the nucleus while the other electron is close to the nucleus, the latter electron may acquire enough energy to escape and ionize. In fact, for eZe collinear motion there is no barrier to prevent ionization and most trajectories ionize, except for the zero measure fractal set of trajectories forming the repeller (and their stable manifolds). All the periodic orbits turn out to be unstable with respect to collinear perturbations. The fractal repeller is therefore fully chaotic and can be described by a binary symbolic dynamics [99]. Note that this is different from the fully chaotic repeller of the transition state of triatomic molecules, where the symbolic dynamics is ternary rather than binary. This difference has its origin in the infinite instability of the Wannier symmetric stretching motion in the case of the Coulomb potential, which becomes singular at short distances during three-body collisions. The long range of the Coulomb potential has the other effect that the orbits of the repeller extend over large distances away from the nucleus. This long-range effect can be observed in Fig. 73, which depicts a combined Poincaré section in the planes $r_1 = r_2$ and $r_2 = 0$ of the phase space. The crosses are the intersections of the periodic orbit with the plane while the lines are the invariant manifolds. Let us mention that, by contrast with collinear perturbations, the eZe motion is stable with respect to perturbations of the bending type, in which the electrons move away from a line.
2. There is another possible collinear configuration, with both elec-

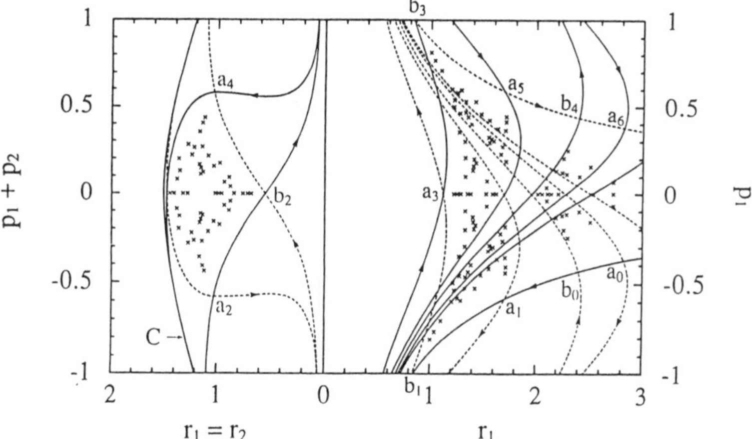

Figure 73. Combined Poincaré sections $r_1 = r_2$ (left-hand side) and $r_2 = 0$ (right-hand side) of the phase space. The crosses are the intersections of the first 22 periodic orbits with the section surfaces. The solid (dashed) line is the unstable manifold of the fixed point at infinity. We see that they delimit the region occupied by the fractal repeller. The sets of points $\{a_k\}$ and $\{b_k\}$ are the successive intersections of two homoclinic orbits with the section surfaces. The curve C is the border of the classically allowed region in the ($r_1 = r_2$, $p_1 + p_2$) plane.

trons on the same side of the nucleus, the Zee configuration. However, no such trapped orbit exists in the case of H^-, contrary to He and positive ions where a stable quasiperiodic island appears in phase space.

3. For H^-, the Langmuir orbit shown in Fig. 72 is stable for in-plane motions but unstable with respect to perturbation in the hyperangle $\alpha = \text{arctg}(r_1/r_2)$.

Next, we also consider the H^- ion in a weak magnetic field, where the orbits are only slightly perturbed with respect to the zero-field situation [101]. Let us remark that the line of collinear eZe motion may point in all possible directions of the sphere since the corresponding quantum state is rotationally invariant and described by an S wave if the angular momentum is assumed to vanish ($\mathbf{L} = 0$).

We now turn to the classical equations of motion for the H^- ion in a magnetic field. As we mentioned in Section X, Coulomb systems obey an important scaling law, which allows us to reduce the study of the classical motion to the energy $E = -1$ according to

$$\mathbf{r}_a \to \alpha \mathbf{r}_a \qquad \mathbf{p}_a \to \beta \mathbf{p}_a \qquad t \to \gamma t \tag{13.3}$$

with

$$\alpha = \frac{Za_0}{|\mathscr{E}|} \qquad \beta = \frac{\hbar|\mathscr{E}|^{1/2}}{a_0} \qquad \gamma = \frac{Zma_0^2}{\hbar|\mathscr{E}|^{3/2}} \tag{13.4}$$

where $\mathscr{E} = (E\hbar^2)/(me^4)$ is the energy expressed in Hartrees and $a_0 = \hbar^2/(me^2)$ is the Bohr radius.

For the eZe configuration of the electrons with respect to the nucleus, the rescaled Hamiltonian is given by

$$\begin{aligned}\mathscr{H}_{\rm cl} &= \frac{p_1^2 + p_2^2}{2} - \frac{1}{r_1} - \frac{1}{r_2} + \frac{1}{Z(r_1 + r_2)} \\ &+ \frac{b}{2} L_z + \frac{b^2}{8} \sin^2\theta(r_1^2 + r_2^2) = -1\end{aligned} \tag{13.5}$$

where L_z is the angular momentum and the parameter b is given by

$$b = \frac{ZB}{|\mathscr{E}|^{3/2} B_0} \qquad \text{with} \qquad B_0 = \frac{\hbar}{ea_0^2} = 2.35 \times 10^5 T \tag{13.6}$$

and θ is the angle between the direction of the magnetic field and the line of motion of the electrons.

For a given periodic orbit, the reduced action S and the period T can now be expressed in terms of the rescaled reduced action $\tilde{S}$ and period $\tilde{T}$ as

$$S = \frac{Z\hbar}{|\mathscr{E}|^{1/2}} \tilde{S} \qquad \text{and} \qquad T = \gamma \tilde{T} \tag{13.7}$$

Since we know that the period of an orbit is given by the derivative of the reduced action with respect to the energy, $T = \partial S/\partial E$, we infer from the scaling law that

$$\tilde{T} = \tfrac{1}{2}\, \tilde{S} \tag{13.8}$$

Note that the semiclassical limit occurs when the energy reaches the double ionization threshold, that is, for $|\mathscr{E}| \to 0$. Since $\mathscr{E}$ is negative, the limit $\mathscr{E} \to 0$ is consistent with the increase of energy that is generally expected for the transition toward classical behavior.

In Coulomb systems, the numerical integration of the equations of motion is plagued with difficulties due to two-body collisions, since then r_1 or $r_2 \sim (t - t_0)^{2/3}$, while $p_1 = \dot{r}_1$ or $p_2 = \dot{r}_2$ diverges as $(t - t_0)^{-1/3}$. This problem can be overcome by use of regularization methods developed in celestial mechanics [264]. Setting

$$r_a = R_a^2 \qquad p_a = \frac{P_a}{2R_a} \qquad a = 1, 2$$

$$\text{with} \qquad dt = R_1^2 R_2^2 \, d\tau \tag{13.9}$$

the new variables have analytic behavior when two-body collisions occur since $\tau - \tau_0 \sim (t - t_0)^{1/3}$, R_1 or $R_2 \sim \tau - \tau_0$, while P_1 or P_2 reach the constant values $\pm\sqrt{8}$.

The motion of the variables r_1 and r_2 is classically chaotic. As compared with helium, the motion in the H^- ion is more unstable since escape to infinity usually occurs more rapidly.

We have carried out a search for periodic orbits and obtained a set that is in 1:1 correspondence with the binary symbolic dynamics $\{+, -\}$ defined by Ezra et al. [99]. In this respect, the H^- ion turns out to behave like helium. In Fig. 74 we show the 23 first periodic orbits of H^- in the plane of the positions r_1 and r_2.

The classical dynamics of ionization can be understood in terms of phase-space transport across the partial separatrices formed by the stable and unstable manifolds of the fixed point at infinity [101]. This fixed point is defined by assuming that the electron at infinity is at rest while the bounded electron carries all the energy with $\varepsilon = -1$. In Fig. 73, we have drawn some periodic orbits of the fractal repeller together with the stable and unstable manifolds of the fixed point at infinity. At the intersections between the stable and the unstable manifolds, we find the homoclinic orbits associated with the fixed point at infinity; two of these are identified in Fig. 73 as $\{a_n\}$ and $\{b_n\}$. The stable manifold (dashed line) is composed of several successive parts, the last of which is the dashed line joining the intersections b_3, a_5, b_4, and a_6 of the homoclinic orbits with the section plane at positive values of the momentum p_1. This last part of the stable manifold constitutes a separatrix for the trajectories in its neighborhood. Trajectories above the separatrix have a momentum p_1 large enough to escape to infinity and ionize. However, during motion below the separatrix, the position r_1 increases for a while at positive p_1 before decreasing until the next collision on the nucleus. Therefore, trajectories below the separatrix do not have enough momentum to escape immediately and they remain bounded till the next return near the nucleus. The stable (as well as the unstable) manifold therefore forms a partial separatrix for the escape dynamics.

The linear stability analysis carried out with the methods of Section III

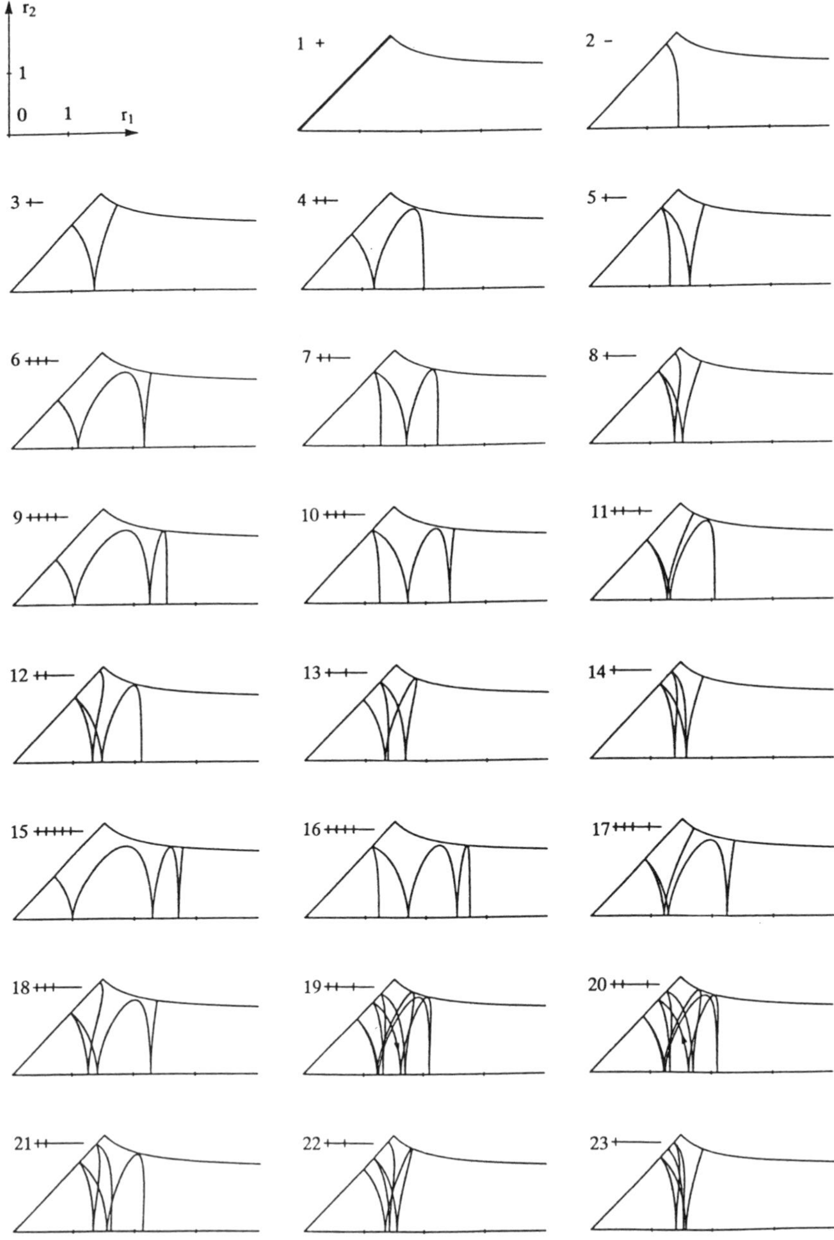

Figure 74. The 23 shortest periodic orbits of H^- for the eZe configuration up to symbolic period 6. They are depicted in the fundamental domain of the position space (r_1, r_2).

shows that the motion of the electrons is unstable with a positive Lyapunov exponent for perturbations that maintain the collinear configuration. On the other hand, perturbations of the bending type are stable with imaginary stability eigenvalues and corresponding rotation numbers. The bending motion can be considered as a doubly degenerate oscillator.

2. *Semiclassical Quantization of* H^-

We have carried out a semiclassical quantization on the basis of the classical collinear eZe orbits. Accordingly, the average of the cosine of the angle θ_{12} between the vectors $\mathbf{r}_1$ and $\mathbf{r}_2$ would take the value $\langle\cos(\theta_{12})\rangle \simeq -1$ for the corresponding electronic eigenfunctions. Since the classical motion is chaotic for such a configuration, the semiclassical quantization can be performed thanks to the zeta function formalism of Section VI.C. Since the linear stability of the eZe orbits shows positive as well as vanishing Lyapunov exponents, we have to use the Zeta function [Eq. (6.24)] with $u=1$ and $s=2$. Indeed, the bending motion is equivalent to a doubly degenerate oscillator of quantum number $v=n_1+n_2$, where each level v has a degeneracy equal to $v+1$. The degenerate levels v are labeled by the quantum number ℓ of angular momentum projected on the line of collinear motion ($\ell=v, v-2, \ldots, -v$).

At the leading semiclassical approximation, the eigenenergies are thus given by the zeros of the following zeta function

$$Z_{Sv\ell}(\mathscr{E}) = \prod_p \prod_{k=0}^{\infty} \left\{1 - \frac{\exp[(i/\hbar)S_p(\mathscr{E}) - i(\pi/2)\mu_{pS} - 2\pi i \rho_p(v+1)]}{|\Lambda_p|^{1/2}\Lambda_p^k}\right\} \tag{13.10}$$

in terms of the reduced actions [Eq. (13.7)] of the periodic orbits and of the Maslov indexes. Since the present system contains two indistinguishable Fermions, the Maslov indexes depend on the spin quantum number S

$$\begin{aligned} \mu_{pS} &= \alpha_p && \text{for} \quad S=1 \\ \mu_{pS} &= \alpha_p + 2\nu_p^{(-)} && \text{for} \quad S=0 \end{aligned} \tag{13.11}$$

where α_p is the Morse index while $\nu_p^{(-)}$ is the number of crossings of the diagonal $r_1=r_2$ during the period. The parameter $\nu_p^{(-)}$ is also equal to the number of symbols $(-)$ in the symbolic dynamical code of the periodic orbit p. The result [Eq. (13.11)] has its origin in the Pauli principle postulating the antisymmetry of the total wave function of an assembly of

Fermions. Consequently, the wave function must be antisymmetric in space for singlet states ($S = 0$) so that there is a Dirichlet boundary condition on $r_1 = r_2$. As a consequence, the phase of the path acquires a minus sign at each crossing of $r_1 = r_2$ besides the phase $(-\pi/2)$ acquired at each new conjugate point. Conversely, the wave function must be symmetric in space for triplet states ($S = 1$) with a Neumann boundary condition on $r_1 = r_2$, so that the phase does not change at crossings of $r_1 = r_2$ and increases only by $(-\pi/2)$ at each new conjugate point.

The effect of a weak magnetic field on the eigenenergies can be studied thanks to the semiclassical perturbation method described in Section VII. The action is modified to take into account the perturbation of the magnetic field according to [101]

$$\frac{S_p}{\hbar} = \frac{Z}{|\mathscr{E}|^{1/2}} \left[\tilde{S}_p - \frac{Z^2 B^2}{12|\mathscr{E}|^3 B_0^2} \tilde{A}_p \right] \tag{13.12}$$

where

$$\tilde{A}_p = \oint_p (r_1^2 + r_2^2)\, dt \tag{13.13}$$

is evaluated in the rescaled variables.

In order to calculate the energy eigenvalues in the presence of a magnetic field, we have to insert these actions in the zeta function [Eq. (13.10)]. For each zero of the zeta function we can then define the susceptibility χ_n according to

$$\chi_n = -B_0^2 \frac{\partial^2 E_n}{\partial B^2}(B = 0) = -\frac{1}{6} \langle r_1^2 + r_2^2 \rangle_n \tag{13.14}$$

which gives an estimate of the size of the doubly excited state (see Fig. 75).

We calculated by periodic-orbit quantization the eigenenergies of the families ${}^1S^e$ and ${}^3S^e$ of autoionizing states restricted to the case $v = 0$, $L = 0$, and $M_L = 0$ [101]. Table VI compares the results of the periodic-orbit quantization with those of conventional wave function methods for the ${}^1S^e$ states. The ground state is the lowest ${}^1S^e$ state and is the only bound state with an energy well below -0.5, which is the single-electron ionization threshold. We see that the H^- ion is stable according to our semiclassical calculation and that the energy is obtained with a precision of 5%. At higher excitations, the precision increases up to 1%. Moreover, the periodic-orbit quantization provides us with new eigenenergies we could not find in the literature [265–269].

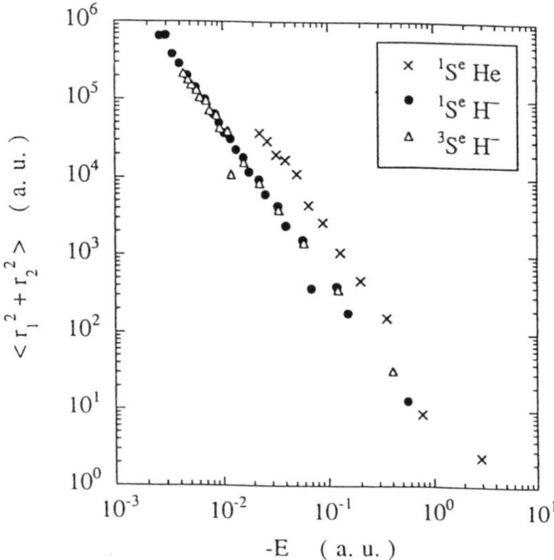

Figure 75. Log–log plot of the semiclassical averages $\langle r_1^2 + r_2^2 \rangle$ versus energy $-E$ for the $^1S^e$ (solid circles) and $^3S^e$ (triangles) states of H^- compared with the $^1S^e$ (crosses) states of He.

The energies of the other $^1S^e$ states are above -0.5, so that these states are all autoionizing and have a finite lifetime. The semiclassical calculation also gives an imaginary part of the order of 10^{-6}–10^{-3} to the zeros of the Selberg zeta function for high quantum numbers. These imaginary parts are very small and are in rough agreement with the known quantum mechanical orders of magnitude. However, they are too small with respect to the error on the real parts to be reliable.

The semiclassical formulation provides us with several important scaling laws. In the presence of a weak magnetic field, Eqs. (13.12)–(13.14) imply that

$$\langle r_1^2 + r_2^2 \rangle_{nSv} \simeq \frac{C_v Z^2}{|\mathscr{E}_n|^2} \tag{13.15}$$

where the constant takes the value $C_v \simeq 4.7$. This value does not seem to depend on whether the states are singlets or triplets. In terms of the principal quantum number N, the energy scales according to $|\mathscr{E}| \sim N^{-2}$, so that we recover the known scaling law $\langle r_1^2 + r_2^2 \rangle \sim N^4$ (see Fig. 75) [267].

Furthermore, in spite of the difficulty in getting reliable numerical results for the lifetime at the leading semiclassical approximation, we can

TABLE VI
Energy Levels of the $^1S^e$ States of H^- (a.u.)

$(N\ell, N'\ell')$	$-\mathrm{Re}\,\mathscr{E}_{WF}$ [a]	$-\mathrm{Re}\,\mathscr{E}_{PO}$ [b]
$1s1s$	0.5259[c]	0.557
$2s2s$	0.14879	0.150
$2s3s$	0.12585	0.119
$3s3s$	0.06960	0.0688
$3s4s$	0.05765	0.0573
$4s4s$	0.039925	0.0399
$4s5s$	0.03353	0.0337
$5s5s$	0.026015	0.0257
$5s6s$	0.02224	0.0222
$6s6s$	0.018205	0.0179
$6s7s$		0.0158
$7s7s$	0.01353[d]	0.0134
$7s8s$	0.01195	0.0118
$8s8s$	0.01042	0.0102
$8s9s$	0.00935	0.00912
$9s9s$	0.008265	0.00832
$9s10s$		0.00808
$10s10s$	0.0066807[e]	0.00670
$11s11s$	0.005537[f]	0.00552
$12s12s$	0.004663	0.00462
$13s13s$		0.00392
$14s14s$		0.00337
$15s15s$		0.00293
$16s16s$		0.00259

[a] $\mathscr{E}_{WF}$: Energies obtained by standard wave function methods.
[b] $\mathscr{E}_{PO}$; Energies obtained by the periodic-orbit method by Gaspard and Rice [101].
[c] Reference [265].
[d] Reference [266].
[e] Reference [267].
[f] Reference [268].

obtain the scaling law of the half-widths of the scattering resonances [270]. For the present Coulomb system the lifetimes are large and correspond to sharp resonances with small widths. Because the actions scale like $S_p \sim |\mathscr{E}|^{-1/2}$, we infer from the general form of the zeta function [Eq. (13.10)] that the widths scale as

$$\Gamma \sim |\mathscr{E}|^{3/2} \tag{13.16}$$

or as $\Gamma \sim N^{-3}$ in terms of the principal quantum number $N \sim |\mathscr{E}|^{-1/2}$. Therefore, the small widths of the resonances are typical of Coulomb systems near the ionization threshold, where $|\mathscr{E}| \to 0$ and $N \to \infty$.

This general behavior of the resonances is in contrast with resonance behavior in other systems which also have a classically chaotic repeller, for example, the hard disk scatterers and the Hamiltonian mappings that we studied elsewhere. For the disk scatterers the actions increase as $S_p \sim E^{1/2}$ with the energy $E \to +\infty$ in the classical limit. The consequence is that the widths increase as $\Gamma \sim E^{1/2}$, while the reduced widths (defined as the widths divided by the velocity v of the outgoing particle) remain constant, $\Gamma^0 \equiv \Gamma/v \sim \Gamma/E^{1/2} \sim \text{Cst}$. The value of this constant is determined mainly by the Lyapunov exponents of the periodic orbits. In these systems a gap in the spectrum is observed, which was related to the absence of sharp resonances with a typical Fano profile in the scattering cross-section (see Section VIII) [63].

C. Mesoscopic Semiconducting Devices

Scattering of electrons in semiconducting devices is an important process at the origin of the property of conductance [45]. In ultrapure nanometric devices at low temperatures, the motion of electrons is essentially ballistic and slightly affected by scattering on impurities or phonons. Under these conditions, the time evolution is governed by the one-electron Schrödinger equation and semiclassical considerations are important. If the potential felt by the electron is sharp at the border of the conducting circuit, the smooth potential may be replaced by a hard wall and the circuit may be modeled as a billiard [40–43, 48].

Assuming that the leads of the circuit are electronic waveguides extending to large distances, the conductance can be formulated as a scattering problem of the electronic wave in the circuit [43]. In this respect, the electronic conductance problem is very similar to the problem of a molecular reaction involving a collinear triatomic complex. Landauer [271] showed that the conductance at energy E is given by

$$g(E) = \frac{e^2}{h} \operatorname{tr} \hat{T}(E)^\dagger \hat{T}(E) \qquad (13.17)$$

in terms of the transmission matrix $\hat{T}(E)$ of a wave through the scattering potential. The transmission matrix is itself directly related to the scattering matrix so that this problem can be reduced to the aforementioned calculation of total cross-sections. Fluctuations in the conductance $g(E)$ are the object of many current studies [40, 41, 43, 45]. In this context, it is worthwhile to mention that similar fluctuations may exist in the cross-sections of molecular collisions, but they are more difficult to observe experimentally. The semiclassical approach is particularly important here because the properties of these fluctuations appear to be related to the chaotic or nonchaotic behavior of the classical motion of the electron [43].

In typical devices, the classical motion is often chaotic so that the question of the consequences of classical chaos on the properties of electronic conductance turns out to be important. In this context, different billiard models have been considered particularly by Smilansky, Stone, and others [42, 43, 272]. Experiments have been carried out on an open stadium billiard as well as on a four-arm circuit closely related to the four-disk scatterer, as shown in Fig. 76 [40, 41].

This circuit is interesting in view of a comparison with molecular reactions, and we would like to describe its symbolic dynamics. Figure 76 shows that a quarter of the four-arm circuit looks like a sharp potential model of a chemical reaction with two exit channels. Moreover, we can use the symmetry by the diagonal to fully reduce the billiard to its fundamental domain. Like the four-disk scatterer, the four-arm billiard has a classically chaotic fractal repeller that is invariant under the C_{4v} symmetry group [273]. When the disks are not too close to each other, all the trapped orbits are in 1:1 correspondence with a symbolic dynamics based on the four symbols $\{a, b, c, d\}$, which are the labels of the four disks. There is a constraint that two successive symbols cannot be equal so that the symbolic dynamics is actually ternary with a topological

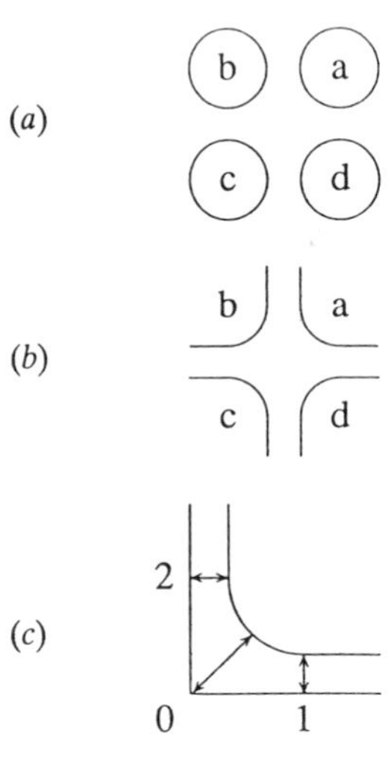

Figure 76. Definitions of the symbolic dynamics in the case of C_{4v} billiards and the corresponding symmetry-reduced billiards: (*a*) Four-disk billiard; (*b*) Four-arm billiard; (*c*) Billiard obtained by reduction of the four-arm billiard using the horizontal and vertical reflections; (*d*) Fundamental domain of the four-arm billiard obtained from (*c*) using a further diagonal reflection.

entropy of ln 3 per symbol. By using horizontal and vertical reflections, we can reduce the symbolic dynamics $\{a, b, c, d\}$ to the symbolic dynamics with the three symbols $\{0, 1, 2\}$ of the molecular transition state of HgI_2 in Section XII, which can equivalently be used in the quarter of the original billiard, as shown in Fig. 76. In these new symbols $\{0, 1, 2\}$, the symbolic dynamics is free of constraint. The translation rules from $\{a, b, c, d\}$ to $\{0, 1, 2\}$ are derived by using horizontal and vertical reflections to fold the full orbit onto an orbit of the quarter of the billiard. A part of the orbit is assigned to a particular symbol of $\{0, 1, 2\}$ if the corresponding path is followed. Table VII gives the translation rules.

We can further reduce the symbolic description to the symbols $\{0, +, -\}$ by using a further reflection through the diagonal to go to the fundamental domain formed by an eighth of the plane. The translation rules are given in Table VIII. In a sequence of symbols $\{0, 1, 2\}$, the symbols 0 remain unchanged. The symbol 1 (resp. 2) is replaced by + if, ignoring the symbols 0, it is preceded by 1 (resp. 2). Conversely, 1 (resp. 2) is replaced by − if it is preceded by 2 (resp. 1), here again ignoring the symbols 0 of the sequence. Accordingly, the three symbolic dynamics schemes are equivalent. We would like to emphasize the generality of this

TABLE VII
Translation Rules between the Symbolic Dynamics $\{a, b, c, d\}$ with the Constraint $\omega_n \neq \omega_{n+1}$ of the Four-Disk Scatterer and the Symbolic Dynamics $\{0, 1, 2\}$ without Constraint of the Quarter of Billiard Reduced by Horizontal and Vertical Reflections As Well As of the Molecular Transition State

$\{a, b, c, d\}$	$\{0, 1, 2\}$
ac, ca, bd, db	0
ad, da, bc, cb	1
ab, ba, cd, dc	2

TABLE VIII
Translation Rules between the Symbolic Dynamics $\{0, 1, 2\}$ without Constraint and the Symbolic Dynamics $\{0, +, -\}$ without Constraint of the Fundamental Domain Reduced by a Further Diagonal Reflection in the Billiard As Well As in the Molecular Transition State

$\{0, 1, 2\}$	$\{0, +, -\}$
$10^n1, 20^n2$	0^n+
$10^n2, 20^n1$	0^n-

symbolic description in two-degree-of-freedom systems: It is encountered in fourfold symmetric billiards [273], in the hydrogen atom in a magnetic field [274], as well as in the molecular transition state [21].

The semiclassical quantization of classically chaotic billiard models of electronic circuits can be carried out with the zeta function formalism presented in Section VI.C along lines that are similar to those developed in Section VIII on quantum billiards [273]. In open circuits, those methods allow us to obtain the scattering resonances. Such metastable states with varying lifetimes also arise here, and can be studied by femtosecond laser excitation of the device. A simple calculation shows that the lifetimes would be in the range of 500–1000 fs for a device with a size of the order of 100 nm [97]. These considerations may be particularly important if we are interested in producing ultrafast devices. In this context, the characteristic time taken by the electronic wave to exit the device is determined by the lifetimes of the scattering resonances.

Here, let us also mention the very recent work on scattering of electronic surface waves by artificially placed surface impurities, which can be observed thanks to the electron tunnel microscope [46]. In this context, semiclassical methods may also be of importance in the modelization of such processes.

Another application of periodic-orbit quantization concerns the magnetic equilibrium properties of mesoscopic systems such as quantum dots considered as tiny billiards for electrons [275–278]. Superlattices with up to 10^5 copies of the same quantum billiard can be built by lithographic methods, which allows the observation of its magnetic susceptibility. The consequences of the transition between regular and irregular energy spectra due to classically chaotic behavior have been studied, particularly by Nakamura and Thomas [275, 276]. The magnetic susceptibility has been expressed within the semiclassical approximation in terms of the periodic orbits of the classical motion of the electron [277]. This periodic-orbit theory explains oscillatory structures observed in the magnetic susceptibility as a function of the Fermi energy controlled by the bias voltage [278].

In a related field, periodic-orbit theory has also been applied to explain nonconventional magnetoresistance oscillations in mesoscopic conductors formed by GaAs/AlGaAs antidot square superlattices (with a lattice period of 200–300 nm) [279]. At low temperatures, spurious oscillations in the behavior of the resistance ρ_{xx}, as a function of the applied magnetic field, have been attributed to electrons trapped on quantized periodic orbits whirling around one antidot or within a unit cell of the lattice [279]. (See also [280].)

These different directions of research show that the number of potential applications of semiclassical effects are large in the field of mesoscopic devices.

XIV. DISCUSSION AND CONCLUSIONS

In this chapter, we reviewed and illustrated several new and recent developments in the semiclassical quantization method. A chart is given in Fig. 77 showing the different pathways we followed in this chapter. The principle of semiclassical quantization is to perform an $\hbar$-expansion of the trace of the resolvent of the quantum Hamiltonian around classical orbits. Accordingly, a knowledge of the classical motion is required. The discovery that classically chaotic dynamics is typical of nonlinear Hamiltonian systems has thus been a challenge for the traditional semiclassical methods. The latter essentially assumed separability between the different degrees of freedom of the system, which is equivalent to integrability of the classical motion. However, separability is precluded in classical chaos where the trajectories are found wandering among unstable periodic orbits over large regions of phase space.

A. Classical Chaos, Periodic Orbits, and Quantum Spectra

This difficulty has been partially overcome thanks to the Gutzwiller trace formula, which expresses the trace of the resolvent as a sum over periodic orbits. The Gutzwiller trace formula has been reformulated in terms of a zeta function, with the great advantage that quantum eigenvalues are given as zeros of the zeta function, as we explained in Section VI. We showed that zeta functions are directly related to the determinant of the semiclassical transfer operator, which contains the quantum probability amplitudes of the transitions between the different semiclassical states of motion. The introduction of the transfer operator is required as soon as the classical motion can be decomposed into a succession of transitions between phase-space cells. This branching process is controlled by a symbolic dynamics defined by an alphabet and grammatical rules. The classical trajectories should be unambiguously associated with infinite sequences of symbols taken from the alphabet and assembled according to the grammatical rules. This symbolic dynamics helps us to list all the unstable periodic orbits of phase space. At the same time, we need to study the phase-space geometry of classical motion in order to establish the correspondence to a symbolic dynamics. The Smale horseshoe is the phase space object that plays the central role in this geometrical construction. We illustrated the central role of the Smale horseshoe and of symbolic dynamics in a number of examples in this chapter, in particular the three- and four-disk scatterers, the H^- ion, as well as other

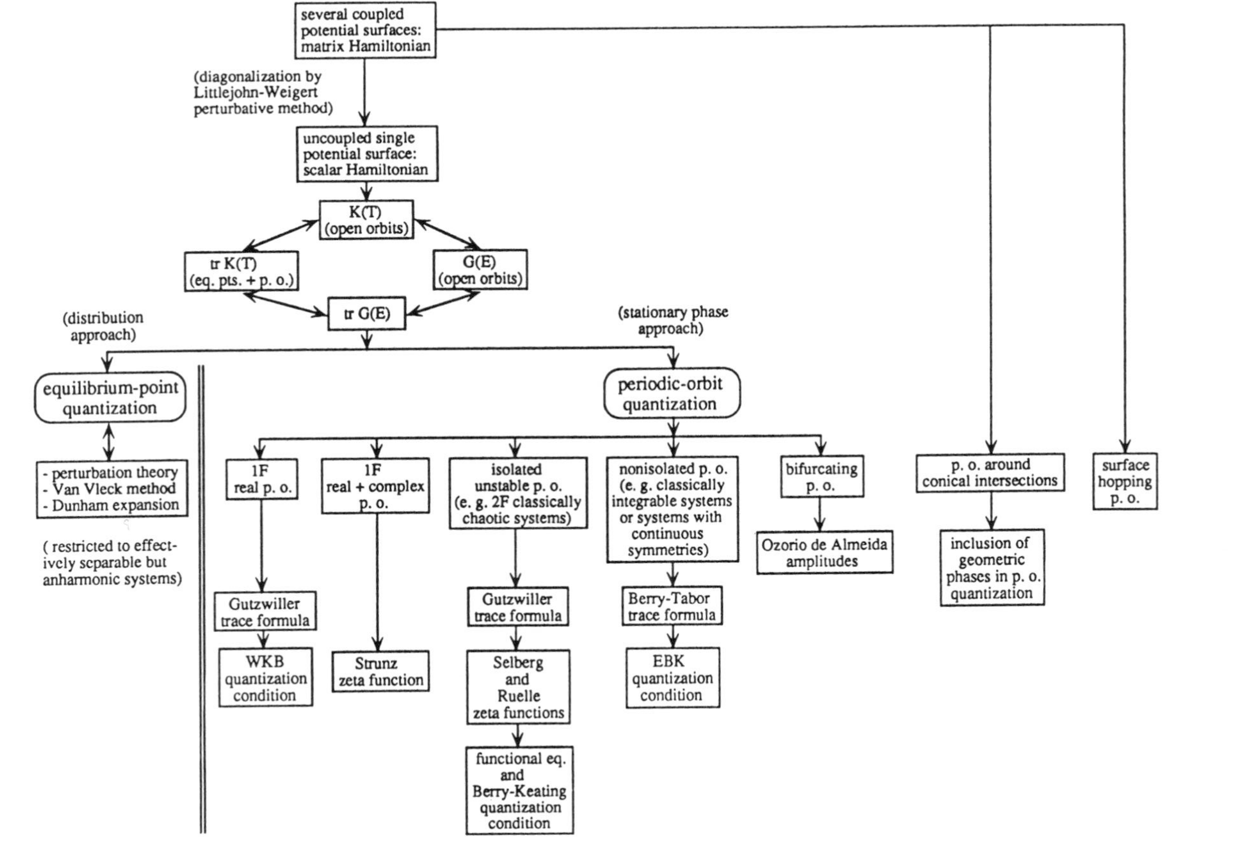

Figure 77. Chart of the relationships between the different Hamiltonians, classical orbits, and corresponding semiclassical formulas, which are discussed in this chapter.

two-electron atomic species, and the dissociating triatomic molecule. Even tunneling in a multiwell potential is described by a semiclassical transfer operator and its related zeta functions, which are constructed on the symbolic dynamics controlling the complex classical orbits.

At the semiclassical level, the zeta function formalism provides a clear answer to the question of the correspondence between the quantum-mechanical eigenvalues and the periodic orbits. In a system with a single unstable periodic orbit, such as the two-disk scatterer or the antibonding molecular potential at an energy just above its saddle equilibrium point, the single periodic orbit (and its repetitions) determines a whole family of complex energy eigenvalues, as shown in Sections VIII and XII. This result is natural in view of the Fourier-type relationship between the energy and time domains, and of the time–energy Heisenberg uncertainty relation $\Delta E\,\Delta t \sim \hbar$. Accordingly, a single periodic orbit with a precisely defined period cannot be associated with a particular energy eigenvalue and vice versa. However, the zeta function shows that a family of energy eigenvalues is associated with a periodic orbit (and its repetitions). Now, classical flows contain a countable family of periodic orbits of increasing periods in chaotic regimes. As we explained, the higher periods turn out to be approximate combinations of a small set of lower periods related to the fundamental periodic orbits. In strongly chaotic systems, these fundamental periodic orbits have the smallest stability eigenvalues and, therefore, dominate the cycle expansion of the zeta function. The energy eigenvalues then basically result from the interference between the semiclassical quantum amplitudes associated with the fundamental periodic orbits. In this way, the periodic-orbit quantization naturally extends to nonseparable systems. The interference phenomenon explains, or at least suggests, that the spectrum of the quantum eigenvalues may become irregular in classically chaotic systems.

It is important to note that the irregularity in the distribution of the eigenvalues results from a dispersion in the periods of the periodic orbits rather than from the instability of the periodic orbits. For instance, the two-disk scatterer has a regular spectrum due to its single unstable periodic orbit, while the spectrum of the many-disk scatterer becomes more and more irregular because the variety of the possible periods increases with the number of fundamental periodic orbits. We think that this remark is relevant to the problem of the origin of the Wigner spacing distributions in irregular discrete energy spectra of bounded quantum systems. It is commonly assumed that a Wigner spacing distribution is a signature of the quantum system being chaotic or, at least, dynamically random in the semiclassical limit, $\hbar \to 0$ [281]. Counterexamples have been found for the reverse statement that classical chaos in the sense of sensitivity to initial conditions would imply a Wigner spacing distribution [282]. Moreover, there exist quantum systems showing a Wigner spacing distribution, but which do not necessarily have a classical counterpart [283]. The delicate and difficult question of the relationship between irregular quantum spectra and classical chaos is still open today in spite of the effort invested over the past years [284]. In this regard, important questions still remain concerning the semiclassical quantization of bounded quantum systems, although definitive advances have been made.

B. Scattering Systems and Applications to Chemical Reaction Dynamics

Actually, the situation is clearer for scattering systems, where the zeta function formalism proved to be an effective method to determine the complex energy eigenvalues that are the scattering resonances. This chapter summarizes several results in this context, specifically, in Section VIII. The semiclassical methods turn out to be extremely powerful not only to calculate the scattering resonances but also to provide an understanding of the distribution of the resonances and, in particular, of semiclassical scaling laws between the half-widths and the energies, or gap laws relating the half-widths and the lifetimes to the Lyapunov exponents of the periodic orbits, as well as to the entropy of the classical repeller. In classically allowed decay processes, the resonant metastable states have short lifetimes (except in Coulomb-type potentials, due to the long range of the interaction, as discussed in Sections X and XIII). Therefore, the semiclassical method turns out to provide a very natural framework to describe the ultrashort molecular dissociation processes that are currently studied in femtosecond laser photochemistry [39].

In Section XII we applied the semiclassical methods to the dynamics of the molecular transition state. The case of HgI_2 has been treated in detail. We presented results recently obtained by Burghardt and Gaspard [21] on the three-branch horseshoe and the ternary symbolic dynamics that controls the classical dynamics of the molecular transition complex of triatomic molecules. We also described the bifurcations of the classical dynamics that lead to the formation of the classically chaotic repeller. This process seems to be very general on antibonding potential surfaces and nicely explains a number of observations dispersed in the literature over several decades. In particular, the three-branch horseshoe contains as fundamental periodic orbits the famous PODS defined by Pollak and Pechukas [236], while the antisymmetric stretching periodic orbit recently emphasized turns out to be a higher period orbit. We illustrated our discussion with simple Hamiltonian mappings describing the dissociation process in symmetric ABA or nonsymmetric ABC molecules. Thanks to the discovery of the three-branch horseshoe, all the periodic orbits of the molecular transition complex can be rigorously listed at high energies, and the periodic-orbit quantization can be carried out.

Connections to the statistical and phase-space theories of chemical reaction rates are possible at this point. In typical chemical reactions, the classical dynamics is characterized not only by chaotic motion, with unstable periodic orbits, but also by quasiperiodic motion [285]. The unstable periodic orbits give rise to homoclinic tangles between their stable and unstable manifolds. At their intersections, turnstiles are formed that also contain several small subsidiary Smale horseshoes according to a mechanism described by Poincaré and, later, by Birkhoff and Smale [286]. The turnstiles of the homoclinic tangles are known to be at the origin of phase-space transport, which determines the chemical reaction rates, as shown by Rice and co-workers [285]. In principle, the unstable periodic orbits of the chaotic zones formed by the homoclinic tangles can be used in the Gutzwiller trace formula to obtain the quantum properties. Nevertheless, the situation is complicated by the presence of the stable periodic orbits of the elliptic type, which are possibly surrounded by elliptic islands. Complex orbit effects of dynamical tunneling out of the vicinity of the stable orbits may be important in such cases. Therefore, the scattering resonances of such systems may acquire long lifetimes. The resulting large fluctuations in the half-widths may be conveniently discussed in the framework of random matrix theories, which open the way to statistical rate theories.

In this context, semiclassical methods are important to treat systems where the dynamics is dominated by particular molecular vibrational or rotational motions in regimes that are intermediate between the statistical regimes of RRKM theories and the regular regimes of simple scattering systems. It is the purpose of the vibrograms introduced in Section XI to extract from photabsorption energy spectra the possible periods of the classical dynamics, which emerge from the quantum dynamics in highly excited molecular systems. The application of vibrograms has been illustrated for several discrete and continuous molecular spectra. Here, we provided a semiclassical explanation of the stroboscopic effect described by Pique et al. [216–218] in terms of the selection of periodic orbits that are used in the Berry–Tabor trace formula for the periodic-orbit quantization of integrable systems [180].

C. Periodic-Orbit Quantization versus Equilibrium Point Quantization

The connection between periodic-orbit quantization and equilibrium point quantization has been discussed in detail. On a fundamental level, the equilibrium point quantization appears as an $\hbar$-expansion around the equilibrium points of classical dynamics. In the time domain, equilibrium points as well as periodic orbits contribute to the semiclassical expression of the trace of the propagator. It is only on going to the energy domain that we face the necessity to follow one or the other of two alternative quantization schemes, that is, either the periodic-orbit quantization based on the stationary phase method, or the equilibrium point quantization, which turns out to be identical with standard perturbation theory, as well as with the Van Vleck contact transformation method. Here, we obtained several results for a systematic development of the equilibrium point quantization. In particular, we showed that there is a difference between the standard perturbation method and the semiclassical quantization based on Birkhoff normal forms at equilibrium points. We explicitly obtained this difference (ΔE_2) at the $\hbar^2$-order. This difference can be interpreted as a zero-point effect due to the anharmonicities of the potential beyond the well-known zero-point effect at the harmonic approximation, which shifts the ground state by $\hbar\omega/2$. Furthermore, the higher order terms of the perturbation series can be obtained as diagrams, which may help to calculate them, at least in particular cases. Moreover, we discussed the extension of the equilibrium point quantization from minima of bonding potentials to saddle equilibrium points of antibonding potentials, where this method can efficiently provide the scattering resonances. For dissociating molecules, equilibrium point quantization is also the method of choice to obtain the rotational structure of the diffuse molecular bands [249], including all the possible vibrational stable or unstable normal modes around the saddle equilibrium points, as discussed in Section XII. However, the equilibrium point quantization is restricted to the effectively separable regimes, where the spectrum remains regular, as discussed in Section V. Indeed, the $\hbar$-expansion of the equilibrium point quantization involves the same kind of transformations of the flow around the equilibrium point as the nonlinear contact transformations used to obtain the classical Birkhoff normal form. In such transformations, the nonseparable classically chaotic behaviors like the splitting of the separatrices and the related homoclinic tangles are removed from the $\hbar$-expansion into the nonanalytical rest of the series resulting from an infinite sequence of contact transformations. Thus, classical chaos and its effect on quantum dynamics cannot be described at the level of equilibrium point quantization, because the local dynamics around the equilibrium point has been reduced to the one of a separable system. Periodic-orbit quantization is required as soon as the dynamics becomes nonseparable and classically chaotic. In periodic-orbit quantization, interferences between several periodic orbits may then result in complex spectral structures related to irregular spectra as aforementioned.

D. $\hbar$-Expansion and the Implications of Anharmonicities

The $\hbar$-expansion extends from the equilibrium point quantization to the periodic-orbit quantization where it allows us to obtain the energy eigenvalues beyond Gutzwiller's leading approximation. In the equilibrium point quantization, the $\hbar$-expansion goes beyond the harmonic approximation around an equilibrium point and takes into account anharmonicities, that is, of the cubic, quartic, and higher degree terms of the potential. As we discussed above, the $\hbar$-expansion applied to equilibrium points treats the system as separable. Nevertheless, strong anharmonicities can still be taken into account in such a framework, as is the case for the Fermi resonance in the model by Pique et al. [216, 217] of $CS_2(\tilde{X}\,^1\Sigma_g^+)$, where the two-degree-of-freedom Hamiltonian remains integrable in spite of the strong 2:1 resonance. On the other hand, the main feature of the periodic-orbit quantization is that nonseparable systems, which typically emerge from a transition between regular and chaotic classical regimes, can now be treated because periodic orbits are global phase-space objects. The zeta function formalism can include the effect of several different periodic orbits, thus capturing the dynamics in the vicinity of different phase-space regions. At the leading order of the $\hbar$-expansion, the periodic-orbit quantization takes into account anharmonicities in the direction along each periodic orbit through the nonlinear dependencies of the actions $S_p(E)$ on energy. However, the leading-order periodic-orbit quantization treats the motion transverse to the periodic orbit as harmonic, since the Gutzwiller trace formula uses the linear stability eigenvalues Λ_p of each periodic orbit. At the next-to-leading orders of the $\hbar$-expansion, the $\hbar$-corrections incorporate in the trace formula or the zeta functions the effects of nonlinear stability of the periodic orbits, that is, of anharmonicities for the motion transverse to the periodic orbits [19–21]. In this way, the inclusion of $\hbar$-corrections can capture dynamical effects around each periodic orbit that extend over a larger phase-space region than at the leading order of the $\hbar$-expansion. We illustrated the application of the $\hbar$-expansion for the theoretical models of the disk scatterers, where we observe very significant improvements of the accuracy of the method, sometimes by more than one order of magnitude [19, 20]. Moreover, new effects of diffraction type are described, which may lead to a lengthening of the resonances' lifetimes, as we showed for the two-disk scatterer in Section VIII.G.2 [19, 20].[7] The

[7] Let us now mention recent works on diffraction effects in billiards, such as the disk scatterers, due to creeping orbits partially circling on the border of the disks, and which are at the origin of very short-lived scattering resonances lying at small Re κ with Im κ deeper in the complex plane than the first row of scattering resonances given by Eq. (8.44) [287].

$\hbar$-expansion can be applied to other systems such as the autoionization states of H^- or He in order to improve the accuracy of the method and, possibly, to obtain the lifetimes that are inaccessible at the leading $\hbar$-order. These results show the importance of anharmonicities around each periodic orbit, an aspect that has been previously overlooked.

In a larger context, we may expect anharmonicities to become essential for periodic orbits undergoing bifurcations, with the linear stability of the orbits passing through neutrality. In these circumstances, the nonlinear stability controls the motion in the vicinity of the bifurcation. It is remarkable that recent experimental works on Rydberg hydrogen-like atoms in electric and magnetic fields have reached the point where such bifurcations can be observed in the Fourier-type transform of the energy spectra [34, 35]. Indeed, bifurcations are typical nonlinear phenomena of classical dynamical systems, while the atoms and molecules obey the linear Schrödinger equation. However, due to the accumulation of energy levels, the classical nonlinear properties start to emerge from the linear wave dynamics. In this sense, it is remarkable that we are today able to see in microscopic systems this emergence of typical nonlinear behaviors like bifurcations, which were long believed to be restricted to the macroscopic world. We may expect important developments in this direction in the future, particularly for electronically excited atomic or molecular species.

In our section on matrix Hamiltonians, we showed that a semiclassical treatment of such systems is possible and that an $\hbar$-expansion is already necessary to perform an effective separation between the different potential energy surfaces. In the same way as chaotic behaviors are globally removed in equilibrium point quantization via the use of an infinite sequence of Van Vleck contact transformations, surface hopping transitions are removed into nonanalytical terms, which shows that surface hopping amplitudes are exponentially small in $\hbar$. As in the Landau–Zener model [201], such amplitudes can be obtained at the superadiabatic level by using complex classical orbits that connect the different adiabatic potential surfaces through excursions into complex phase space, as in tunneling. As a consequence, the decay of the metastable states is classically forbidden, and they may acquire long lifetimes due to the exponentially small hopping amplitudes. Fundamental phenomena, such as gauge structures and geometric phases, appear when two or more potential surfaces form a conical intersection. A new phase must be taken into account in the semiclassical treatment for periodic orbits encircling the conical intersection, which gives rise to the assignment of half-integer quantum numbers to energy eigenstates as we discussed in Section IX.C. Moreover, we showed that the upper surface of a conical intersection can sustain long-lived scattering resonances, which is a result we have not seen mentioned in the literature. At the classical level, another new result is that the dynamics on the upper cone may become weakly chaotic if the conical intersection becomes anisotropic.

E. Other Problems Related to the Semiclassical Limit

Despite the extension of this chapter, there are many topics closely related to semiclassical methods that we did not mention. In particular, the semiclassical methods are very successful in the calculation of atomic and molecular scattering cross-sections, as in rainbow scattering [288, 289]. Of more recent interest, there is the problem of cross-sections in irregular scattering; here important new results have been obtained on the statistical properties of the S-matrix elements and eigenvalues, particularly by Smilansky and co-workers [290, 291]. Besides the observation of irregular caustic-like structures in the angular dependency of the cross-sections in irregular scattering, the S-matrix elements and eigenvalues have been shown to fluctuate according to the predictions of random matrix theories on circular unitary ensembles. We were surprised [291] that the eigenvalues show a Wigner repulsion even for scattering systems that do not have a classically chaotic set of trapped orbits, such as the two-disk scatterer. The reason for this apparently paradoxical result is that the S-matrix is determined by scattering orbits rather than by periodic orbits. The eigenvalues of the S-matrix are determined by iterating the scattering Poincaré map rather than by iterating the internal dynamics on the classical repeller. Consequently, classical chaos may arise by this iterating process of the transient scattering orbits, leading to statistical properties of the quantum S-matrix [291]. Moreover, important work is in progress concerning the energy dependence of irregular fluctuations in the scattering cross-sections [40–45]. The interest in the phenomenon of Ericson fluctuations, originally observed in neutron–nuclei scattering at medium and high energies, has been revived by the numerical observation of similar fluctuations in classically chaotic scattering [42]. Although the main activity focuses on electronic conductance in semiconductor devices and in nuclear reactions, molecular scattering analogs of this phenomenon should also become manifest.

Among the problems that we did not discuss, are the applications of semiclassical methods to many-body systems of statistical mechanics [50]. The time evolution of such systems is ruled by the Landau–von Neumann equation for the density matrix, $\rho(\mathbf{x}, \mathbf{x}') = \psi^*(\mathbf{x}')\psi(\mathbf{x})$ or $\Sigma_i\, p_i\psi_i^*(\mathbf{x}')\psi_i(\mathbf{x})$. In the semiclassical limit ($\hbar \to 0$) and using the Weyl–Wigner representation of Section II.D, the Landau–von Neumann equation is known to reduce to the classical Liouville equation of nonequilibrium statistical mechanics [292]. Recent works have been concerned with finding the eigenvalues and the spectral decompositions of the classical Liouville operator [121, 293–296]. In chaotic classical systems, the Liouville eigenvalue equation also involves a zeta function, which we briefly mentioned in Section III.D. This other zeta function is related to the classical probabilities rather than to the quantum probability amplitudes associated with each unstable periodic orbit. It is important to understand that the classical zeta function is also the eigenvalue equation of the semiclassical limit ($\hbar \to 0$) of the quantum Landau–von Neumann operator. Indeed, we have seen that the probability amplitude associated with an unstable periodic orbit is given by

$$\varphi_p(E) = \frac{1}{|\Lambda_p|^{1/2}} e^{(i/\hbar)S_p(E) - i(\pi/2)\mu_p} \tag{14.1}$$

which corresponds to the deterministic level of the Schrödinger equation. At the statistical level of the Landau–von Neumann equation, the probability associated with the unstable periodic orbit is given by an expression that is quadratic in the amplitudes as for a density matrix [297]

$$\mathcal{P}_p(E', E) = \varphi_p^*(E')\varphi_p(E) = \frac{1}{|\Lambda_p|} e^{(i/\hbar)[S_p(E) - S_p(E')]} \tag{14.2}$$

for two different energies, E and E', separated by the eigenvalue of the Landau–von Neumann equation. These eigenvalues are the Bohr frequencies, or the corresponding rates in the case where the frequencies become complex, since we have for $\hat{\rho} = \exp(st)|E\rangle\langle E'|$

$$\partial_t \hat{\rho} = \frac{1}{i\hbar}[\hat{\mathcal{H}}, \hat{\rho}] = s\hat{\rho} \tag{14.3}$$

with $s = (E - E')/i\hbar$ as the eigenvalue. Therefore, the two energies, E and E', are separated by a quantity that is proportional to $\hbar$: $E' = E - i\hbar s$. Using $T_p = \partial_E S_p$, we get the expansion

$$S_p(E') = S_p(E - i\hbar s) = S_p(E) - i\hbar s T_p(E) + \mathcal{O}(\hbar^2 s^2) \tag{14.4}$$

and the probability associated with the unstable periodic orbit becomes

$$\mathcal{P}_p(E', E) = \mathcal{P}_p(s; E) = \frac{1}{|\Lambda_p|} e^{-sT_p(E) + \mathcal{O}(\hbar)} \tag{14.5}$$

which is precisely the weight of the periodic orbit in the classical zeta function [Eq. (3.58)] for the classical exponent $\beta = 1$. The preceding reasoning establishes the connection between the level of the quantum wave function and the statistical level of transport and rate processes, which are the subject of nonequilibrium statistical mechanics. At the classical level, highly significant results have been obtained, which show that irreversibility in transport phenomena like diffusion can be rigorously understood in terms of classical Ruelle resonances, which are the zeros of the classical zeta function, and in terms of the associated spectral decomposition [121, 293–296]. The preceding semiclassical reasoning suggests important connections between quantum and classical statistical mechanics, the understanding of which is still at the beginning.

To conclude, the new developments in the semiclassical methods that we exposed and illustrated in this chapter highlight their ability to provide a unified approach to the understanding of many phenomena. This applies particularly to reaction processes, and thus responds to the need for such a unified approach, which has been recently emphasized by Marcus [51]. We expect further progress in these directions in this rapidly advancing field of research.

ACKNOWLEDGMENTS

It is our pleasure to thank Professors G. Nicolis and I. Prigogine for support in this research as well as to Professor S. A. Rice for his invitation to write this chapter. We also thank Professors E. Heller, M. Herman, R. Jost, and J.-P. Pique for fruitful discussions and contacts. The authors are financially supported, respectively, by the National Fund for Scientific Research (FNRS Belgium) for P. G., by a NATO postdoctoral grant obtained from the German Academic Exchange Service (DAAD) as well as by a EEC postdoctoral fellowship No. ERBCHBICT 930852 for I. B., and by a EEC doctoral fellowship No. ERBCHBICT920097 for D. A. The research is financially supported by the SADOVEM Project under EEC contract No. SC1-CT91-0711, as well as by the Quantum Keys for Reactivity ARC Project of the Communauté Française de Belgique.

NOTE ADDED IN PROOF

Since this manuscript was completed, several works have come to our knowledge, and new results have been obtained.

In Section II.C.2, we should add that the distinction between overlapping and nonoverlapping scattering resonances has been discussed many times, in particular in Refs. [42, 114, 253], as well as in Refs. [298, 299]. We believe that the distinction is a delicate problem, especially at the level of the consequences on the cross-sections or in the time domain. Furthermore, the relationship between the scattering resonances and the energy dependence of the reaction probabilities has been discussed by Truhlar and coworkers [300].

Concerning the $\hbar$-expansion of the propagator in Section II.F.5, we would like to mention the recent work of Ref. [301] where a $\hbar$-expansion of the propagator has been obtained within a phase-space formulation rather than a position-space formulation.

About the shift in the zero-point energy due to the anharmonicities in Section V.A, its presence has recently been pointed out in the context of spectroscopic analysis [302]. We remark that our calculation [Eq. (5.7)] thus confirms previous works [137, 299, 302].

Concerning the vibrograms in Section XI, let us mention that Fourier transforms with a scanning Blackmann–Harris window function have previously been considered in Ref. [303], which came to our knowledge after the completion of the present work. Very recently, we constructed and analyzed the vibrogram of C_2HD, which has been modeled by a pure Dunham expansion leading to a classically integrable Hamiltonian [304]. In this analysis, we used the function

$$f(T, E; \Delta E) = \frac{|S(T, E; \Delta E)|^2}{|S(0, E; \Delta E)|},$$

instead of Eq. (11.6), with the advantage that the amplitudes of the peaks remain approximately constant with energy. Moreover, the role of periodic orbits embedded in lower-dimensional subsystems has been clarified both in C_2HD and in CS_2, as will be reported elsewhere [305, 306].

In our analysis of the CS_2 molecule in Section XI.B.2, we would further like to point out that the transformation between the Hamiltonians (11.20) and (11.21) introduces a factor of four in the Jacobian of the angle variables. Consequently, the Hamiltonian (11.21) generates four times more periodic orbits than the original one. This extra feature has also to be taken into account in the interpretation of the vibrogram of Fig. 50.

In the present chapter, we did not mention the numerous current works on the properties of the eigenfunctions of classically chaotic systems. We refer the reader to the literature on this fundamental problem [6, 15, 307]. The question of the limit of validity of semiclassical propagation in time is also of importance, which has been discussed by Tomsović and Heller [308] (see also [309]). We think that the inclusion of $\hbar$-corrections that we described in this chapter can significantly enlarge the time interval of validity of the semiclassical approximation. In a related context, Berry and Howls showed very recently how the asymptotic series in $\hbar$ representing the Weyl–Wigner level density is controlled by a Stokes mechanism involving the shortest periodic orbits [310].

REFERENCES

1. R. D. Levine and R. B. Bernstein, *Molecular Reaction Dynamics and Chemical Reactivity*, Oxford University Press, New York, 1987.
2. A. J. Lieberman and A. J. Lichtenberg, *Regular and Stochastic Motion*, Springer, New York, 1983.
3. R. S. MacKay and J. D. Meiss, Eds., *Hamiltonian Dynamical Systems: A Reprint Selection*, Adam Hilger, Bristol, 1987.
4. J.-P. Eckmann and D. Ruelle, *Rev. Mod. Phys.* **57**, 617 (1985).
5. G. Casati and J. Ford, Eds., *Stochastic Behavior in Classical and Quantum Hamiltonian Systems*, Lecture Notes in Physics, Vol. 93, Springer, Berlin, 1979.
6. M.-J. Giannoni, A. Voros, and J. Zinn-Justin, Eds., *Chaos and Quantum Physics*, North-Holland, Amsterdam, 1991.
7. A. M. Ozorio de Almeida, *Hamiltonian Systems: Chaos and Quantization*, Cambridge University Press, 1988.

8. M. Jammer, *The Conceptual Development of Quantum Mechanics*, McGraw-Hill, New York, 1973.
9. B. L. van der Waerden, *Sources of Quantum Mechanics*, Dover, New York, 1967.
10. J. Dalibard, Y. Castin, and K. Mölmer, *Phys. Rev. Lett.* **68**, 580 (1992); N. Gisin and I. C. Percival, *J. Phys. A: Math. Gen.* **25**, 5677 (1992); H. J. Carmichael, *Phys. Rev. Lett.* **70**, 2273 (1993).
11. M. C. Gutzwiller, *J. Math. Phys.* **8**, 1979 (1967); **10**, 1004 (1969); **11**, 1791 (1970); **12**, 343 (1971); **18**, 806 (1977).
12. R. Balian and C. Bloch, *Ann. Phys.* **60**, 401 (1970); **63**, 592 (1971); **64**, 271 (1971); *Errata* **84**, 559 (1974).
13. R. Balian and C. Bloch, *Ann. Phys.* **69**, 76 (1972); **85**, 514 (1974).
14. M. C. Gutzwiller, *Phys. Rev. Lett.* **45**, 150 (1980); M. C. Gutzwiller, in R. L. Devaney and Z. H. Nitecki, Eds., *Classical Mechanics and Dynamical Systems*, Marcel-Dekker, New York, 1981; M. C. Gutzwiller, *J. Phys. Chem.* **92**, 3154 (1988).
15. M. C. Gutzwiller, *Chaos in Classical and Quantum Mechanics*, Springer, New York, 1990.
16. P. Cvitanović, Ed., *Periodic Orbit Theory*, Chaos, Vol. 2, Nos. 1–4 (1992).
17. G. Casati, I. Guarneri, and U. Smilansky, Eds., *Quantum Chaos*, Proceedings of the International School of Physics "Enrico Fermi," Varenna, Italy, 1991, North-Holland, Amsterdam, 1993.
18. D. Wintgen, *Phys. Rev. Lett.* **58**, 1589 (1987).
19. P. Gaspard and D. Alonso, *Phys. Rev. A* **47**, R3468 (1993); P. Gaspard, in G. Casati and B. V. Chirikov, Eds., *Quantum Chaos*, Cambridge University Press, UK, 1994.
20. D. Alonso and P. Gaspard, *Chaos* **3**, 601 (1993); Erratum, ibid. **4**, 105 (1994).
21. I. Burghardt and P. Gaspard, *J. Chem. Phys.* **100**(9), 6395 (1994); I. Burghardt and P. Gaspard, *J. Phys. Chem.* (1995) in press; I. Burghardt and P. Gaspard, in preparation.
22. V. P. Maslov and M. V. Fedoriuk, *Semiclassical Approximation in Quantum Mechanics*, Reidel, Boston, MA, 1981.
23. D. Wintgen, K. Richter, and G. Tanner, *Chaos* **2**, 19 (1992).
24. H. Poincaré, *Les Méthodes Nouvelles de la Mécanique Céleste*, 3 Vols., Paris, Gauthier-Villars, 1899, Dover, New York, 1957, AIP, New York, 1993; J. Moser, *Stable and Random Motions in Dynamical Systems*, Princeton University Press, New Jersey, 1973.
25. A. Einstein, *Verh. Deut. Phys. Ges.* **19**, 82 (1917).
26. P. Ehrenfest, *Verh. Deut. Phys. Ges.* **15**, 451 (1913); *Proc. Kon. Akad. Amsterdam* **16**, 591 (1913); *Ann. der Phys.* **51**, 327 (1916).
27. W. Pauli, *General Principles of Quantum Mechanics*, Springer, Berlin, 1980.
28. G. Wentzel, *Z. Physik* **38**, 518 (1926); H. A. Kramers, *Z. Physik* **39**, 828 (1926); L. Brillouin, *C. R. Acad. Sci. Paris* **183**, 24 (1926).
29. E. P. Wigner, *Phys. Rev.* **40**, 749 (1932).
30. L. H. Thomas, *Proc. Cambridge Philos. Soc.* **23**, 542 (1927); E. Fermi, *Rend. Accad. Naz. Lincei* **6**, 602 (1927); E. H. Lieb, *Rev. Mod. Phys.* **53**, 603 (1981).
31. J. K. G. Watson, M. Herman, J. C. Van Craen, and R. Colin, *J. Mol. Spectrosc.* **95**, 101 (1982); J. Vander Auwera, T. R. Huet, M. Herman, C. D. Hamilton, J. L. Kinsey, and R. W. Field, *ibid.* **137**, 381 (1989); E. Abramson, R. W. Field, D. Imre, K. K. Innes, and J. L. Kinsey, *J. Chem. Phys.* **80**, 2298 (1984).
32. G. E. Ewing, 1978, *Chem. Phys.* **29**, 253 (1978); G. E. Ewing, *J. Chem. Phys.* **71**,

3143 (1979); K. E. Johnson, L. Wharton, and D. H. Levy, *ibid.* **69**, 2719 (1978); J. E. Kenny, K. E. Johnson, W. Sharfin, and D. H. Levy, *ibid.* **72**, 1109 (1980); R. J. Le Roy and J. S. Carley, *Adv. Chem. Phys.* **XLII**, 353 (1980).

33. M. G. Littman, M. L. Zimmerman, T. W. Ducas, R. R. Freeman, and D. Kleppner, *Phys. Rev. Lett.* **36**, 788 (1976); M. L. Zimmerman, M. G. Littman, M. M. Kash, and D. Kleppner, *Phys. Rev. A* **20**, 2251 (1979); C. Iu, G. R. Welch, M. M. Kash, D. Kleppner, D. Delande, and J. C. Gay, *Phys. Rev. Lett.* **66**, 145 (1991); A. Buchleitner, B. Grémaud, and D. Delande, *J. Phys. B: At. Mol. Opt. Phys.* **27**, 2663 (1994).

34. A. Holle, J. Main, G. Wiebusch, H. Rottke, and K. H. Welge, *Phys. Rev. Lett.* **61**, 161 (1988); J.-M. Mao and J. B. Delos, *Phys. Rev. A* **45**, 1746 (1992); T. van der Velt, W. Vassen, and W. Hogervorst, *Europhys. Lett.* **21**, 903 (1993); J. Main, G. Wiebush, K. Welge, J. Shaw, and J. B. Delos, *Phys. Rev. A* **49**, 847 (1994).

35. J.-M. Mao, K. A. Rapelje, S. J. Blodgett-Ford, J. B. Delos, A. König, and H. Rinneberg, *Phys. Rev. A* **48**, 2117 (1993).

36. U. Eichmann, V. Lange, and W. Sandner, *Phys. Rev. Lett.* **68**, 21 (1992); R. P. Wood and C. H. Greene, *Phys Rev. A* **49**, 1029 (1994).

37. M. Lombardi, P. Labastie, M. C. Bordas, and M. Broyer, *J. Chem. Phys.* **89**, 3479 (1988); M. Lombardi and T. H. Seligman, *Phys. Rev. A* **47**, 3571 (1993).

38. R. Blümel et al., *Nature (London)* **334**, 309 (1988); J. Hoffnagle and R. G. Brewer, *Phys. Rev. Lett.* **71**, 1828 (1993); D. Farrelly and J. E. Howard, *Phys. Rev. A* **49**, 1494 (1994); and references cited therein.

39. A. H. Zewail, *Science* **242**, 1645 (1988); M. Gruebele and A. H. Zewail, *Phys. Today*, 24–33 (1990); L. R. Khundkar and A. H. Zewail, *Annu. Rev. Phys. Chem.* **41**, 15 (1990); A. H. Zewail, *J. Phys. Chem.* **97**, 12427 (1993).

40. C. W. J. Beenakker and H. van Houten, *Phys. Rev. Lett.* **63**, 1857 (1989); M. L. Roukes and O. L. Alerhand, *Phys. Rev. Lett.* **65**, 1651 (1990).

41. C. M. Marcus, R. M. Westervelt, P. F. Hopkins, and A. C. Gossard, *Chaos* **3**, 643 (1993); and references cited therein.

42. R. Blümel and U. Smilansky, *Phys. Rev. Lett.* **60**, 477 (1988); **64**, 241 (1990); B. Eckhardt, *Chaos* **3**, 613 (1993).

43. R. A. Jalabert, H. U. Baranger, and A. D. Stone, *Phys. Rev. Lett.* **65**, 2442–2445 (1990); H. U. Baranger, R. A. Jalabert, and A. D. Stone, *Chaos* **3**, 665 (1993).

44. J. Main and G. Wunner, *Phys. Rev. Lett.* **69**, 586 (1992).

45. H. A. Weidenmüller, *Physica A* **167**, 28 (1990).

46. M. F. Crommie, C. P. Lutz, and D. M. Eigler, *Science* **262**, 218 (1993); G. P. Collins, *Phys. Today*, Nov. 1993, pp. 16–19; E. J. Heller, M. F. Crommie, C. P. Lutz, and D. M. Eigler, *Nature* **369**, 464 (1994).

47. K. E. Drexler, *Nanosystems*, Wiley, New York, 1992.

48. E. Doron, U. Smilansky, and A. Frenkel, *Phys. Rev. Lett.* **65**, 3072 (1990); *Physica D* **50**, 367 (1992); H. D. Graf, H. L. Harney, H. Lengeler, C. H. Lewenkopf, C. Rangacharyulu, A. Richter, P. Schardt, and H. A. Weidenmüller, *Phys. Rev. Lett.* **69**, 1296 (1992).

49. J. W. Young and J. C. Bertrand, *J. Acoust. Soc. Am.* **58**, 1190 (1975); F. Mortessagne, O. Legrand, and D. Sornette, *Chaos* **3**, 529 (1993).

50. A. J. Leggett, S. Chakravarty, A. T. Dorsey, M. P. A. Fisher, A. Garg, and W. Zwerger, *Rev. Mod. Phys.* **59**, 1 (1987); G. A. Voth, D. Chandler, and W. H. Miller, *J. Phys. Chem.* **93**, 7009 (1989); G. A. Voth, D. Chandler, and W. H. Miller, *J. Chem. Phys.* **91**, 7749 (1989); G. A. Voth, *J. Phys. Chem.* **97**, 8365 (1993).
51. R. M. Marcus, *Faraday Discuss. Chem. Soc.* **91**, 479 (1991).
52. L. I. Schiff, *Quantum Mechanics*, McGraw-Hill, Tokyo, 1968.
53. C. Cohen-Tannoudji, J. Dupont-Roc, and G. Grynberg, *Processus d'interaction entre photons et atomes*, InterEditions/Editions du CNRS, Paris, 1988.
54. C. J. Joachain, *Quantum Collision Theory*, North-Holland, Amsterdam, 1975.
55. J. Humblet and L. Rosenfeld, *Nucl. Phys.* **70**, 1 (1965).
56. A. Böhm, *Quantum Mechanics: Foundations and Applications*, 3rd ed., Springer, New York, 1993.
57. A. Böhm, *The Rigged Hilbert Space and Quantum Mechanics*, Lecture Notes in Physics, Vol. 78, Springer, Berlin, 1978; A. Böhm and M. Gadella, *Dirac kets, Gamow vectors, and the Gel'fand triplets*, Lecture Notes in Physics, Vol. 348, Springer, Berlin, 1989.
58. W. Thirring, *Quantum Mechanics of Atoms and Molecules*, Springer, New York, 1981.
59. R. Loudon, *The Quantum Theory of Light*, Clarendon Press, Oxford, 1983; W. H. Louisell, *Quantum Statistical Properties of Radiation*, Wiley, New York, 1973.
60. D. A. McQuarrie, *Statistical Mechanics*, Harper & Row, New York, 1976; J. H. Van Vleck and D. L. Huber, *Rev. Mod. Phys.* **49**, 939 (1977).
61. E. J. Heller, *J. Chem. Phys.* **68**, 2066 (1978); S.-Y. Lee and E. Heller, *J. Chem. Phys.* **71**, 4777 (1979); E. J. Heller and S. Tomsovic, *Phys. Today*, Jul. 1993, pp. 38–46.
62. R. Schinke, *Photodissociation Dynamics*, Cambridge University Press, UK, 1993.
63. U. Fano and A. R. P. Rau, *Atomic Collisions and Spectra*, Academic, Orlando, FL, 1986.
64. E. P. Wigner, *Phys. Rev.* **98**, 145 (1955); F. T. Smith, *ibid.* **118**, 349 (1960); H. Narnhofer, *Phys. Rev. D* **22**, 2387 (1980); H. Narnhofer and W. Thirring, *Phys. Rev. A* **23**, 1688 (1981); E. H. Hauge and J. A. Stövneng, *Rev. Mod. Phys.* **61**, 917 (1989).
65. P. J. Robinson and K. A. Holbrook, *Unimolecular Reactions*, Wiley-Interscience, London, 1972; W. Forst, *Theory of Unimolecular Reactions*, Academic, New York, 1973; H. O. Pritchard, *The Quantum Theory of Unimolecular Reactions*, Cambridge University Press, Cambridge, 1984; H. O. Pritchard, *J. Phys. Chem.* **89**, 3970 (1985).
66. W. H. Miller, *Adv. Chem. Phys.* **XXV,** 69 (1974); **XXX,** 77 (1975).
67. H. Weyl, *Z. Phys.* **46**, 1 (1927); K. Imre, E. Özizmir, M. Rosenbaum, and P. F. Zweifel, *J. Math. Phys.* (*N.Y.*) **8**, 1097 (1967).
68. J. L. Dahl, *Phys. Scripta* **25**, 499 (1982); N. L. Balazs and B. K. Jennings, *Phys. Rep.* **104**, 347 (1984).
69. P. Carruthers and F. Zachariasen, *Rev. Mod. Phys.* **55**, 245 (1983).
70. H. Goldstein, *Classical Mechanics*, Addison-Wesley, Reading, MA 1950.
71. V. I. Arnold, *Mathematical Methods of Classical Mechanics*, Springer, New York, 1978.
72. M. Morse, *The Calculus of Variations in the Large*, Colloquium Publication, Vol. 18, American Mathematical Society, New York, 1934.
73. J. B. Delos, *J. Chem. Phys.* **86**, 425 (1987).
74. V. P. Maslov, *Théorie des Perturbations et Méthodes Asymptotiques*, Dunod/Gauthier-Villars, Paris, 1972.

75. M. V. Berry and K. E. Mount, *Rep. Prog. Phys.* **35**, 315–397 (1972).
76. J. H. Van Vleck, *Proc. Natl. Acad. Sci. USA* **14**, 178 (1928).
77. C. Morette, *Phys. Rev.* **81**, 848 (1951).
78. S. Levit and U. Smilansky, *Ann. Phys.* **103**, 198 (1977); W. Dittrich and M. Reuter, *Classical and Quantum Dynamics: From Classical Paths to Path Integrals*, Springer, Berlin, 1992.
79. R. P. Feynman and A. R. Hibbs, *Quantum Mechanics and Path Integrals*, McGraw-Hill, New York, 1965.
80. R. Kosloff, *J. Phys. Chem.* **92**, 2087 (1988).
81. N. Bleistein and R. A. Handelsman, *Asymptotic Expansion of Integrals*, Dover, New York, 1986.
82. L. S. Schulman, *Techniques and Applications of Path Integration*, Wiley, New York, 1981.
83. C. DeWitt-Morette, *Ann. Phys.* **97**, 367 (1976); C. DeWitt-Morette, A. Maheshwari, and B. Nelson, *Phys. Rep.* **50**, 255 (1979).
84. B. DeWitt, *Supermanifolds*, 2nd ed., Cambridge University Press, 1992.
85. J. Iliopoulos, C. Itzykson, and A. Martin, *Rev. Mod. Phys.* **47**, 165 (1975).
86. R. F. Dashen, B. Hasslacher, and A. Neveu, *Phys. Rev. D* **10**, 4114 (1974).
87. G. Parisi, *Statistical Field Theory*, Addison-Wesley, Redwood City, CA, 1988.
88. G. Nicolis and I. Prigogine, *Exploring Complexity: An Introduction*, Freeman, New York, 1989.
89. V. I. Oseledec, *Trans. Moscow Math. Soc.* **19**, 197 (1968).
90. D. Ruelle, *Ann. N.Y. Acad. Sci.* **316**, 408 (1978).
91. V. I. Arnold, Ed., *Dynamical Systems III*, Springer, Berlin, 1988.
92. B. Eckhardt and D. Wintgen, *J. Phys. A: Math. Gen.* **24**, 4335 (1991).
93. S. A. Rice, P. Gaspard, and K. Nakamura, *Adv. Class. Traj. Meth.* **1**, 215 (1992).
94. K. R. Meyer, *Trans. Am. Math. Soc.* **149**, 95 (1970); J. M. Greene, R. S. MacKay, F. Vivaldi and M. J. Feigenbaum, *Physica D* **3**, 468 (1981); both reprinted in [3].
95. S. Wiggins, *Global Bifurcations and Chaos*, Springer, New York, 1988; S. Wiggins, *Chaotic Transport in Dynamical Systems*, Springer, New York, 1992.
96. Ya. G. Sinai, Ed., *Dynamical Systems: Collection of Papers*, World Scientific, Singapore, 1991.
97. P. Gaspard, in *Quantum Chaos*, G. Casati, I. Guarneri, and U. Smilansky, Eds., Proceedings of the International School of Physics "Enrico Fermi," Varenna, Italy 1991, North-Holland, Amsterdam, 1993, pp. 307–383.
98. J. P. Dahl and M. Springborg, *J. Chem. Phys.* **88**, 4535 (1988); and references cited therein.
99. G. S. Ezra, K. Richter, G. Tanner, and D. Wintgen, *J. Phys. B: At. Mol. Opt. Phys.* **24**, L413–L420 (1991).
100. K. Richter, G. Tanner, and D. Wintgen, *Phys. Rev. A* **48**, 4182 (1993).
101. P. Gaspard and S. A. Rice, *Phys. Rev. A* **48**, 54 (1993).
102. M. Hénon and C. Heiles, *Astronom. J.* **69**, 73 (1964); R. C. Churchill, G. Pecelli, and D. L. Rod, in [5], pp. 76–136.
103. B. Barbanis, *Astronom. J.* **71**, 415 (1966); G. Contopoulos, *Physica D* **8**, 142 (1983); G. Contopoulos, *Astron. Astrophys.* **161**, 244 (1986).
104. M. V. Berry, N. L. Balazs, M. Tabor, and A. Voros, *Ann. Phys.* **122**, 26 (1979).

105. S. H. Tersigni, P. Gaspard, and S. A. Rice, *J. Chem. Phys.* **92**, 1775 (1990).
106. B. V. Chirikov, Phys. Rep. **52**, 265 (1979).
107. M. Hénon, *Q. J. Appl. Math.* **27**, 291 (1969).
108. P. Gaspard and S. A. Rice, *J. Phys. Chem.* **93**, 6947 (1989).
109. L. A. Bunimovich and Ya. G. Sinai, *Commun. Math. Phys.* **78**, 247, 479 (1980); J. Machta and R. Zwanzig, *Phys. Rev. Lett.* **50**, 1959 (1983).
110. G. D. Birkhoff, *Acta Math.* **50**, 359 (1927); in [3] pp. 154–174.
111. D. Alonso and P. Gaspard, *Quantization of the Circle Billiard*, in preparation.
112. L. A. Bunimovich, *Commun. Math. Phys.* **65**, 295 (1979).
113. B. Eckhardt, *J. Phys. A: Math. Gen.* **20**, 5971–5979 (1987); B. Eckhardt, *Physica D* **33**, 89 (1988).
114. P. Gaspard and S. A. Rice, *J. Chem. Phys.* **90**, 2225, 2242, 2255 (1989); **91**, E3279 (1989).
115. P. Cvitanović and B. Eckhardt, *Phys. Rev. Lett.* **63**, 823 (1989).
116. S. Smale, *The Mathematics of Time*, Springer, New York, 1980.
117. E. Ott, *Chaos in Dynamical Systems*, Cambridge University Press, UK, 1993.
118. R. Artuso, E. Aurell, and P. Cvitanović, *Nonlinearity* **3**, 325, 361 (1990).
119. P. Walters, *An Introduction to Ergodic Theory*, Springer, Berlin, 1981.
120. S. Bleher, C. Grebogi, and E. Ott, *Phys. Rev. Lett.* **63**, 919 (1989); *Physica D* **46**, 87 (1990).
121. P. Gaspard and D. Alonso Ramirez, *Phys. Rev. A* **45**, 8383 (1992).
122. Ya. G. Sinai, *Russ. Math. Surveys* **27**, 21 (1972).
123. R. Bowen, *Equilibrium States and the Ergodic Theory of Anosov Diffeomorphisms*, Lecture Notes in Mathematics, Vol. 470, Springer, Berlin, 1975.
124. D. Ruelle, *Thermodynamic Formalism*, Addison-Wesley, Reading, MA, 1978.
125. Ya. B. Pesin, *Math. USSR Izv.* **10**(6), 1261 (1976); Ya. B. Pesin, *Russ. Math. Surveys* **32**(4), 55 (1977); H. Kantz and P. Grassberger, *Physica D* **17**, 75 (1985).
126. P. Szepfalusy and T. Tél, *Phys. Rev. A* **34**, 2520 (1986); T. Tél, *Phys. Lett. A* **119**, 65 (1986); *Phys. Rev. A* **36**, 1502 (1987); T. Bohr and D. Rand, *Physica D* **25**, 387 (1987); D. Bessis, G. Paladin, G. Turchetti, and S. Vaienti, *J. Stat. Phys.* **51**, 109 (1988).
127. D. Ruelle, *Invent. Math.* **34**, 231 (1976); *J. Stat. Phys.* **44**, 281 (1986); *Commun. Math. Phys.* **125**, 239 (1989).
128. J. Chazarain, *Invent. Math.* **24**, 65 (1974); A. Voros, *Suppl. Prog. Theor. Phys.* **116**, 17 (1994).
129. R. Balian, G. Parisi, and A. Voros, *Phys. Rev. Lett.* **41**, 1141 (1978); in S. Albeverio et al., Eds., *Feynman Path Integrals*, Lecture Notes in Physics, Vol. 106, Springer, Berlin, 1979, pp. 337–360; A. Voros, *Nucl. Phys. B* **165**, 209 (1980).
130. B. Grammaticos and A. Voros, *Ann. Phys.* **123**, 359 (1979).
131. R. G. Littlejohn, *J. Math. Phys.* **31**, 2952 (1990).
132. M. V. Berry, *Ann. Phys.* 131, 163 (1981).
133. D. Alonso and P. Gaspard, *J. Phys. A: Math. Gen.* **27**, 1599 (1994).
134. D. Papousek and M. R. Aliev, *Molecular Vibrational–Rotational Spectra*, Elsevier, Amsterdam, 1982.
135. I. M. Mills, in *Molecular Spectroscopy: Modern Research*, K. N. Rao and C. W. Mathews, Eds., Academic, New York, 1972, pp. 115–140.

136. H. H. Nielsen, *Rev. Mod. Phys.* **23**, 90 (1951).
137. W. H. Miller, R. Hernandez, N. C. Handy, D. Jayatilaka, and A. Willetts, *Chem. Phys. Lett.* **172**, 62 (1990).
138. T. Seideman and W. H. Miller, *J. Chem. Phys.* **95**, 1768 (1991).
139. J. L. Dunham, *Phys. Rev.* **41**, 721 (1932).
140. G. D. Birkhoff, *Dynamical Systems*, Colloquium Publication, Vol. 9, American Mathematical Society, New York, 1927; F. G. Gustavson, *Astronom. J.* **71**, 670 (1966).
141. V. I. Arnold, *Geometric Methods in the Theory of Ordinary Differential Equations*, Springer, New York, 1983.
142. C. E. Porter, *Statistical Theories of Spectra: Fluctuations*, Academic, New York, 1965.
143. R. T. Swimm and J. B. Delos, *J. Chem. Phys.* **71**, 1706 (1979); C. Jaffé and W. P. Reinhardt, *J. Chem. Phys.* **71**, 1862 (1979); **77**, 5191 (1982).
144. T. Uzer and R. A. Marcus, *J. Chem. Phys.* **81**, 5013 (1984).
145. L. E. Fried and G. S. Ezra, *J. Phys. Chem.* **92**, 3144 (1988).
146. M. Robnik, *J. Phys. A: Math. Gen.* **17**, 109 (1984); *ibid.* **25**, 5311 (1992).
147. H. A. Weidenmüller, *Phys. Rev. A* **48**, 1819 (1993).
148. H. P. Baltes and E. R. Hilf, *Spectra of Finite Systems*, Bibliographisches Institut-Wissenschaftsverlag, Mannheim/Wien/Zürich, 1976.
149. M. C. Gutzwiller, in [6], pp. 201–249.
150. J. M. Robbins, *Nonlinearity* **4**, 343 (1991).
151. S. C. Creagh, J. M. Robbins, and R. G. Littlejohn, *Phys. Rev. A* **42**, 1907 (1990).
152. A. Selberg, *J. Indian Math. Soc.* **20**, 47 (1956).
153. W. H. Miller, *J. Chem. Phys.* **56**, 38 (1972); **63**, 996 (1975); **83**, 960 (1979).
154. A. Voros, *J. Phys. A: Math. Gen.* **21**, 685 (1988).
155. B. Eckhardt and E. Aurell, *Europhys. Lett.* **9**, 509 (1989).
156. C. Gérard and J. Sjöstrand, *Commun. Math. Phys.* **108**, 391 (1987); **116**, 193 (1988); J. Sjöstrand, *Duke Math. J.* **60**, 1 (1990).
157. M. Ikawa, *Osaka J. Math.* **22**, 657 (1985); **27**, 281 (1990); *Ann. Inst. Fourier*, **38**(2), 113 (1988).
158. B. Eckhardt and G. Russberg, *Phys. Rev. E* **47**, 1578 (1993).
159. D. Wintgen, A. Bürgers, K. Richter, and G. Tanner, *Suppl. Prog. Theor. Phys.* **116**, 121 (1994).
160. P. Cvitanović, P. E. Rosenqvist, G. Vattay, and H. H. Rugh, *Chaos* **3**, 619 (1993); P. Cvitanović and G. Vattay, *Phys. Rev. Lett.* **71**, 4138 (1993).
161. P. Cvitanović, *Phys. Rev. Lett.* **61**, 2729 (1988).
162. E. Bogomolny, *Chaos* **2**, 5 (1992).
163. T. Szeredi, J. H. Lefebvre, and D. A. Goodings, *Phys. Rev. Lett.* **71**, 2891 (1993); T. Szeredi and D. A. Goodings, *Phys. Rev. E* **48**, 3518, 3529 (1993); G. Tanner and D. Wintgen, *Classical and semiclassical zeta functions in terms of transition probabilities*, to be published in *Chaos, Solitons, and Fractals* (1994).
164. M. V. Berry and J. P. Keating, *J. Phys. A* **23**, 4839 (1990); *Proc. R. Soc. Lond. A* **437**, 151 (1992); O. Agam and S. Fishman, *J. Phys. A: Math. Gen.* **26**, 2113 (1993).
165. M. Sieber and F. Steiner, *Phys. Rev. Lett.* **67**, 1941 (1991); C. Matthies and F. Steiner, *Phys. Rev. A* **44**, R7877 (1991); R. Aurich, C. Matthies, M. Sieber, and F. Steiner, *Phys. Rev. Lett.* **68**, 1629 (1992).

166. E. Bogomolny, *Comm. At. Mol. Phys.* **25**, 63 (1990).

167. G. Tanner and D. Wintgen, *Chaos* **2**, 53 (1992).

168. G. Tanner, P. Scherer, E. B. Bogomolny, B. Eckhardt, and D. Wintgen, *Phys. Rev. Lett.* **67**, 2410 (1991).

169. P. Gaspard, *Suppl. Prog. Theor. Phys.* **116**, 59 (1994).

170. M. S. Child, *Semiclassical Mechanics with Molecular Applications*, Clarendon, Oxford, 1991.

171. W. G. Harter and C. W. Paterson, *J. Chem. Phys.* **80**, 9 (1984); J. M. Robbins, S. C. Creagh, and R. G. Littlejohn, *Phys. Rev. A* **39**, 2838 (1989); **41**, 6052 (1990).

172. W. T. Strunz, *J. Phys. A: Math. Gen.* **25**, 3855 (1992); W. T. Strunz, G. Alber, and J. S. Briggs, *J. Phys. B: Math. Gen.* **24**, 5091 (1991); *J. Phys. A: Math. Gen.* **26**, 5157 (1993).

173. T. C. Bountis, *Int. J. Bif. Chaos*, **2**, 217 (1992).

174. S. Destrain, *Quantification Semiclassique d'Applications Hamiltoniennes presque Intégrables et Faiblement Chaotiques: Etude Numérique des Quasi-Énergies*, Mémoire de Licence, ULB, 1990.

175. S. Adachi, *Ann. Phys. (N.Y.)* **195**, 45 (1989).

176. A. Shudo and K. Ikeda, *Suppl. Prog. Theor. Phys.* **116**, 283 (1994).

177. J. M. Greene and I. C. Percival, *Physica D* **3**, 530 (1981); I. C. Percival, *Physica D* **6**, 67 (1982).

178. V. G. Gelfreich, V. F. Lazutkin, C. Simó, and M. B. Tabanov, *Int. J. Bif. Chaos* **2**, 353 (1992).

179. W. H. Miller, *J. Chem. Phys.* **62**, 1899 (1975); W. H. Miller, S. D. Schwartz, and J. W. Tromp, *J. Chem. Phys.* **79**, 4889 (1983).

180. M. V. Berry and M. Tabor, *Proc. R. Soc. Lond. A* **349**, 101 (1976); *J. Phys. A: Math. Gen.* **10**, 371 (1977).

181. S. C. Creagh and R. G. Littlejohn, *Phys. Rev. A* **44**, 836 (1991); S. C. Creagh, *Semiclassical Mechanics of Symmetry Reduction*, preprint.

182. P. Gaspard, in *Quantum Chaos-Quantum Measurement*, P. Cvitanović, I. Percival and A. Wirzba, Eds., Kluwer, Dordrecht, 1992, pp. 19–42.

183. M. Kuś, F. Haake, and D. Delande, *Phys. Rev. Lett.* **71**, 2167 (1993).

184. A. M. Ozorio de Almeida and J. H. Hannay, *J. Phys. A: Math. Gen.* **20**, 5873 (1987).

185. M. V. Berry and C. Upstill, *Progress in Optics*, Vol., 18, E. Wolf, Ed., North-Holland, Amsterdam, 1980, pp. 257–348.

186. T. Poston and I. Stewart, *Catastrophe Theory and its Applications*, Pitman, London, 1978.

187. L. Landau and E. Lifshitz, *Quantum Mechanics*, Pergamon Press, New York, 1958.

188. M. L. Du and J. B. Delos, *Phys. Rev. A* **38**, 1896, 1913 (1988).

189. B. Eckhardt, S. Fishman, K. Müller, and D. Wintgen, *Phys. Rev. A* **45**, 3531 (1992); B. Eckhardt, in [17], pp. 77–112.

190. E. Heller, *Adv. Class. Traj. Method.* **1** 165 (1992).

191. M. V. Berry and M. Wilkinson, *Proc. R. Soc. London A* **392**, 15 (1984).

192. T. Harayama and A. Shudo, *Phys. Lett. A* **165**, 417 (1992).

193. B. Lauritzen, *Phys. Rev. A* **43**, 603 (1991); P. Cvitanović and B. Eckhardt, *Nonlinearity* **6**, 277 (1992).

194. C. E. Porter and R. G. Thomas, *Phys. Rev.* **104**, 483 (1956); P. A. Moldauer, *Phys. Rev. B* **136,** 947 (1964); T. A. Brody, J. Flores, J. B. French, P. A. Mello, A. Pandey, and S. S. M. Wong, *Rev. Mod. Phys.* **53**, 385 (1981).

195. A. Wirzba, *Chaos* **2**, 77 (1992); A. Wirzba, private communication.

196. J.-W. Turner, unpublished results.

197. H. A. Bethe and R. W. Jackiw, *Intermediate Quantum Mechanics*, Benjamin, New York, 1968; G. Baym, *Lectures on Quantum Mechanics*, Benjamin, New York, 1969; C. J. Ballhausen and A. E. Hansen, *Annu. Rev. Phys. Chem.* **23**, 15 (1972).

198. S. Weigert and R. G. Littlejohn, *Phys. Rev. A* **47**, 3506 (1993).

199. R. G. Littlejohn and S. Weigert, *Phys. Rev. A* **48**, 924 (1993).

200. T. A. Stephenson and S. A. Rice, *J. Chem. Phys.* **81**, 1083 (1984); W. D. Sands, L. F. Jones, and R. Moore, *J. Phys. Chem.* **93**, 6601 (1989); A. R. Tiller, A. C. Peet, and D. C. Clary, *Chem. Phys.* **129**, 125 (1989).

201. C. Zener, *Proc. R. Soc. London A* **137**, 696 (1932); L. Landau, *Sov. Phys.* **1**, 89 (1932); E. G. C. Stueckelberg, *Helv. Phys. Acta* **5**, 369 (1932).

202. A. Shapere and F. Wilczek, eds., *Geometric Phases in Physics*, World Scientific, Singapore, 1989.

203. C. A. Mead and D. G. Truhlar, *J. Chem. Phys.* **70**, 2284 (1979); in [202], pp. 90–103.

204. C. A. Mead, *Rev. Mod. Phys.* **64**, 51 (1992).

205. M. V. Berry, *Proc. R. Soc. London A* **392**, 45 (1984); in [202] pp. 124–137.

206. M. Rosker, T. S. Rose, and A. H. Zewail, *Chem. Phys. Lett.* **146**, 175 (1988); T. S. Rose, M. Rosker, and A. H. Zewail, *J. Chem. Phys.* **88**, 6672 (1988).

207. J. von Neumann and E. Wigner, *Phys. Z.* **30**, 467 (1929).

208. G. Delacrétaz, E. R. Grant, R. L. Whetten, L. Wöste, and J. W. Zwanziger, *Phys. Rev. Lett.* **56**, 2598 (1986); in [202], pp. 240–246.

209. G. Stock and W. Domcke, *J. Phys. Chem.* **97**, 12466 (1993).

210. D. Delande and J. C. Gay, *Phys. Rev. Lett.* **57**, 2006 (1986); H. Friedrich and D. Wintgen, *Phys. Rep.* **183**, 37 (1989).

211. J. K. G. Watson, *Mol. Phys.* **15**, 479 (1968); *ibid.* **19**, 465 (1970).

212. P. A. Braun, *Rev. Mod. Phys.* **65**, 115 (1993).

213. G. Herzberg, *Molecular Spectra and Molecular Structure*, *Volume III*, Krieger, Malabar, 1966.

214. E. Heller and P. Gaspard, unpublished results.

215. M. Gruebele, G. Roberts, M. Dantus, R. M. Bowman, and A. H. Zewail, *Chem. Phys. Lett.* **166**, 459 (1990).

216. J.-P. Pique, M. Joyeux, J. Manners, and G. Sitja, *J. Chem. Phys.* **95**, 8744 (1991).

217. J.-P. Pique, J. Manners, G. Sitja, and M. Joyeux, *J. Chem. Phys.* **96**, 6495 (1992).

218. J.-P. Pique, private communication.

219. M. E. Kellman and E. D. Lynch, *J. Chem. Phys.* **85**, 7216 (1986); **88**, 2205 (1988); Z. Li, X. Lin, and M. E. Kellman, *ibid.* **92**, 2251 (1990).

220. R. S. Nakata, K. Watanabe, and F. M. Matsunaga, *Sci. Light* **14**, 54 (1965).

221. S. Sato, *J. Chem. Phys.* **23**, 592, 2465 (1955).

222. K. C. Kulander and J. C. Light, *J. Chem. Phys.* **73**, 4337 (1980).

223. R. Schinke and V. Engel, *J. Chem. Phys.* **93**, 3252 (1990).
224. R. T. Pack, *J. Chem. Phys.* **65**, 4765 (1976).
225. L. C. Lee, X. Wang, and M. Suto, *J. Chem. Phys.* **86**, 4353 (1987).
226. L. Butler, *Chem. Phys. Lett.* **182**, 393 (1991).
227. B. Heumann, R. Düren, and R. Schinke, *Chem. Phys. Lett.* **180**, 583 (1991).
228. H. Eyring, *J. Chem. Phys.* **3**, 107 (1935); K. J. Laidler, *Chemical Kinetics*, McGraw-Hill, New York, 1965.
229. B. Simon, *Commun. Math. Phys.* **27**, 1 (1972); W. P. Reinhardt, *Annu. Rev. Phys. Chem.* **33**, 223 (1982); Y. K. Ho, *Phys. Rep.* **99**, 1 (1983); N. Moiseyev, in *Resonances–Models and Phenomena*, S. Albeverio, L. S. Ferreira, and L. Streit, Eds., Lecture Notes in Physics, Vol. 211, Springer, Berlin, 1984, pp. 235–256; E. Brändas and N. Elander, Eds., in *Resonances*, Lecture Notes in Physics, Vol. 325, Springer, Berlin, 1987, pp. 459–474.
230. W. A. Lester, Jr., and R. B. Bernstein, *J. Chem. Phys.* **48**, 4896 (1968); W. A. Lester, Jr., *Adv. Quantum Chem.* **9**, 199 (1975).
231. E. P. Wigner and L. Eisenbud, *Phys. Rev.* **72**, 29 (1947); A. M. Lane and R. G. Thomas, *Rev. Mod. Phys.* **30**, 257 (1958); E. P. Wigner, in *Dispersion Relations and their Connection with Causality*, E. P. Wigner, Ed., Academic, New York, 1964, pp. 40–67; P. G. Burke, I. Mackey, and I. Shimamura, *J. Phys. B: At. Mol. Phys.* **10**, 2497 (1977); cf. [63].
232. A. H. Zewail, *Faraday Discuss. Chem. Soc.* **91**, 207 (1991).
233. M. Dantus. R. M. Bowman, M. Gruebele, and A. H. Zewail, *J. Chem. Phys.* **91**, 7437 (1989).
234. M. Gruebele, G. Roberts, and A. H. Zewail, *Philos. Trans. R. Soc. London A* **332**, 223 (1990).
235. E. Pollak, *J. Chem. Phys.* **74**, 5586 (1981).
236. E. Pollak and P. Pechukas, *J. Chem. Phys.* **69**, 1218 (1978); E. Pollak, M. S. Child, and P. Pechukas, *ibid.* **72**, 1669 (1980); E. Pollak and M. S. Child, *Ibid.* **73**, 4373 (1980).
237. E. Pollak and M. S. Child, *Chem. Phys.* **60**, 23 (1981).
238. J. Guckenheimer and P. Holmes, *Nonlinear Oscillations, Dynamical Systems, and Bifurcations of Vector Fields*, Springer, New York, 1983.
239. B. R. Johnson and J. L. Kinsey, *Phys. Rev. Lett.* **62**, 1607 (1989).
240. O. Hahn, J. M. Gomez-Llorente, and H. S. Taylor, *J. Chem. Phys.* **94**, 2608 (1991).
241. R. T. Skodje and M. J. Davis, *J. Chem. Phys.* **88**, 2429 (1988).
242. J. Manz, E. Pollak, and J. Römelt, *Chem. Phys. Lett.* **86**, 26 (1982); E. Pollak, *J. Chem. Phys.* **76**, 5843 (1982).
243. M. J. Davis, *J. Chem. Phys.* **86**, 3978 (1987); R. Sadeghi and R. T. Skodje, *ibid.* **99**, 5126 (1993); D. C. Chatfield, R. S. Friedman, G. C. Lynch, D. G. Truhlar, and D. W. Schwenke, *J. Chem. Phys.* **98**, 342 (1993).
244. J. M. Ziman, *Models of Disorder*, Clarendon, Oxford, 1979.
245. W. H. Press, B. P. Flannery, S. A. Teukolsky, and V. T. Vetterling, *Numerical Recipes*, Cambridge University Press, UK, 1986, p. 85 (subroutine RATINT).
246. M. D. Feit, J. A. Fleck, and A. Steiger, *J. Comput. Phys.* **47**, 412 (1982).
247. D. Kosloff and R. Kosloff, *J. Comput. Phys.* **52**, 35 (1983); **63**, 363 (1986).
248. J. Maya, *J. Chem. Phys.* **67**, 4976 (1977); H. Hofmann and S. R. Leone, *ibid.* **69**, 3819 (1978).

249. J.-A. Sepulchre and P. Gaspard, *Rotational effect on the lifetime of molecular resonances*, to appear in *J. Chem. Phys.* (1995).

250. G. Herzberg, *Molecular Spectra and Structure, Volume II*, Krieger, Malabar, 1991.

251. G. Breit and E. P. Wigner, *Phys. Rev.* **49**, 519, 642 (1936); cf. [187].

252. W. F. Polik, C. B. Moore, and W. H. Miller, *J. Chem. Phys.* **89**, 3584 (1988); W. F. Polik, D. R. Guyer, W. H. Miller, and C. B. Moore, *J. Chem. Phys.* **92**, 3471 (1990); W. H. Miller, R. Hernandez, and C. B. Moore, *ibid.* **93**, 5657 (1990).

253. M. Desouter-Lecomte and F. Culot, *J. Chem. Phys.* **98**, 7819 (1993); F. M. Izrailev, D. Saher, and V. V. Sokolov, *Phys. Rev. E* **49**, 130 (1994); and references cited therein.

254. J. A. Joens, *J. Chem. Phys.* **100**, 3407 (1994).

255. M. J. Seaton, *Rep. Prog. Phys.* **46**, 167 (1983); C. H. Greene, A. R. P. Rau, and U. Fano, *Phys. Rev. A.* **26**, 2441 (1982); cf. [63].

256. J. G. Leopold, I. C. Percival, and A. S. Tworkowski, *J. Phys. B: At. Mol. Phys.* **13**, 1025 (1980); J. G. Leopold and I.C. Percival, **13**, 1037 (1980).

257. R. Blümel and W. P. Reinhardt, in *Directions in Chaos*, B.-L. Hao, D. H. Feng, and J.-M. Yuan, Eds., World Scientific, Singapore, 1991.

258. J. Müller, J. Burgdörfer, and D. Noid, *Phys. Rev. A* **45**, 1471 (1992); J. Müller and J. Burgdörfer, *Phys. Rev. Lett.* **70**, 2375 (1993).

259. K. Richter and and D. Wintgen, *J. Phys. B: At. Mol. Opt. Phys.* **23**, L197 (1990); **24**, L565 (1991).

260. K. Richter, J. M. Rost, R. Thürwächter, J. S. Briggs, D. Wintgen, and E. A. Solov'ev, *Phys. Rev. Lett.* **66**, 149 (1991).

261. G. Ferrante, R. Geracitano, and L. Lo Cascio, *Phys. Rev. A* **14**, 558 (1976).

262. H. Bethe, *Z. Phys.* **57**, 815 (1929).

263. E. Hylleraas, *Z. Phys.* **60**, 624 (1930).

264. V. Szebehely, *Theory of Orbits*, Academic, New York, 1967.

265. N. Koyama, H. Fukuda, T. Motoyama, and M. Matsuzawa, *J. Phys. B: At. Mol. Phys.* **19**, L331 (1986).

266. H. Fukuda, N. Koyama, and M. Matsuzawa, *J. Phys. B: At. Mol. Phys.* **20**, 2959 (1987).

267. J. M. Rost and J. S. Briggs, *J. Phys. B: At. Mol. Phys.* **22**, 3587 (1989).

268. J. M. Rost and J. S. Briggs, *J. Phys. B: At. Mol. Phys.* **24**, 4293 (1991).

269. Y. K. Ho and J. Callaway, *Phys. Rev. A* **34**, 130 (1986); Y. K. Ho, *Phys. Rev. A* **41**, 1492 (1990).

270. U. Fano and J. W. Cooper, *Phys. Rev. A.* **137**, 1364 (1965).

271. R. Landauer, *Philos. Mag.* **21**, 863 (1970).

272. K. Nakamura and H. Ishio, *J. Phys. Soc. Jpn.* **61**, 2649, 3939 (1992).

273. P. Gaspard, D. Alonso, T. Okuda, and K. Nakamura, *Phys. Rev. E* **50**, 2591 (1994).

274. B. Eckhardt and D. Wintgen, *J. Phys. B: At. Mol. Phys.* **23**, 355 (1990).

275. K. Nakamura and H. Thomas, *Phys. Rev. Lett.* **61**, 247 (1988).

276. K. Nakamura, *Quantum Chaos*, Cambridge University Press, 1993.

277. O. Agam, *The magnetic response of chaotic mesoscopic systems*, preprint Technion (1993).

278. D. Ullmo, K. Richter, and R. A. Jalabert, *Orbital Magnetism in Ensembles of Ballistic Billiards*, preprint Orsay (1993).

279. D. Weiss, K. Richter, A. Menschig, R. Bergmann, H. Schweizer, K. von Klitzing, and G. Weimann, *Phys. Rev. Lett.* **70**, 4118 (1993).
280. R. Fleischmann, T. Geisel, and R. Ketzmerick, *Phys. Rev. Lett.* **68**, 1367 (1992); T. Geisel, R. Ketzmerick, and O. Schedletzky, *Phys. Rev. Lett.* **69**, 1680 (1992).
281. O. Bohigas, M. J. Giannoni, and C. Schmidt, *Phys. Rev. Lett.* **52**, 1 (1984).
282. J. Bolte, G. Steil, and F. Steiner, *Phys. Rev. Lett.* **69**, 2188 (1992).
283. P. van Ede van der Pals and P. Gaspard, *Phys. Rev. E* **49**, 79 (1994).
284. J. Keating, in [17], pp. 145–185; N. Argaman, F.-M. Dittes, E. Doron, J. P. Keating, A. Yu. Kitaev, M. Sieber, and U. Smilansky, *Phys. Rev. Lett.* **71**, 4326 (1993).
285. S. K. Gray, S. A. Rice, and D. W. Noid, *J. Chem. Phys.* **84**, 3745 (1986); M. Zhao and S. A. Rice, *ibid.* **96**, 3542, 6654, 7483 (1992); **97**, 943 (1992).
286. R. S. MacKay, J. D. Meiss, and I. C. Percival, *Physica D* **13**, 55 (1984), reprinted in [3], pp. 606–632; D. Bensimon and L. P. Kadanoff, *Physica D* **13**, 82 (1984).
287. G. Vattay, A. Wirzba, and P. E. Rosenqvist, *Phys. Rev. Lett.* **73**, 2304 (1994).
288. H. M. Nussenzveig, *Diffraction Effects in Semiclassical Scattering*, Cambridge University Press, UK, 1992.
289. D. Beck, *J. Chem. Phys.* **37**, 2884 (1962); cf. Chap. 3 in [1]; cf. [288].
290. D. W. Noid, S. K. Gray, and S. A. Rice, *J. Chem. Phys.* **84**, 2649 (1986); U. Smilansky, in [6], pp. 371–441; V. Balasubramanian, B. K. Misra, A. Bahel, S. Kumar, and N. Sathyamurthy, *J. Chem. Phys.* **95**, 4160 (1991); K. Someda, R. Ramaswamy, and H. Nakamura, *ibid.* **98**, 1156 (1993); T. Tél and E. Ott, Eds., *Chaotic Scattering*, *Chaos* **3**(4) (1993); and references cited therein.
291. R. Blümel, B. Dietz, C. Jung, and U. Smilansky, *J. Phys. A: Math. Gen.* **25**, 1483 (1992).
292. I. Prigogine, *Nonequilibrium Statistical Mechanics*, Wiley, New York, 1962.
293. P. Gaspard and G. Nicolis, *Phys. Rev. Lett.* **65**, 1693 (1990); P. Gaspard, *J. Stat. Phys.* **68** (1992) 673.
294. P. Gaspard, *J. Phys. A: Math. Gen.* **25**, L483 (1992); P. Gaspard, *Phys. Lett. A* **168**, 13 (1992); P. Gaspard, *Chaos* **3**, 427 (1993).
295. H. H. Hasegawa and W. C. Saphir, *Phys. Rev. A* **46**, 7401 (1992).
296. S. Tasaki and P. Gaspard, in M. Yamaguti, ed., *Towards the Harnessing of Chaos*, Proc. 7th Toyota Conference, Mikkabi, Japan, Nov. 1993, Elsevier Science B.V., Amsterdam, 1994, pp. 273–288.
297. P. Gaspard, *Lecture Notes ULB* (1991–1992, 1992–1993).
298. F. H. Mies and M. Krauss, *J. Chem. Phys.* **45**, 4455 (1966).
299. R. Hernandez, W. H. Miller, C. B. Moore, and W. F. Polik, *J. Chem. Phys.* **99**, 950 (1993).
300. R. S. Friedman and D. G. Truhlar, *Chem. Phys. Lett.* **183**, 539 (1991); D. C. Chatfield, R. S. Friedman, D. W. Schwenke, and D. G. Truhlar, *J. Phys. Chem.* **96**, 2414 (1992); D. C. Chatfield, R. S. Friedman, G. C. Lynch, D. G. Truhlar, and D. W. Schwenke, *J. Chem. Phys.* **98**, 342 (1993).
301. F. H. Molzahn and T. A. Osborn, *Ann. Phys.* **230**, 343 (1994).
302. M. J. Cohen, N. C. Handy, R. Hernandez, and W. H. Miller, *Chem. Phys. Lett.* **192**, 407 (1992).
303. B. R. Johnson and J. L. Kinsey, *J. Chem. Phys.* **91**, 7638 (1989).
304. J. Liévin, M. Abbouti Temsamani, P. Gaspard, and M. Herman, *Chem. Phys.* **190**, 419 (1995).

305. J.-P. Pique, G. Sitja, H. Ring, L. Michaille, and P. Gaspard, in preparation.

306. P. Gaspard, in preparation.

307. M. Srednicki, *Phys. Rev. E* **50**, 888 (1994).

308. S. Tomsović and E. J. Heller, *Phys. Rev. Lett.* **67**, 664 (1991); *Phys. Rev. E* **47**, 282 (1993).

309. L. S. Schulman, *J. Phys. A: Math. Gen.* **27**, 1703 (1994).

310. M. V. Berry and C. J. Howls, *High orders of the Weyl expansion for quantum billiards: Resurgence of periodic orbits, and the Stokes phenomenon*, preprint submitted to *Proc. Roy. Soc. Lond.* (1994).

AUTHOR INDEX

Numbers in parentheses are reference numbers and indicate that the author's work is referred to although his name is not mentioned in the text. Numbers in *italic* show the pages on which the complete references are listed.

SUBJECT INDEX